Rabin Raut and M. N. S. Swamy

Modern Analog Filter Analysis and Design

Related Titles

Shenoi, B. A.

Introduction to Digital Signal Processing and Filter Design

2005
ISBN: 978-0-471-46482-2

Madsen, C. K., Zhao, J. H.

Optical Filter Design and Analysis
A Signal Processing Approach

1999
ISBN: 978-0-471-18373-0

Rabin Raut and M. N. S. Swamy

Modern Analog Filter Analysis and Design

A Practical Approach

WILEY-VCH Verlag GmbH & Co. KGaA

The Authors

Dr. Rabin Raut
Concordia University
Electrical and Computer Engineering
Montreal, Canada
rabinr@ece.concordia.ca

Dr. M. N. S. Swamy
Concordia University
Electrical and Computer Engineering
Montreal, Canada
swamy@ece.concordia.ca

Cover
Spieszdesign, Neu-Ulm, Germany

All books published by Wiley-VCH are carefully produced. Nevertheless, authors, editors, and publisher do not warrant the information contained in these books, including this book, to be free of errors. Readers are advised to keep in mind that statements, data, illustrations, procedural details or other items may inadvertently be inaccurate.

Library of Congress Card No.: applied for

British Library Cataloguing-in-Publication Data
A catalogue record for this book is available from the British Library.

Bibliographic information published by the Deutsche Nationalbibliothek
The Deutsche Nationalbibliothek lists this publication in the Deutsche Nationalbibliografie; detailed bibliographic data are available on the Internet at <http://dnb.d-nb.de>.

© 2010 WILEY-VCH Verlag & Co. KGaA, Boschstr. 12, 69469 Weinheim, Germany

All rights reserved (including those of translation into other languages). No part of this book may be reproduced in any form – by photoprinting, microfilm, or any other means – nor transmitted or translated into a machine language without written permission from the publishers. Registered names, trademarks, etc. used in this book, even when not specifically marked as such, are not to be considered unprotected by law.

Cover Design Adam Design, Weinheim
Typesetting Laserwords Private Ltd., Chennai, India
Printing and Binding Fabulous Printers Pte Ltd, Singapore

Printed in Singapore
Printed on acid-free paper

ISBN: 978-3-527-40766-8

आचार्यात्पादमादत्ते पादं शिष्यस्वमेधया ।
पादं सब्रह्मचारिभ्यः पादं कालक्रमेण तु ॥

A student acquires a quarter of knowledge from the teacher, a quarter from self study, a quarter from class mates, and the final quarter in course of time.
From Neeti Sara

To
Our parents, teachers, and our wives Sucheta & Leela
and
Late Prof. B. B. Bhattacharyya
(Ph.D. supervisor of R. Raut and
First Ph.D. student of M. N. S Swamy)

Contents

Preface *XV*
Abbreviations *XIX*

1 **Introduction** *1*

2 **A Review of Network Analysis Techniques** *7*
2.1 Transformed Impedances *7*
2.2 Nodal Analysis *9*
2.3 Loop (Mesh) Analysis *9*
2.4 Network Functions *11*
2.5 One-Port and Two-Port Networks *12*
2.5.1 One-Port Networks *12*
2.5.2 Two-Port Networks *13*
2.5.2.1 Admittance Matrix Parameters *13*
2.5.2.2 Impedance Matrix Parameters *14*
2.5.2.3 Chain Parameters (Transmission Parameters) *14*
2.5.2.4 Interrelationships *15*
2.5.2.5 Three-Terminal Two-Port Network *16*
2.5.2.6 Equivalent Networks *16*
2.5.2.7 Some Commonly Used Nonreciprocal Two-Ports *16*
2.6 Indefinite Admittance Matrix *18*
2.6.1 Network Functions of a Multiterminal Network *20*
2.7 Analysis of Constrained Networks *24*
2.8 Active Building Blocks for Implementing Analog Filters *28*
2.8.1 Operational Amplifier *28*
2.8.2 Operational Transconductance Amplifier *30*
2.8.3 Current Conveyor *31*
Practice Problems *33*

3 **Network Theorems and Approximation of Filter Functions** *41*
3.1 Impedance Scaling *41*
3.2 Impedance Transformation *42*
3.3 Dual and Inverse Networks *44*

Modern Analog Filter Analysis and Design: A Practical Approach. Rabin Raut and M. N. S. Swamy
Copyright © 2010 WILEY-VCH Verlag GmbH & Co. KGaA, Weinheim
ISBN: 978-3-527-40766-8

3.3.1	Dual and Inverse One-Port Networks	44
3.3.2	Dual Two-Port Networks	45
3.4	Reversed Networks	47
3.5	Transposed Network	48
3.6	Applications to Terminated Networks	50
3.7	Frequency Scaling	52
3.8	Types of Filters	52
3.9	Magnitude Approximation	54
3.9.1	Maximally Flat Magnitude (MFM) Approximation	55
3.9.1.1	MFM Filter Transfer Function	56
3.9.2	Chebyshev (CHEB) Magnitude Approximation	60
3.9.2.1	CHEB Filter Transfer Function	63
3.9.3	Elliptic (ELLIP) Magnitude Approximation	65
3.9.4	Inverse-Chebyshev (ICHEB) Magnitude Approximation	68
3.10	Frequency Transformations	69
3.10.1	LP to HP Transformation	69
3.10.2	LP to BP Transformation	71
3.10.3	LP to BR Transformation	73
3.11	Phase Approximation	73
3.11.1	Phase Characteristics of a Transfer Function	74
3.11.2	The Case of Ideal Transmission	74
3.11.3	Constant Delay (Linear Phase) Approximation	75
3.11.4	Graphical Method to Determine the BT Filter Function	76
3.12	Delay Equalizers	77
	Practice Problems	78
4	**Basics of Passive Filter Design**	**83**
4.1	Singly Terminated Networks	83
4.2	Some Properties of Reactance Functions	85
4.3	Singly Terminated Ladder Filters	88
4.4	Doubly Terminated LC Ladder Realization	92
	Practice Problems	100
5	**Second-Order Active-RC Filters**	**103**
5.1	Some Basic Building Blocks using an OA	104
5.2	Standard Biquadratic Filters or Biquads	104
5.3	Realization of Single-Amplifier Biquadratic Filters	109
5.4	Positive Gain SAB Filters (Sallen and Key Structures)	111
5.4.1	Low-Pass SAB Filter	111
5.4.2	RC:CR Transformation	113
5.4.3	High-Pass Filter	115
5.4.4	Band-Pass Filter	115
5.5	Infinite-Gain Multiple Feedback SAB Filters	115
5.6	Infinite-Gain Multiple Voltage Amplifier Biquad Filters	117
5.6.1	KHN State-Variable Filter	119

5.6.2	Tow–Thomas Biquad	121
5.6.3	Fleischer–Tow Universal Biquad Structure	123
5.7	Sensitivity	124
5.7.1	Basic Definition and Related Expressions	124
5.7.2	Comparative Results for ω_p and Q_p Sensitivities	126
5.7.3	A Low-Sensitivity Multi-OA Biquad with Small Spread in Element Values	126
5.7.4	Sensitivity Analysis Using Network Simulation Tools	129
5.8	Effect of Frequency-Dependent Gain of the OA on the Filter Performance	130
5.8.1	Cases of Inverting, Noninverting, and Integrating Amplifiers Using an OA with Frequency-Dependent Gain	130
5.8.1.1	Inverting Amplifier	130
5.8.1.2	Noninverting Amplifier	131
5.8.1.3	Inverting Integrating Amplifier	132
5.8.2	Case of Tow–Thomas Biquad Realized with OA Having Frequency-Dependent Gain	132
5.9	Second-Order Filter Realization Using Operational Transconductance Amplifier (OTA)	135
5.9.1	Realization of a Filter Using OTAs	138
5.9.2	An OTA-C Band-Pass Filter	138
5.9.3	A General Biquadratic Filter Structure	139
5.10	Technological Implementation Considerations	140
5.10.1	Resistances in IC Technology	141
5.10.1.1	Diffused Resistor	141
5.10.1.2	Pinched Resistor	142
5.10.1.3	Epitaxial and Ion-Implanted Resistors	142
5.10.1.4	Active Resistors	143
5.10.2	Capacitors in IC Technology	144
5.10.2.1	Junction Capacitors	145
5.10.2.2	MOS Capacitors	145
5.10.2.3	Polysilicon Capacitor	146
5.10.3	Inductors	146
5.10.4	Active Building Blocks	147
5.10.4.1	Operational Amplifier (OA)	147
5.10.4.2	Operational Transconductance Amplifier (OTA)	148
5.10.4.3	Transconductance Amplifiers (TCAs)	150
5.10.4.4	Current Conveyor (CC)	151
	Practice Problems	152
6	**Switched-Capacitor Filters**	**161**
6.1	Switched C and R Equivalence	162
6.2	Discrete-Time and Frequency Domain Characterization	163
6.2.1	SC Integrators: $s \leftrightarrow z$ Transformations	164

6.2.2	Frequency Domain Characteristics of Sampled-Data Transfer Function 167
6.3	Bilinear $s \leftrightarrow z$ Transformation 169
6.4	Parasitic-Insensitive Structures 173
6.4.1	Parasitic-Insensitive-SC Integrators 176
6.4.1.1	Lossless Integrators 176
6.4.1.2	Lossy Integrators 177
6.5	Analysis of SC Networks Using PI-SC Integrators 177
6.5.1	Lossless and Lossy Integrators 177
6.5.1.1	Inverting Lossless Integrator 177
6.5.1.2	Noninverting Lossless Integrator 179
6.5.1.3	Inverting and Noninverting Lossless Integration Combined 179
6.5.1.4	Lossy PI-SC Integrator 181
6.5.2	Application of the Analysis Technique to a PI-SC Integrator-Based Second-Order Filter 181
6.5.3	Signals Switched to the Input of the OA during Both Phases of the Clock Signal 183
6.6	Analysis of SC Networks Using Network Simulation Tools 184
6.6.1	Use of VCVS and Transmission Line for Simulating an SC Filter 184
6.7	Design of SC Biquadratic Filters 187
6.7.1	Fleischer–Laker Biquad 188
6.7.2	Dynamic Range Equalization Technique 191
6.8	Modular Approach toward Implementation of Second-Order Filters 191
6.9	SC Filter Realization Using Unity-Gain Amplifiers 199
6.9.1	Delay-and-add Blocks Using UGA 200
6.9.2	Delay Network Using UGA 201
6.9.3	Second-Order Transfer Function Using DA1, DA2 and D Networks 202
6.9.4	UGA-Based Filter with Reduced Number of Capacitances 202
	Practice Problems 204
7	**Higher-Order Active Filters** 207
7.1	Component Simulation Technique 207
7.1.1	Inductance Simulation Using Positive Impedance Inverters or Gyrators 208
7.1.2	Inductance Simulation Using a Generalized Immittance Converter 210
7.1.2.1	Sensitivity Considerations 213
7.1.3	FDNR or Super-Capacitor in Higher-Order Filter Realization 213
7.1.3.1	Sensitivity Considerations 216
7.2	Operational Simulation Technique for High-Order Active RC Filters 217
7.2.1	Operational Simulation of All-Pole Filters 217
7.2.2	Leapfrog Low-Pass Filters 219

7.2.3	Systematic Steps for Designing Low-Pass Leapfrog Filters	220
7.2.4	Leapfrog Band-Pass Filters	222
7.2.5	Operational Simulation of a General Ladder Structure	223
7.3	Cascade Technique for High-Order Active Filter Implementation	225
7.3.1	Sensitivity Considerations	227
7.3.2	Sequencing of the Biquads	228
7.3.3	Dynamic Range Considerations	228
7.4	Multiloop Feedback (and Feed-Forward) System	229
7.4.1	Follow the Leader Feedback Structure	229
7.4.1.1	$T_i = (1/s)$, a Lossless Integrator	230
7.4.1.2	$T_i = 1/(s + \alpha)$, a Lossy Integrator	231
7.4.2	FLF Structure with Feed-Forward Paths	233
7.4.3	Shifted Companion Feedback Structure	234
7.4.4	Primary Resonator Block Structure	237
7.5	High-Order Filters Using Operational Transconductance Amplifiers	239
7.5.1	Cascade of OTA-Based Filters	239
7.5.2	Inductance Simulation	239
7.5.3	Operational Simulation Technique	239
7.5.4	Leapfrog Structure for a General Ladder	242
7.6	High-Order Filters Using Switched-Capacitor (SC) Networks	245
7.6.1	Parasitic-Insensitive Toggle-Switched-Capacitor (TSC) Integrator	245
7.6.2	A Stray-Insensitive Bilinear SC Integrator Using Biphase Clock Signals	247
7.6.3	A Stray-Insensitive Bilinear Integrator with Sample-and-Hold Input Signal	247
7.6.4	Cascade of SC Filter Sections for High-Order Filter Realization	248
7.6.5	Ladder Filter Realization Using the SC Technique	250
	Practice Problems	251
8	**Current-Mode Filters**	255
8.1	Basic Operations in Current-Mode	255
8.1.1	Multiplication of a Current Signal	255
8.1.1.1	Use of a Current Mirror	256
8.1.1.2	Use of a Current Conveyor	257
8.1.1.3	Use of Current Operational Amplifier	258
8.1.2	Current Addition (or Subtraction)	259
8.1.3	Integration and Differentiation of a Current Signal	259
8.2	Current Conveyors in Current-Mode Signal Processing	264
8.2.1	Some Basic Building Blocks Using CCII	264
8.2.2	Realization of Second-Order Current-Mode Filters	264
8.2.2.1	Universal Filter Implementation	264
8.2.2.2	All-Pass/Notch and Band-Pass Filters Using a Single CCII	265
8.2.2.3	Universal Biquadratic Filter Using Dual-Output CCII	266
8.3	Current-Mode Filters Derived from Voltage-Mode Structures	267

8.4	Transformation of a VM Circuit to a CM Circuit Using the Generalized Dual	269
8.5	Transformation of VM Circuits to CM Circuits Using Transposition	271
8.5.1	CM Circuits from VM Circuits Employing Single-Ended OAs	272
8.5.1.1	CM Biquads Derived from VM Biquads Employing Finite Gain Amplifiers	272
8.5.1.2	CM Biquads Derived from VM Biquads Employing Infinite-Gain Amplifiers	273
8.5.2	CM Circuits from VM Circuits Employing OTAs	274
8.5.2.1	VM Circuits Using Single-Ended OTAs	274
8.5.2.2	VM Circuits Using Differential-Input OTAs	277
8.6	Derivation of CTF Structures Employing Infinite-Gain Single-Ended OAs	279
8.6.1	Illustrative Examples	280
8.6.1.1	Single-Amplifier Second-Order Filter Network	280
8.6.1.2	Tow-Thomas Biquad	281
8.6.1.3	Ackerberg and Mossberg LP and BP Filters	282
8.6.2	Effect of Finite Gain and Bandwidth of the OA on the Pole Frequency, and Pole Q	283
8.7	Switched-Current Techniques	285
8.7.1	Add, Subtract, and Multiply Operations	286
8.7.2	Switched-Current Memory Cell	286
8.7.3	Switched-Current Delay Cell	288
8.7.4	Switched-Current Integrators	288
8.7.5	Universal Switched-Current Integrator	290
8.8	Switched-Current Filters	291
	Practice Problems	294
9	**Implementation of Analog Integrated Circuit Filters**	**299**
9.1	Active Devices for Analog IC Filters	300
9.2	Passive Devices for IC Filters	300
9.2.1	Resistance	300
9.2.2	Switch	302
9.3	Preferred Architecture for IC Filters	303
9.3.1	OA-Based Filters with Differential Structure	303
9.3.1.1	First-Order Filter Transfer Functions	304
9.3.1.2	Second-Order Filter Transfer Functions	304
9.3.2	OTA-Based Filters with Differential Structures	310
9.3.2.1	First-Order Filter Transfer Functions	310
9.3.2.2	Second-Order Filter Transfer Functions	312
9.4	Examples of Integrated Circuit Filters	314
9.4.1	A Low-Voltage, Very Wideband OTA-C Filter in CMOS Technology	314
9.4.2	A Current-Mode Filter for Mobile Communication Application	318

9.4.2.1	Filter Synthesis	*318*
9.4.2.2	Basic Building Block	*319*
9.4.2.3	Inductance and Negative Resistance	*321*
9.4.2.4	Second-Order Elementary Band-Pass Filter Cell	*321*
	Practice Problems	*323*

Appendices *325*

Appendix A *327*

A.1 Denominator Polynomial $D(s)$ for the Butterworth Filter Function of Order n, with Passband from 0 to 1 rad s^{-1} *327*

A.2 Denominator Polynomial $D(s)$ for the Chebyshev Filter Function of Order n, with Passband from 0 to 1 rad s^{-1} *328*

A.3 Denominator Polynomial $D(s)$ for the Bessel Thomson Filter Function of Order n *328*

A.4 Transfer Functions for Several Second-, Third-, and Fourth-Order Elliptic Filters *330*

Appendix B *333*

B.1 Bessel Thomson Filter Magnitude Error Calculations (MATLAB Program) *333*

B.2 Bessel Thomson Filter Delay Error Calculations (MATLAB Program) *334*

Appendix C *337*

C.1 Element Values for All-Pole Single-Resistance-Terminated Low-Pass Lossless Ladder Filters *337*

C.2 Element Values for All-Pole Double-Resistance-Terminated Low-Pass Lossless Ladder Filters *337*

C.3 Element Values for Elliptic Double-Resistance-Terminated Low-Pass Lossless Ladder Filters *340*

References *345*

Index *351*

Preface

Filters, especially analog filters, are employed in many different systems that electrical engineers embark upon to design. Even many signal processing systems that are apparently digital, often contain one or more analog continuous-time filters either internally or as interface with the real-time world, which is analog in nature. This book on analog filters is intended as an intermediate-level text for a senior undergraduate and/or an entry-level graduate class in an electrical/electronic engineering curriculum. The book principally covers the subject of analog active filters with brief introductions to passive filters and integrated circuit filters. In the class of active filters, both continuous-time and sampled-data filters are covered. Further, both voltage-mode and current-mode filters are considered. The book is targeted at students and engineers engaged in signal processing, communications, electronics, controls, and so on.

The book is not intended to be an extensive treatise on the subject of analog filters. The subject of (analog) electrical filters is very vast and numerous authors have written excellent books on this subject in the past. Therefore, the question that naturally arises pertains to the need for yet another book on analog filters.

The subject of analog filters is so fascinating that there is always room to introduce the subject with slightly different orientation, especially one that is directed toward certain class of practitioners in the field of electrical engineering. This book exploits the existing wealth of knowledge to illustrate practical ways to implement an analog filter, both for voltage and current signals. Use of currents for signal processing has been a popular subject during the last two decades, and in this respect the book touches on a modern viewpoint of signal processing, relevant to analog filters. In particular, the concept of transposition and its usefulness in obtaining in a very simple manner a current-mode filter from a voltage-mode filter, or vice versa, is presented for the first time in this book. Even though this concept was developed in 1971 itself, its practical use came only after the advent of IC technology, and hence this concept did not receive much attention in earlier books.

This book has been written with a young practicing engineer in mind. Most of the engineers now have to work between deadline dates and have very little time to plunge into the details of theoretical work to scoop up the practical outcome; namely, the usable device such as the needed filter. Thus, the subject of filters has been introduced in this book in a systematic manner with as much theoretical

Modern Analog Filter Analysis and Design: A Practical Approach. Rabin Raut and M. N. S. Swamy
Copyright © 2010 WILEY-VCH Verlag GmbH & Co. KGaA, Weinheim
ISBN: 978-3-527-40766-8

exposure as is essential to be able to build a filter in question. Ample references have been cited to aid the reader in further exploration of the detailed theoretical matter, if the reader is interested. Most of the theoretical material presented in the book has been immediately illustrated via practical examples of synthesis and design, using modern numerical and circuit simulation tools such as MATLAB and SPICE. These tools are now easily available to an electrical engineer (either a student or a practitioner), so the user of the book will feel very close to the practical world of building the filter at hand.

In the era of computers, building analog filters could involve simple use of several software programs downloaded from the Internet and obtaining the required hardware components for the filter to be designed. The authors, however, expect that there are some inquisitive minds who want to know the *why* and *how* behind the working and implementation of the filters. Thus, the book attempts to infuse some understanding of the elegant mathematical methods behind the synthesis of a filter, and ingenious applications of these methods toward the implementation of the filter. The expected background knowledge on the part of the reader of this book is some basics related to circuit theory, electronics, Laplace transform, and z-transform. These topics are covered in most of the modern electrical engineering curricula within the span of the first two years of study.

Although this book is more compact than many other books on analog filters in the market, we still feel that the material in this text book cannot be covered satisfactorily over the span of the usual three-and-a-half month-long session pursued by most academic institutions in the Western world. For a one-term undergraduate course, material in Chapters 2–6 can be taught at a reasonable pace. Similarly, for a graduate class over a similar term, Chapters 5–9 may be covered. It is expected that a graduate student would be able to learn the materials in Chapters 2–4 by himself/herself, or that he/she has the required background earned previously from an undergraduate course in analog filters. In those schools where a two-semester course is available at the undergraduate level, the material in the whole book can be easily covered in detail.

Plenty of exercise problems have been added at the end of each chapter. The problems are carefully coordinated with the subject matter dealt with in the body of the pertinent chapter. These could be used by the students to profitably enhance their understanding of the subject. Some of the more challenging problems could be assigned as projects, by the instructor. The authors strongly feel that a course given by using this text book must be accompanied by one or more projects, so that the student develops the practical skill involved in designing and implementing an analog filter.

The authors wish to gratefully acknowledge the contributions made by numerous students upon whom the material has been tested over the past several years of teaching at Concordia University. The authors would like to thank their respective wives, Sucheta and Leela, for their patience and understanding during the course of writing this book. They also sincerely extend their thanks to Anja Tschörtner for her patience and cooperation while waiting for the final manuscript.

Montreal, Canada R. Raut
January 2010 M. N. S. Swamy

Abbreviations

AP	all-pass
BDI	backward digital integrator
BJT	bipolar junction transistor
BLI	bilinear integrator
BLT	bilinear transformation
BP	band-pass
BR	band-reject
BS	band-stop
BT	Bessel-Thomson
CA	current amplifier
CC	current conveyor
CCCII+	controlled CCII+
CCCS	current controlled current source
CCE	charge conservation equation
CCII	current-conveyor type 2
CCII−	negative CCII
CCII+	positive CCII
CCVS	current controlled voltage source
CDA	composite delay and add
CDTA	current differencing TCA
CHEB	Chebyshev
CM	current mode
CMOS	complementary MOS
CMRR	common-mode rejection ratio
CNIC	current-inversion type NIC
COA	current operational amplifier
CTF	current transfer function
DA-1	delay and add type 1
DA-2	delay and add type 2
dB	deci-bel
DC	direct current
DDCCII	dual differential CCII
DIOTA	differential-input OTA

Modern Analog Filter Analysis and Design: A Practical Approach. Rabin Raut and M. N. S. Swamy
Copyright © 2010 WILEY-VCH Verlag GmbH & Co. KGaA, Weinheim
ISBN: 978-3-527-40766-8

DISO	dual input single output
DISO-OTA	differential-input single-output OTA
DOCCII	dual output CCII
DPA	driving point admittance
DPI	driving point impedance
DSP	digital signal processing
ELLIP	elliptic
FDCCII	fully differential CCII
FDI	forward digital integrator
FDNR	frequency dependent negative resistance
FIR	finite impulse response
FLF	follow the leader feedback
GD	generalized dual
GDT	generalized dual transpose
GIC	generalized immittance converter
GSM	global system mobile (communication)
HP	high-pass
IAM	indefinite admittance matrix
IC	integrated circuit(s)
ICHEB	inverse Chebyshev
IF	intermediate frequency
IFLF	inverse FLF
IGMFB	infinite gain multiple feedback
II	impedance inverter
KCL	Kirchoff's current law
KHN	Kerwin, Huelsman, Newcomb
LDI	lossless digital integrator
LH	left-half
LHP	left-half plane
LP	low-pass
LTI	linear time invariant
MFM	maximally flat magnitude
MISO	multi-input single output
MLF	multiloop feedback
MOS	metal-oxide semiconductor
MOSFET	MOS field effect transistor
MOS-R	MOSFET resistor
MSF	modified shifted companion feedback
NAM	nodal admittance matrix
NIC	negative immittance converter
NII	negative impedance inverter
NMOS	N-type MOSFET
OA	operational amplifier
OTA	operational transconductance amplifier
OTA-C	OTA-capacitor

OTRA	operational transresistance amplifier
PB	pass-band
PCM	pulse code modulation
PI	parasitic insensitive
PIC	positive immittance converter
PII	positive impedance inverter
PMOS	P-type MOSFET
PRB	primary resonator block
PSRR	power supply rejection ratio
RHS	right-hand side
SAA	systolic array architecture
SAB	single amplifier biquad
SB	stop-band
SC	switched capacitor
SCF	shifted companion feedback
SFG	signal flow graph
SI	switched current
SIDO	single input dual output
SIMO	single input multi output
SK	Sallen & Key
TAF	transadmittance function
TB	transition-band
TCA	transconductance amplifier
TF	transfer function
TIF	transimpedanace function
TSC	toggle-SC
TT	Tow-Thomas
UCNIC	unity gain CNIC
UGA	unity gain amplifier
UVNIC	unity gain VNIC
VA	voltage amplifier
VCCS	voltage controlled current source
VCVS	voltage controlled voltage source
VLSI	very large scale integrated circuit/system
VM	voltage mode
VNIC	voltage inversion-type NIC
VTF	voltage transfer function

1
Introduction

Electrical filters permeate modern electronic systems so much that it is imperative for an electronic circuit or system designer to have at least some basic understanding of these filters. The electronic systems that employ filtering process are varied, such as communications, radar, consumer electronics, military, medical instrumentation, and space exploration. An electrical filter is a network that transforms an electrical signal applied to its input such that the signal at the output has specified characteristics, which may be stated in the frequency or the time domain, depending upon the application. Thus, in some cases the filter exhibits a frequency-selective property, such as passing some frequency components in the input signal, while rejecting (stopping) signals at other frequencies.

The developments of filters started around 1915 with the advent of the electric wave filter by Campbell and Wagner, in connection with telephone communication. The early design advanced by Campbell, Zobel, and others made use of passive lumped elements, namely, resistors, inductors, and capacitors, and was based on *image parameters* (see for example, Ruston and Bordogna, 1971). This is known as the *classical filter theory* and it yields reasonably good filters without very sophisticated mathematical techniques.

Modern filter theory owes its origin to Cauer, Darlington, and others, and the development of the theory started in the 1930s. Major advancements in filter theory took place in the 1930s and 1940s. However, the filters were still passive structures using R, L, and C elements. One of the most important applications of passive filters has been in the design of channel bank filters in frequency division multiplex telephone systems.

Introduction of silicon integrated circuit (IC) technology together with the development of operational amplifiers (OAs) shifted the focus of filter designers in the 1960s to realize inductorless filters for low-frequency (voice band 300–3400 Hz) applications. Thus ensued the era of active-RC filters, with OA being the active element. With computer-controlled laser trimming, the values of the resistances in thick and thin film technologies could be controlled accurately and this led to widespread use of such low-frequency (up to about 4 kHz) active-RC filters in the pulse code modulation (PCM) system in telephonic communication.

Owing to the difficulty in fabricating large-valued resistors in the same process as the OA, low-frequency filters could not be built as monolithic devices. However, the observation that certain configurations of capacitors and

Modern Analog Filter Analysis and Design: A Practical Approach. Rabin Raut and M. N. S. Swamy
Copyright © 2010 WILEY-VCH Verlag GmbH & Co. KGaA, Weinheim
ISBN: 978-3-527-40766-8

periodically operated switches could function approximately as resistors led to the introduction of completely monolithic low-frequency filters. The advent of complementary metal-oxide semiconductor (CMOS) transistors facilitated this alternative with monolithic capacitors, CMOS OAs, and CMOS transistor switches. The switched-capacitor (SC) filters were soon recognized as being in the class of sampled-data filters, since the switching introduced sampling of the signals. In contrast, the active-RC filters are in the category of continuous-time filters, since the signal processed could theoretically take on any possible value at a given time. In the SC technique, signal voltages sampled and held on capacitors are processed via voltage amplifiers and integrators. Following the SC filters, researchers soon invented the complementary technique where current signals sampled and transferred on to parasitic capacitances at the terminals of metal-oxide semiconductor (MOS) transistors could be processed further via current mirrors and dynamic memory storage (to produce the effect of integration). This led to switched-current (SI) filtering techniques, which have become popular in all-digital CMOS technology, where no capacitors are needed for the filtering process.

In recent times, several microelectronic technologies (such as Bipolar, CMOS, and BiCMOS), filter architectures, and design techniques have emerged leading to high-quality fully integrated active filters. Moreover, sophisticated digital and analog functions (including filtering) can coexist on the same very large-scale integrated (VLSI) circuit chip. An example of the existence of several integrated active filters in a VLSI chip is illustrated in Figure 1.1. This depicts the floor plan of a typical PCM codec chip (Laker and Sansen, 1994).

Together with the progress in semiconductor technology, new types of semiconductor amplifiers, such as the operational transconductance amplifier (OTA),

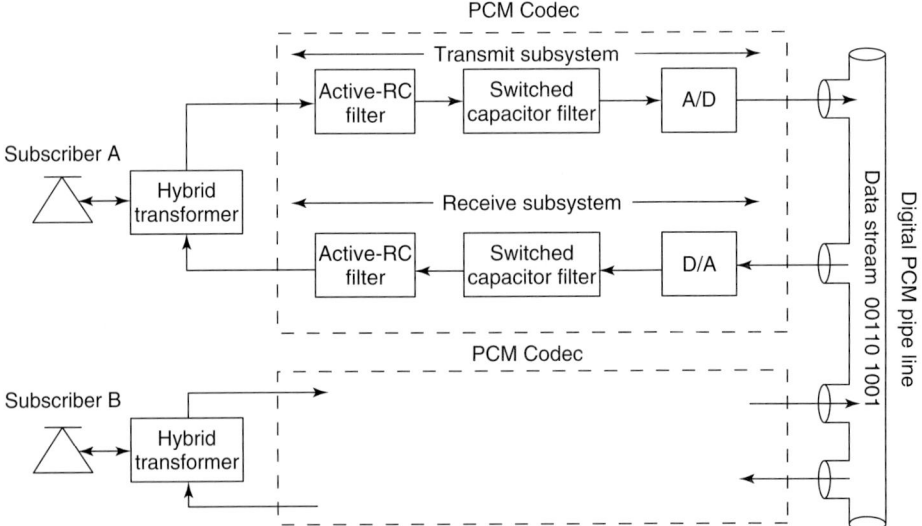

Figure 1.1 A typical VLSI analog/digital system floor plan.

and current conveyor (CC) became realizable in the late 1970s and onwards. This opened up the possibility for implementation of high-frequency filters (50 kHz to $\sim$300 MHz) in monolithic IC technology. An OTA can be conveniently configured to produce the function of a resistor and an inductor, so that usual high-frequency passive LCR filters can be easily replaced by suitable combinations of monolithic OTAs and capacitors leading to operational transconductance amplifier capacitor (OTA-C) (or g_m-C) filters. Introduction of CCs in the 1990s encouraged researchers to investigate signal processing in terms of signal currents rather than signal voltages. This initiated activities in the area of current-mode (CM) signal processing and hence CM filtering, even though the idea of realizing current transfer functions goes back to the late 1950s and the 1960s (Thomas, 1959; Hakim, 1965; Bobrow, 1965; Mitra, 1967, 1969; Daggett and Vlach, 1969). In fact, a very simple and direct method of obtaining a current transfer function realization from that of a voltage transfer function employing the concept of transposition was advanced as early as 1971 by Bhattacharyya and Swamy (1971). Since for CM signal processing, the impedances at the input and output ports are supposed to be very low, the attendant bandwidth can be very large. Modern CMOS devices can operate at very low voltages (around 1 V direct current (DC)) with small currents (0.1 mA or less). Thus, CM signal processing using CMOS technology entails low-voltage high-frequency operation. The intermediate frequency (IF) ($f_o \sim 100$ MHz) filter in a modern mobile communication (global system mobile, GSM) system has typical specifications as presented in Table 1.1. The required filters can be implemented as monolithic IC filters in the CM, using several CC building blocks and integrated capacitors.

Considering applications in ultra wideband ($\sim$10–30 GHz) communication systems, monolithic inductors ($\sim$1–10 nH) can be conveniently realized in modern submicron CMOS technology. Thus, passive LCR filter structures can be utilized for completely monolithic very wideband electronic filters. Advances in IC technology have also led to the introduction of several kinds of digital ICs. These could be used to process an analog signal after sampling and quantization. This has led to digital techniques for implementing an electronic filter (i.e., digital filters), and the area falls under the general category of digital signal processing (DSP).

As the subject of electrical/electronic filter is quite mature, there are a large number of books on this subject contributed by many eminent teachers and researchers. The current book is presented with a practical consideration, namely, that with the advent of computers and the abundance of computer-oriented courses in the electrical engineering curricula, there is insufficient time for a very exhaustive book on analog filters to be used for teaching over the span of one semester or

Table 1.1 Magnitude response characteristics of an IF filter.

Frequency	$f_o \pm 100$ kHz	$f_o \pm 800$ kHz	$f_o \pm 1.6$ MHz	$f_o \pm 3$ MHz	$f_o \pm 6$ MHz
Attenuation (dB)	≤ 0.5	≥ 5	≥ 10	≥ 15	≥ 30

two quarters. The present book is, therefore, relatively concise and is dedicated to current concepts and techniques that are basic and essential to acquire a good initial grasp of the subject of analog filters. Recognizing the popularity of courses that are amenable to the use of computer-aided tools, many circuit analysis (i.e., SPICE) and numerical simulation (i.e., MATLAB) program codes are provided in the body of the book to reinforce computer-aided design and analysis skills. The present book is very close to the practical need of a text book that can be covered over the limited span of time that present-day electrical engineering curricula in different academic institutions in the world can afford to the subject of analog filters. Toward this, the subject matter is presented through several chapters as follows.

Chapter 2 presents a review of several network analysis methods, such as the nodal, loop, and indefinite matrix techniques, as well as a method for analyzing constrained networks. One- and two-port networks are defined and various methods of representing a two-port and the interrelationships between the parameters representing a two-port are also detailed. The analysis methods are illustrated by considering several examples from known filter networks.

Chapter 3 introduces several concepts such as impedance and frequency scaling, impedance transformation, dual (and inverse) two-port networks, reversed two-ports, and transposed networks. Some useful network theorems concerning dual two-ports and transposed two-ports are established, and their applications to singly and doubly terminated networks are considered. Also, the transposes of commonly used active elements are given. Various approximation techniques for both the magnitude and phase of a filter transfer function, as well as frequency transformations to transform a low-pass filter to a high-pass, band-pass, or band-reject filter are also presented in this chapter. Several MATLAB simulation codes are presented.

Chapter 4 presents passive filter realization using singly terminated as well as doubly terminated LC ladder structures. Synthesis of all-pole transfer functions using such ladders is considered in detail.

Chapter 5 introduces the subject of designing second-order filters with active devices and RC elements. The active devices employed are the OAs and the OTAs. Both the single-amplifier and multiamplifier designs are presented. The sensitivity aspect is also discussed. The chapter concludes with a brief introduction to the devices and passive elements that are available in typical microelectronic manufacturing environments. The objective is to provide a modest orientation to the designers of active-RC filters toward IC filter implementation.

Chapter 6 deals with the subject of SC filters. The concept of the equivalence of R and the classical switched-C is refined by introducing the notion of sampled-data sequence and z-transformed equations. Parasitic-insensitive second-order filters are discussed. Filters based on unity-gain buffer amplifiers are also presented. Techniques to utilize the common continuous-time circuit elements (i.e., transmission lines) to simulate the operation of an SC network are introduced. The principles are illustrated using SPICE simulation.

High-order filter realization using active devices and RC elements is presented in Chapter 7. The knowledge base developed through Chapters 3–6 is now integrated to illustrate several well-known techniques for high-order active filter implementation. Inductance simulation, frequency-dependent negative resistance technique, operational simulation, cascade method, and multiloop feedback methods are discussed. Implementations of high-order continuous-time filters using OAs and OTAs, as well as SC high-order filters using OAs are illustrated.

Chapter 8 deals with the subject matter of CM filters. This technique of filtering has been of considerable interest to researchers in the past two decades. The basic difference between, voltage-mode (VM) and CM transfer functions is highlighted and several active devices that can process current signals introduced. Derivation of CM filter structures from a given VM filter structure using the principles of dual networks and network transposition, are illustrated. In particular, the usefulness of the transposition operation in obtaining, in a very simple manner, a CM realization for a given VM realization (or vice versa) is brought out through a number of examples. Implementations of CM transfer functions using OAs, OTAs, and CCs are presented. SI filtering technique is also introduced in this chapter.

Chapter 9 introduces the concepts and techniques relevant to implementation of IC continuous-time filters. The cases of linear resistance simulation using MOS transistors, and integrator implementation using differential architecture are illustrated. Second-order integrated filter implementations using OAs and OTAs are considered. The chapter ends with two design examples for IC implementation: (i) a low-voltage differential wideband OTA-C filter in CMOS technology and (ii) an approach toward an IF filter for a modern mobile communication (GSM) handset.

The book ends with three appendices that contain several tables for the approximation of filter functions, as well as for implementation of the filter functions using LCR elements. It is expected that once the filter transfer function is known, or the specific LCR values for a high-order filter are known, the designer can use the knowledge disseminated throughout the book to implement the required filter using either discrete RC elements and active devices, or using the devices available in a given IC technology. A MATLAB program for deriving the design curves for Bessel–Thomson (BT) filters up to order 15 is also included.

2
A Review of Network Analysis Techniques

In this chapter we deal with some basic background material related to linear network analysis and its applications with reference to realization of continuous-time analog filters. The guidelines for writing network equations using nodal and loop analysis methods are presented. The concept of network function is introduced, followed by the basic theory of two-port networks. Various nonreciprocal two-port elements such as the controlled sources, impedance converters, and impedance inverters are introduced. Analysis of general multinode networks using indefinite admittance matrices (IAMs), as well as the analysis of constrained networks, is presented in detail. Several applications of these network theoretic concepts are discussed. Finally, realization of a few second-order filters is illustrated using popularly known building blocks, such as the OAs, OTAs, and CCs.

2.1
Transformed Impedances

It is well known that the time domain integro-differential equations for lumped linear time-invariant networks (i.e., networks containing linear circuit elements like R, L, and C, linear-controlled sources such as voltage-controlled voltage source (VCVS), voltage-controlled current source (VCCS), current-controlled current source (CCCS), and current-controlled voltage source (CCVS)) can be arranged in a matrix form: $w(p)x(t) = f(t)$ (Chen, 1990) where $w(p)$ is an impedance/admittance matrix operator containing integro-differential elements, $x(t)$ is the unknown current/voltage variables (vectors), and $f(t)$ are the source voltage/current variables (vectors). The p-operator implies $p = d/dt$ and $1/p = \int dt$. If we take the Laplace transform on both sides, we would obtain a matrix equation $W(s)X(s) = F(s) + h(s)$, where $W(s), X(s)$, and $F(s)$ are the Laplace transforms of $w(t), x(t)$, and $f(t)$, and $h(s)$ contains the contributions due to the *initial values*. On inserting $s = j\omega$ one can get the *frequency domain* characterization of the system. The above sequence of operations is, however, rather lengthy and impractical. A more efficient technique is to characterize each network component (R, L, and C) in the *s*-domain including the contribution of the

Modern Analog Filter Analysis and Design: A Practical Approach. Rabin Raut and M. N. S. Swamy
Copyright © 2010 WILEY-VCH Verlag GmbH & Co. KGaA, Weinheim
ISBN: 978-3-527-40766-8

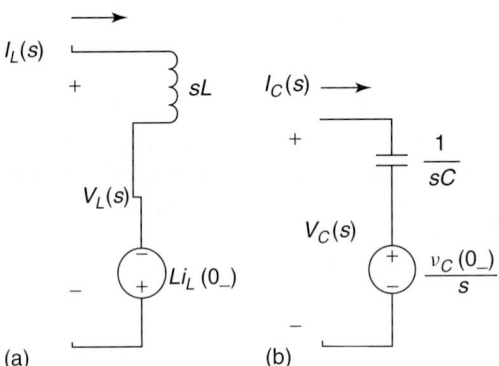

Figure 2.1 Transform representation of (a) an inductor and (b) a capacitor.

initial conditions and formulate the network equations. Impedance elements so expressed are *transformed impedances* and the network becomes a *transformed network*. Characterizations for transformed basic network elements are discussed below.

For ideal voltage or current sources, the transformed quantities are simply the Laplace transforms (e.g., $V_g(s)$, $I_g(s)$). For the i–v relation across a resistor, one can write either $V_R(s) = I_R(s)\, R$ or its dual $I_R(s) = V_R(s)/R$. Thus, there are two characterizations (viz., an I mode and a V mode) for each element. The particular choice depends upon which of these, $I(s)$ and $V(s)$, is the independent variable. In loop analysis, $I(s)$ is considered the independent variable. Similarly, in nodal analysis, the nodal voltage $V(s)$ is considered the independent variable.

Figures 2.1a and 2.1b give the transform representations of an inductor and a capacitor, respectively. These representations are derived from the i–v relationships for an inductor L and a capacitor C. For an inductor L, since $v_L(t) = L\frac{di_L}{dt}$, after taking Laplace transform, one obtains $V_L(s) = sLI_L(s) - Li_L(0_-)$. Similarly, for a capacitor C, since $v_C(t) = \frac{1}{C}\int_0^t i_C dt + v_C(0_-)$, on taking Laplace transform, one gets $V_C(s) = \frac{I_C(s)}{sC} + \frac{v_C(0_-)}{s}$. An alternate set of representations may be obtained by rewriting the above equations as $I_L(s) = \frac{1}{sL}V_L(s) + \frac{i_L(0_-)}{s}$ and $I_C(s) = sCV_C(s) - Cv_C(0_-)$. The corresponding representations are shown in Figures 2.2a and 2.2b, respectively. In loop analysis, the models given in Figures 2.1a and 2.1b should be used, while in nodal analysis, the representations given in Figures 2.2a and 2.2b should be used.

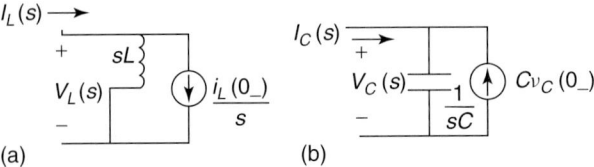

Figure 2.2 Alternate transform representations of (a) an inductor and (b) a capacitor.

2.2
Nodal Analysis

In nodal analysis, a voltage source in series with an element should be transformed into a current source with a shunt element by employing the source transformation. This will reduce the number of nodes and also make the analysis more homogeneous in that we have to deal with only node voltages and current sources (independent or dependent), which are the principal variables in nodal analysis. It may be recalled that the nodal system of equations is represented by the matrix equation $Y(S)V(S) = J(S)$, where $Y(S)$ is the admittance matrix, $V(S)$ is the node voltage vector, and $J(S)$ is the current source vector. The given network has to be converted to a network with transformed impedances with model representation for inductances and capacitances as shown in Figures 2.2a and 2.2b.

If a voltage source feeds several impedances in a parallel connection, E-shift technique (Chen, 1990) is to be used before embarking on the source transformation operation. In the following steps, a systematic procedure to set up the nodal matrix equation is given.

Step 1. **Identification of the sources:** Identify all dependent and independent sources. These are to be included initially as elements of the vector $J(s)$.

Step 2. **Set up the matrix elements:**
  a. y_{ii} is the sum of the admittances connected to the node i.
  b. y_{ij} is the negative of the admittances connected between the node pair (i, j).

  The above two sets of elements are to be included in the $Y(s)$ part of the matrix equation $Y(s)V(s) = J(s)$.

  c. The kth row element in $J(s)$ is the sum of the current sources connected to node k, taken to be positive if the source is directed toward and negative when directed away from the node. These are to be included in the $J(s)$ part of the matrix equation $Y(s)V(s) = J(s)$.

Step 3. In $J(s)$, decode the dependent current sources in terms of the node (voltage) variables, that is, elements of $V(s)$.

Step 4. Transpose the quantities obtained in Step 3 to the other side and allocate them to appropriate locations in the $Y(s)$ matrix.

2.3
Loop (Mesh) Analysis

In this technique, one has to begin with the equivalent circuit by using series model versions of the transform impedances for the inductors and capacitors. Further, all current sources are to be converted to equivalent voltage sources with series impedances using the source transformation (Thevenin's theorem). If a current source exists with no impedance in parallel, I-shift technique (Chen, 1990) is to be used before applying the source transformation. It may be recalled that the loop

2 A Review of Network Analysis Techniques

system matrix equation has the form $Z(s)I(s) = E(s)$, where $Z(s)$ is the impedance matrix, $I(s)$ is the loop current vector, and $E(s)$ is the loop voltage vector.

In the following steps, a systematic procedure to set up the loop matrix equation is given.

Step 1. **Identification of sources:** Identify all the voltage sources (independent and dependent) to be initially included as elements of the $E(s)$ vector.

Step 2. **Identify the loop impedance matrix operator elements and loop voltage source vector:**

  a. *Self-loop impedance* z_{ii} is the sum of all impedances in the loop i.
  b. *Mutual-loop impedance* z_{ij} is the impedance shared by loop i and loop j. If the currents in loops i and j are in the same direction, z_{ij} is taken with a positive sign. On the other hand, if the currents in loops i and j are in opposite directions, it is taken with a negative sign.
  c. The element e_i in the *loop source vector* $E(s)$ is the algebraic sum of all the voltage sources in loop i. The components are taken with a positive sign if a potential rise occurs in the direction of the loop current. If a potential drop takes place in the direction of the loop current, the voltage element is taken with a negative sign. We thus have the preliminary form $Z(s)I(s) = E(s)$.

Step 3. In $E(s)$ found above, express the dependent sources in terms of the loop current variables (i.e., elements of $I(s)$).

Step 4. Transpose the dependent components of $E(s)$ to the other side and allocate the associated coefficients to proper location of the $Z(s)$ matrix.

Example 2.1. Figure 2.3a shows the AC equivalent circuit of a typical semiconductor device such as a bipolar junction transistor (BJT). Develop the loop matrix formulation for the network.

Figure 2.3b shows the equivalent circuit redrawn after taking Laplace transformation and applying source transformation to the VCCS of value $g_m V(s)$. The currents I_1 and I_2 are the loop currents. Applying the suggested steps, one gets

$$\begin{bmatrix} R_1 + R_2 & 0 \\ 0 & R_3 \end{bmatrix} \begin{bmatrix} I_1(s) \\ I_2(s) \end{bmatrix} = \begin{bmatrix} V_s(s) \\ -g_m V(s) R_3 \end{bmatrix} \qquad (2.1)$$

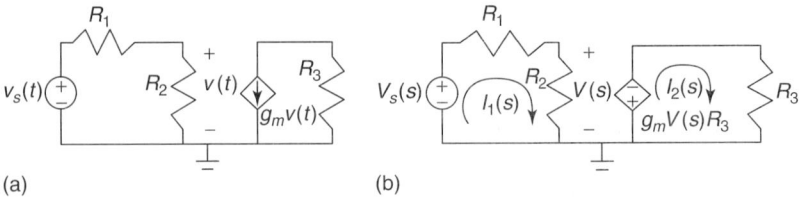

Figure 2.3 (a) Circuit with a voltage source for loop analysis. (b) Circuit reconfigured for loop analysis using matrix formulation.

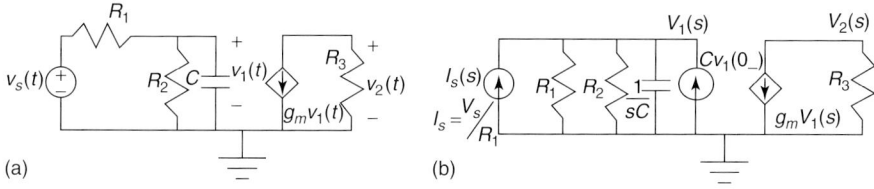

Figure 2.4 (a) Circuit with a voltage source for nodal analysis. (b) Transformed network for using nodal analysis using matrix formulation.

But, $V(s) = I_1(s) R_2$. On substituting and bringing it to the left side, one gets

$$\begin{bmatrix} R_1 + R_2 & 0 \\ g_m R_2 R_3 & R_3 \end{bmatrix} \begin{bmatrix} I_1(s) \\ I_2(s) \end{bmatrix} = \begin{bmatrix} V_s(s) \\ 0 \end{bmatrix} \quad (2.2)$$

Equation 2.2 is the loop matrix formulation for Figure 2.3a.

Example 2.2. Figure 2.4a shows a circuit with a voltage source. Develop the nodal matrix formulation for the network.

Figure 2.4b shows the equivalent circuit redrawn after applying source transformation to $v_S(t)$ and transforming the network using the representation for a capacitor as shown in Figure 2.2b. Since admittances are to be used, letting $G = 1/R$ for the conductance, one can write

$$\begin{bmatrix} G_1 + G_2 + sC & 0 \\ 0 & G_3 \end{bmatrix} \begin{bmatrix} V_1(s) \\ V_2(s) \end{bmatrix} = \begin{bmatrix} I_s(s) + Cv_1(0_-) \\ -g_m V_1(s) \end{bmatrix} \quad (2.3)$$

where $I_s(s) = G_1 V_S(s)$. On substituting and bringing $V_1(s)$ to the left side, the final formulation becomes

$$\begin{bmatrix} G_1 + G_2 + sC & 0 \\ g_m & G_3 \end{bmatrix} \begin{bmatrix} V_1(s) \\ V_2(s) \end{bmatrix} = \begin{bmatrix} V_s(s) G_1 \\ 0 \end{bmatrix} + \begin{bmatrix} Cv_1(0_-) \\ 0 \end{bmatrix} \quad (2.4)$$

It is observed that Eq. (2.4) is in the form $W(s)X(s) = F(s) + h(s)$.

2.4 Network Functions

If we study the relationships developed in connection with nodal and loop analyses, we discover a general format, namely, $W(s)X(s) = F(s) + h(s)$, where $W(s)$ can be either an admittance or an impedance matrix, $X(s)$ a nodal voltage vector or a loop current vector, $F(s)$ the vector of independent sources, and $h(s)$ the vector of initial conditions. Using this equation, one can easily arrive at $X(s) = W^{-1}(s)F(s) + W^{-1}(s)h(s)$. The *first* part of the solution on the right-hand side (RHS) is the complete solution if the initial values were zero (i.e., $h(s) = 0$). This is called the *zero (initial)-state response*. The *second* part of the solution on the RHS is the complete solution if the forcing functions were zero (i.e., $F(s) = 0$). This is known as the *zero-input* or *natural response*.

A network function is defined with regard to the zero-state response in a network when there is only one independent voltage/current forcing function (driving function) in the network. It is the ratio of the Laplace transform of the zero-state response in a network to the Laplace transform of the input. The input could be a voltage (or current) across (or through), and similarly the output response could be a voltage (or current) across (or through) a pair of nodes.

Consider a linear time-invariant network under zero–initial state conditions. Then, the network function (also called *system function*) is defined as

$$\frac{\text{Laplace transform of the output response}}{\text{Laplace transform of the input}}$$

Depending upon the location of the pair of nodes, we have different network functions. If the two pairs of nodes corresponding to the input and the response are physically coincident, we talk of *driving point impedance* (DPI) or *driving point admittance* (DPA). If the pairs of nodes are distinct (one of the nodes may be common between the pairs), then we can define (i) a *voltage transfer function* (VTF), (ii) a *current transfer function* (CTF), (iii) a *transfer impedance (transimpedance) function* (TIF), and (iv) a *transfer admittance (transadmittance) function* (TAF).

2.5
One-Port and Two-Port Networks

2.5.1
One-Port Networks

A pair of terminals such that the current entering one of the terminals is the same as the current leaving the other is called a *one-port*. Figure 2.5 shows a one-port or a two-terminal network. This one-port can be characterized by two network functions depending on whether the input is a current or a voltage, in which case the response is a voltage or a current, respectively. The ratio of the response to the input are respectively called the *driving point impedance* and the *driving point admittance*. Symbolically, these are given by

$$Z_{in}(s) = \frac{V(s)}{I(s)} \text{ and } Y_{in}(s) = \frac{I(s)}{V(s)} \quad (2.5)$$

Of course, $Z_{in}(s) = 1/Y_{in}(s)$.

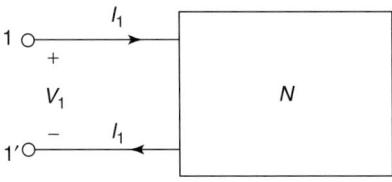

Figure 2.5 A one-port network.

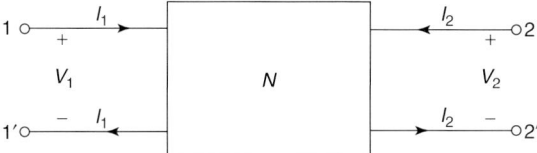

Figure 2.6 A two-port network.

2.5.2
Two-Port Networks

A two-port network has two accessible terminal pairs; that is, there are two pairs of terminals that can be used as input and output ports. The study of two-port network is very important, as they form the building blocks of most of the electrical or electronic systems. Figure 2.6 shows a general two-port and the standard convention adopted in designating the terminal voltages and currents.

There are a number of ways of characterizing a two-port. Of the four variables V_1, I_1, V_2, and I_2, only two can be considered independent variables and the other two as dependent variables. Hence, in all, we have six different ways of choosing the independent variables (there is a seventh set of choice, leading to *scattering parameters* characterization, which is defined analogous to those in transmission line theory, and is not considered here). Here, we define three of these sets of choices, leading to three sets of matrix parameter descriptions. These will be of great utility in the context of this book.

2.5.2.1 Admittance Matrix Parameters

If I_1 and I_2 are chosen as the independent variables, then we characterize the network N by the equations

$$\begin{bmatrix} y_{11} & y_{12} \\ y_{21} & y_{22} \end{bmatrix} \begin{bmatrix} V_1 \\ V_2 \end{bmatrix} = \begin{bmatrix} I_1 \\ I_2 \end{bmatrix}$$

or

$$[y] \begin{bmatrix} V_1 \\ V_2 \end{bmatrix} = \begin{bmatrix} I_1 \\ I_2 \end{bmatrix} \tag{2.6}$$

The four admittance parameters may be determined by using the relations $y_{11} = [I_1/V_1]|_{V_2=0}$, $y_{12} = [I_1/V_2]|_{V_1=0}$, $y_{21} = [I_2/V_1]|_{V_2=0}$, and $y_{22} = [I_2/V_2]|_{V_1=0}$. Since $V_1 = 0$, $V_2 = 0$ implies AC short circuit conditions, the parameters y_{11}, y_{12}, y_{21}, and y_{22} are often referred to as *short circuit admittance parameters*, and the matrix $[y]$ is called the *short circuit admittance* (or simply the admittance) matrix of the two-port. The AC equivalent circuit model (i.e., equivalent to the network inside the black box) associated with the above set of equations is shown in Figure 2.7.

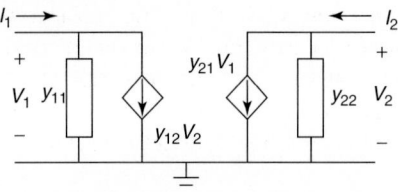

Figure 2.7 Equivalent circuit model for two-port admittance matrix.

2.5.2.2 Impedance Matrix Parameters

If V_1 and V_2 are chosen as the independent variables, then the two-port may be characterized by

$$\begin{bmatrix} z_{11} & z_{12} \\ z_{21} & z_{22} \end{bmatrix} \begin{bmatrix} I_1 \\ I_2 \end{bmatrix} = \begin{bmatrix} V_1 \\ V_2 \end{bmatrix}$$

or

$$[z] \begin{bmatrix} I_1 \\ I_2 \end{bmatrix} = \begin{bmatrix} V_1 \\ V_2 \end{bmatrix} \qquad (2.7)$$

where the parameters z_{11}, z_{12}, z_{21}, and z_{22} may be determined by the equations $z_{11} = [V_1/I_1]|_{I_2=0}$, $z_{12} = [V_1/I_2]|_{I_1=0}$, $z_{21} = [V_2/I_1]|_{I_2=0}$, and $z_{22} = [V_2/I_2]|_{I_1=0}$. Since $I_1 = 0$, $I_2 = 0$ implies open (to AC) circuit conditions, the parameters z_{11}, z_{12}, z_{21}, and z_{22} are called *open circuit impedance parameters*, and $[z]$ the *open circuit impedance matrix* (or simply, impedance matrix) of the two-port. It is obvious from Eqs. (2.6) and (2.7) that

$$[y] = [z]^{-1} \qquad (2.8)$$

The AC equivalent circuit model is shown in Figure 2.8.

2.5.2.3 Chain Parameters (Transmission Parameters)

If we consider V_2 and I_2 as the independent variables, then we may write

$$\begin{bmatrix} V_1 \\ I_1 \end{bmatrix} = \begin{bmatrix} A & B \\ C & D \end{bmatrix} \begin{bmatrix} V_2 \\ -I_2 \end{bmatrix} \text{ or } \begin{bmatrix} V_1 \\ I_1 \end{bmatrix} = [a] \begin{bmatrix} V_2 \\ -I_2 \end{bmatrix} \qquad (2.9)$$

The matrix $[a]$ is called the *chain matrix* or *transmission matrix* of the two-port. The parameters A, B, C, and D are called the *chain or transmission parameters*. Note that $-I_2$ instead of I_2 is used to imply a current flowing outward at port 2. It is seen that the parameters A, B, C, and D may be defined as

$$\frac{1}{A} = \frac{V_2}{V_1}\bigg|_{I_2=0}, \quad \frac{1}{B} = \frac{-I_2}{V_2}\bigg|_{V_2=0}, \quad \frac{1}{C} = \frac{V_2}{I_1}\bigg|_{I_2=0}, \quad \frac{1}{D} = \frac{-I_2}{I_1}\bigg|_{V_2=0} \qquad (2.10)$$

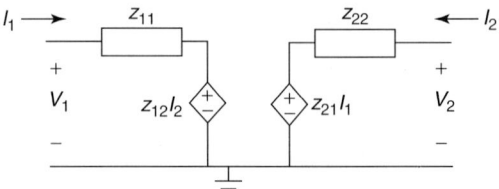

Figure 2.8 Equivalent circuit model for two-port impedance matrix.

A and C are measured under open circuit conditions, while B and D are measured under short circuit conditions. Further, A and D are ratios of similar physical quantities, while B has the unit of an impedance and C has the unit of an admittance. In fact, $1/A$ is the forward open circuit voltage gain, $1/D$ is the forward short circuit current gain, $1/B$ represents the forward transconductance, and $1/C$ is the forward transimpedance function.

Similarly, we may define hybrid matrices $[h]$ and $[g]$, as well as reverse chain matrix $[\tilde{a}]$ by the following relations:

$$\begin{bmatrix} V_1 \\ I_2 \end{bmatrix} = \begin{bmatrix} h_{11} & h_{12} \\ h_{21} & h_{22} \end{bmatrix} \begin{bmatrix} I_1 \\ V_2 \end{bmatrix} = [h] \begin{bmatrix} I_1 \\ V_2 \end{bmatrix} \tag{2.11}$$

$$\begin{bmatrix} I_1 \\ V_2 \end{bmatrix} = \begin{bmatrix} g_{11} & g_{12} \\ g_{21} & g_{22} \end{bmatrix} \begin{bmatrix} V_1 \\ I_2 \end{bmatrix} = [g] \begin{bmatrix} V_1 \\ I_2 \end{bmatrix} \tag{2.12}$$

and

$$\begin{bmatrix} V_2 \\ I_2 \end{bmatrix} = \begin{bmatrix} A & B \\ C & D \end{bmatrix} \begin{bmatrix} V_1 \\ -I_1 \end{bmatrix} = [\alpha] \begin{bmatrix} V_1 \\ -I_1 \end{bmatrix} \tag{2.13}$$

It can be easily established from Eqs. (2.9) and (2.13) that

$$[\alpha] = \frac{1}{AD-BC} \begin{bmatrix} D & B \\ C & A \end{bmatrix} \begin{bmatrix} V_1 \\ -I_1 \end{bmatrix} \tag{2.14}$$

2.5.2.4 Interrelationships

Since the $[y]$, $[z]$, $[a]$, $[g]$, $[h]$, and $[\alpha]$ matrices are all different characteristics of the same network, it is obvious that these parameters are related. We give in Table 2.1 these relationships for $[z]$, $[y]$, and $[a]$ matrices, since these are more commonly used in practice. The network N is said to be reciprocal if $y_{12} = y_{21}$ (hence, $z_{12} = z_{21}$ and determinant $[a] = \Delta_a = AD - BC = 1$); otherwise it is said to be nonreciprocal.

Table 2.1 Interrelationships among $[z]$, $[y]$, and $[a]$ parameters of a two-port network.

Matrix	[y]		[z]		[a]	
$[y]$	y_{11}	y_{12}	$\frac{y_{22}}{\Delta_y}$	$\frac{-y_{12}}{\Delta_y}$	$\frac{-y_{22}}{y_{21}}$	$\frac{-1}{y_{21}}$
	y_{21}	y_{22}	$\frac{-y_{21}}{\Delta_y}$	$\frac{y_{11}}{\Delta_y}$	$\frac{-\Delta_y}{y_{21}}$	$\frac{-y_{11}}{y_{21}}$
$[z]$	$\frac{z_{22}}{\Delta_z}$	$\frac{-z_{12}}{\Delta_z}$	z_{11}	z_{12}	$\frac{z_{11}}{z_{21}}$	$\frac{\Delta_z}{z_{21}}$
	$\frac{-z_{21}}{\Delta_z}$	$\frac{z_{11}}{\Delta_z}$	z_{21}	z_{22}	$\frac{1}{z_{21}}$	$\frac{z_{22}}{z_{21}}$
$[a]$	$\frac{D}{B}$	$\frac{-\Delta_a}{B}$	$\frac{A}{C}$	$\frac{\Delta_a}{C}$	A	B
	$\frac{-1}{B}$	$\frac{A}{B}$	$\frac{1}{C}$	$\frac{D}{C}$	C	D

$\Delta_z = z_{11}z_{22} - z_{12}z_{21}$, $\Delta_y = y_{11}y_{22} - y_{21}y_{12}$, $\Delta_a = AD - BC$.

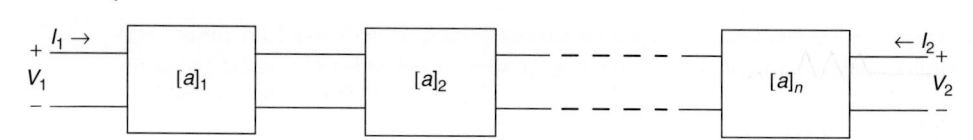

Figure 2.9 A cascade of two-port networks.

The different two-port characterizations are useful to obtain the overall description of a complicated two-port which may be made up of several two-ports. For example, when two two-ports are connected in series, [z] matrices are useful, since the overall [z] matrix becomes equal to the sum of the constituent [z] matrices. Similarly, [y] matrices are useful when two-ports are connected in parallel, and chain matrices are employed when two two-ports are connected in cascade. Readers interested in more details may refer to Moschytz (1974). Here, we consider an example of a cascade connection since it is quite common in practice.

Consider a number of two-ports connected in cascade, as shown in Figure 2.9. By identifying the terminal voltage and current variables in adjacent blocks, it becomes immediately clear that the overall chain matrix of the cascade is given by $[a] = [a]_1[a]_2 \ldots [a]_n$.

2.5.2.5 Three-Terminal Two-Port Network

One of the most important types of two-ports is the three-terminal two-port, wherein the input and output ports have a common terminal as shown in Figure 2.10. A practical example is an OA used as an inverting amplifier.

2.5.2.6 Equivalent Networks

Two one-port networks are said to be equivalent if their DPIs are the same. Similarly, two two-ports are said to be equivalent if their [y] (or [z], or [a]) matrices are the same. The concept of equivalent networks can be extended to networks with more than two-ports. An example follows.

The two one-ports shown in Figure 2.11a,b are equivalent, since the DPI of either of the networks is equal to unity.

2.5.2.7 Some Commonly Used Nonreciprocal Two-Ports

A nonreciprocal two-port can be modeled using the four basic controlled sources, namely, the VCVS, CCCS, VCCS, and CCVS. The corresponding equivalent circuit models for ideal controlled sources are shown in Figures 2.12a–2.12d.

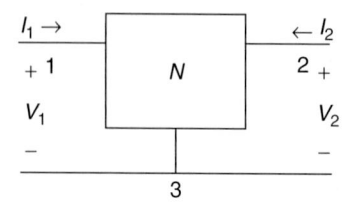

Figure 2.10 A three-terminal two-port network.

2.5 One-Port and Two-Port Networks

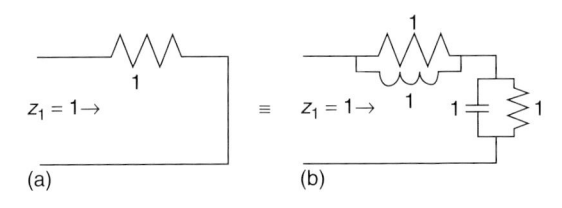

Figure 2.11 (a) A one-port with DPI of 1 Ω. (b) An equivalent one-port with same DPI as in (a).

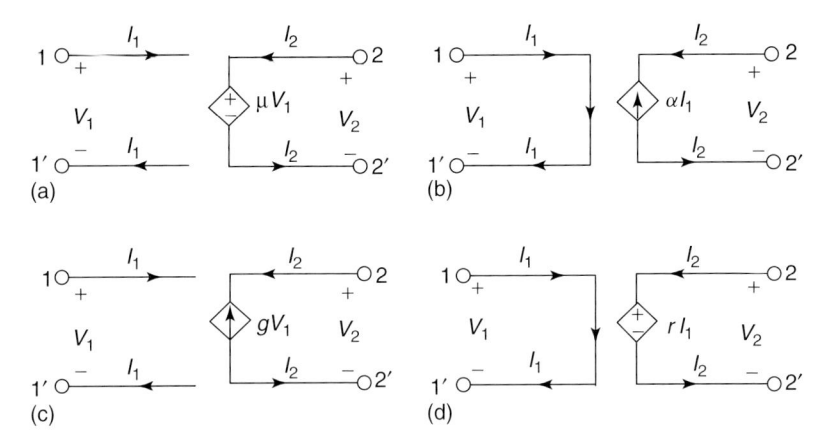

Figure 2.12 Controlled sources: (a) VCVS ($I_1 = 0, V_2 = \mu V_1$), (b) CCCS ($V_1 = 0, I_2 = -\alpha I_1$), (c) VCCS ($I_1 = 0, I_2 = -gV_1$), and (d) CCVS ($V_1 = 0, V_2 = rI_1$).

The above models are all ideal, since all the input, output, and feedback impedances are assumed to be either zero or infinite. The two-port parameters of the ideal controlled sources are listed in Table 2.2. It is observed that except for the VCCS, the admittance matrix does not exist for the other ideal sources, even though the chain matrices do exist for all of them. However, it should be mentioned that the [y] matrices do exist for all these sources if they are nonideal (i.e., when the input and output impedances are added), which anyway is the case in practice.

Some of the common active two-ports that are used in many complex electronic circuits are the (i) impedance converters (both positive and negative types) and (ii) impedance inverters (both positive and negative types). The two-port parameters of these devices are also listed in Table 2.2. The impedance converter is said to be a *positive impedance converter* (PIC) if $K_1 K_2 > 0$, and a *negative impedance converter* (NIC) if $K_1 K_2 < 0$. Further, the NIC is called a *current-inverting negative impedance converter* (CNIC) if $K_1 > 0$ and $K_2 < 0$, while it is called a *voltage-inverting negative impedance converter* (VNIC) if $K_1 < 0$ and $K_2 > 0$. In addition, if $K_1 = 1$ and $K_2 = -1$, the NIC is called a *unity-gain current-inverting negative impedance converter* (UCNIC), and when $K_1 = -1$ and $K_2 = 1$, the NIC is called a *unity-gain voltage-inverting negative impedance converter* (UVNIC). Similarly, if $G_2/G_1 > 0$, the

Table 2.2 [z], [y], and [a] parameters for some nonreciprocal two-ports.

Nonreciprocal two-port	[z]	[y]	[a]
VCVS	–	–	$\begin{bmatrix} 1/\mu & 0 \\ 0 & 0 \end{bmatrix}$
VCCS	–	$\begin{bmatrix} 0 & 0 \\ g & 0 \end{bmatrix}$	$\begin{bmatrix} 0 & 1/g \\ 0 & 0 \end{bmatrix}$
CCVS	$\begin{bmatrix} 0 & 0 \\ r & 0 \end{bmatrix}$	–	$\begin{bmatrix} 0 & 0 \\ 1/r & 0 \end{bmatrix}$
CCCS	–	–	$\begin{bmatrix} 0 & 0 \\ 0 & 1/\alpha \end{bmatrix}$
Impedance converter	–	–	$\begin{bmatrix} K_1 & 0 \\ 0 & K_2 \end{bmatrix}$
Impedance inverter (II)	$\begin{bmatrix} 0 & -1/G_1 \\ 1/G_2 & 0 \end{bmatrix}$	$\begin{bmatrix} 0 & G_2 \\ -G_1 & 0 \end{bmatrix}$	$\begin{bmatrix} 0 & 1/G_1 \\ G_2 & 0 \end{bmatrix}$

impedance inverter is called a *positive impedance inverter* (PII) and if $G_2/G_1 < 0$, it is called a *negative impedance inverter* (NII).

2.6 Indefinite Admittance Matrix

The IAM presents a powerful method of analyzing electrical networks. The advantage of such a method is that it does not require the setting up of network equations to obtain a network function.

Consider the *n*-terminal *N* of Figure 2.13 where *N* is a linear time-invariant network with zero initial conditions.

Let $I_k(k = 1, 2, \ldots, n)$ be the current injected into node k, whose voltage with reference to some external node, assumed to be the ground node, is V_k. Then, such a network has the property (by Kirchoff's current law, KCL) that $\sum_{k=1}^{n} I_k = 0$. The

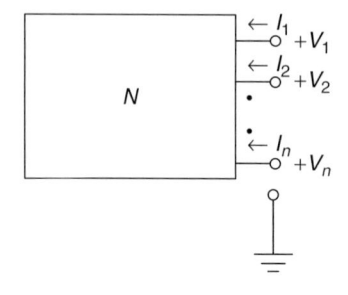

Figure 2.13 An *n*-terminal network.

network N can be characterized by the nodal matrix equation

$$\begin{bmatrix} I_1 \\ . \\ . \\ . \\ I_n \end{bmatrix} = \begin{bmatrix} y_{11} & . & . & y_{1n} \\ . & & & \\ . & & & \\ . & & & \\ y_{n1} & . & . & y_{nn} \end{bmatrix} \begin{bmatrix} V_1 \\ . \\ . \\ . \\ V_n \end{bmatrix} \quad \text{or } [I] = [Y]_N [V] \quad (2.15)$$

The matrix $[Y]_N$ is called the *indefinite admittance matrices of N*, since no definite node in the whole network is considered as the reference node. The matrix elements y_{ij} can be calculated using the relation $y_{ij} = I_i|_{V_j=1, V_m=0, m\neq j}$. Hence, the coefficient y_{ij} is the current that flows into the node i when all the nodes except for jth node are grounded and a unit voltage (AC signal) is applied at node j.

The IAM has the following properties:

1) The IAM $[Y]_N$ is singular, and the sum of the elements of any row or any column is zero.
2) If any terminal is grounded, then the admittance matrix of order $(n-1)$ of the resulting network is obtained by deleting the corresponding row and column in $[Y]_N$. This no longer a singular matrix. This now becomes a definite admittance matrix or the conventional nodal admittance matrix (NAM).
3) When two terminals (i.e., nodes) are shorted, the elements of the corresponding rows are added together, and so also the elements of the corresponding columns to form a new row and a new column. The dimension of the matrix is reduced by unity and the resulting matrix becomes the IAM of the new network.
4) The IAM of order n can be obtained from the NAM of order $(n-1)$ by adding a row and a column such that in the new matrix, the sum of the elements of every row and of every column is zero.
5) A row and a column of zeros correspond to an isolated node. Hence, any n-node network may be treated as an m-node $(m > n)$ network by adding $(m-n)$ rows and columns of zeros to $[Y]_N$.
6) If the mth terminal of an n-node network, with an IAM $[Y]_N$, is suppressed (i.e., $I_m = 0$), then the new network N^* has an IAM $[\hat{Y}]_{N^*}$ which is of the order $(n-1)$ and whose elements are given by

$$\hat{y}_{kl} = \frac{\begin{vmatrix} y_{kl} & y_{km} \\ y_{ml} & y_{mm} \end{vmatrix}}{y_{mm}} \quad (2.16)$$

7) It may be noted that if $[Y]_N$ is an NAM, then $[\hat{Y}]_{N^*}$ is also an NAM.
8) The IAM of a number of n-node networks connected in parallel is obtained by adding the IAMs of the individual n-node networks.
9) For an n-node network consisting of only one-port elements, the IAM is obtained in a simple manner. The element y_{ii} is obtained as the sum of the admittances connected to node i, while the element y_{ij} is obtained as the negative of the sum of all the admittances connected exclusively between nodes i and j.

For details one may refer to Moschytz (1974).

2.6.1
Network Functions of a Multiterminal Network

Consider a multiterminal network, where we assume that the current I_{kl} flows into node k and gets out of node l and that all the other currents are zero. Hence,

$$I_{kl} = I_k - I_l \tag{2.17}$$

Also, let V_{ij} be the voltage between nodes i and j, that is,

$$V_{ij} = V_i - V_j \tag{2.18}$$

Then, it can be shown that the transfer impedance z_{kl}^{ij} is given by (Moschytz, 1974)

$$z_{kl}^{ij} = \frac{V_{ij}}{I_{kl}} = \text{sgn}(k-l) \times \text{sgn}(i-j) \frac{Y_{ij}^{kl}}{Y_l^l} \tag{2.19}$$

where

$Y_{ij}^{kl} = (-1)^{i+j+k+l} M_{ij}^{kl}$,

$Y_l^l = M_l^l$,

M_{ij}^{kl} = minor obtained from the IAM $[Y]_N$ by deleting the kth and lth rows, and ith and lth columns,

M_l^l = minor obtained from the IAM by deleting the lth row and lth column, and

$\text{Sgn}(x) = 1$, if $x > 0$, and -1, if $x < 0$.

Also the DPI between the nodes k and l is given by

$$z_{kl} = \frac{V_{kl}}{I_{kl}} = \frac{Y_{kl}^{kl}}{Y_l^l} \tag{2.20}$$

Finally, the VTF between the nodes (i, j) and (k, l) is given by

$$T_{kl}^{ij} = \frac{V_{ij}}{V_{kl}} = \text{sgn}(k-l) \times \text{sgn}(i-j) \frac{Y_{ij}^{kl}}{Y_{kl}^{kl}} \tag{2.21}$$

Example 2.3. Obtain the admittance matrix of the bridged-T circuit shown in Figure 2.14, using the method of IAM.

Since the given circuit consists of only one-ports, it is easy to write the IAM directly using property number 8 of an IAM.

$$[Y]_N = \begin{array}{c} \\ \\ \end{array} \begin{array}{cccc} 1 & 2 & 3 & 4 \end{array} \\ \left[\begin{array}{cccc} Y_1 + Y_4 & -Y_4 & -Y_1 & 0 \\ -Y_4 & Y_2 + Y_4 & -Y_2 & 0 \\ -Y_1 & -Y_2 & Y_1 + Y_2 + Y_3 & -Y_3 \\ 0 & 0 & -Y_3 & Y_3 \end{array} \right]$$

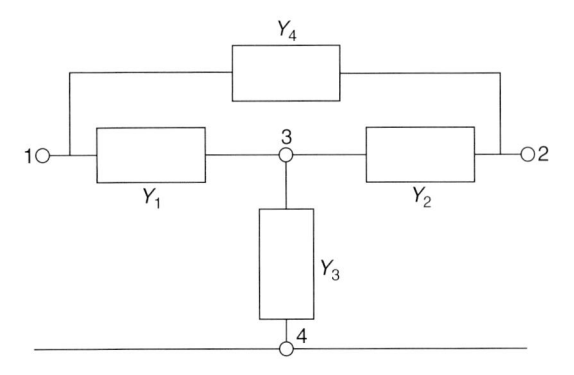

Figure 2.14 A bridged-T network.

Since node 4 is taken as reference node, we may cancel the fourth row and fourth column. Then the corresponding definite admittance matrix of the circuit with node 4 as ground is

$$[Y]_{N*} = \begin{matrix} & 1 & 2 & 3 \\ & \begin{bmatrix} Y_1 + Y_4 & -Y_4 & -Y_1 \\ -Y_4 & Y_2 + Y_4 & -Y_2 \\ -Y_1 & -Y_2 & Y_1 + Y_2 + Y_3 \end{bmatrix} \end{matrix}$$

Since we need the short circuit admittance matrix of the bridged-T two-port with terminals (1, 4) as input and terminals (2, 4) as output, node 3 is to be suppressed. Hence, we use Eq. (2.16) to obtain the two-port $[y]$ matrix description

$$[y] = \begin{bmatrix} \begin{vmatrix} Y_1 + Y_4 & -Y_1 \\ -Y_1 & Y_S \end{vmatrix} & \begin{vmatrix} -Y_4 & -Y_1 \\ -Y_2 & Y_S \end{vmatrix} \\ \begin{vmatrix} -Y_4 & -Y_2 \\ -Y_1 & Y_S \end{vmatrix} & \begin{vmatrix} Y_2 + Y_4 & -Y_2 \\ -Y_2 & Y_S \end{vmatrix} \end{bmatrix} \frac{1}{Y_S}$$

$$= \begin{bmatrix} Y_1(Y_2 + Y_3) + Y_S Y_4 & -(Y_1 Y_2 + Y_4 Y_S) \\ -(Y_1 Y_2 + Y_4 Y_S) & Y_2(Y_1 + Y_3) + Y_4 Y_S \end{bmatrix} \frac{1}{Y_S}$$

where

$$Y_S = Y_1 + Y_2 + Y_3$$

Example 2.4. In this example, we consider a circuit which consists not only of one-ports but also a two-port, which does not possess a $[y]$ matrix. The technique of IAM can still be applied to such a case. Consider the network in Figure 2.15a which makes use of a CNIC to realize the transfer function V_2/V_1.

This structure was proposed by Yanagisawa (1957) to realize VTFs. It is seen that the CNIC has no $[y]$ matrix, but has only the chain matrix $[a]$, and hence an

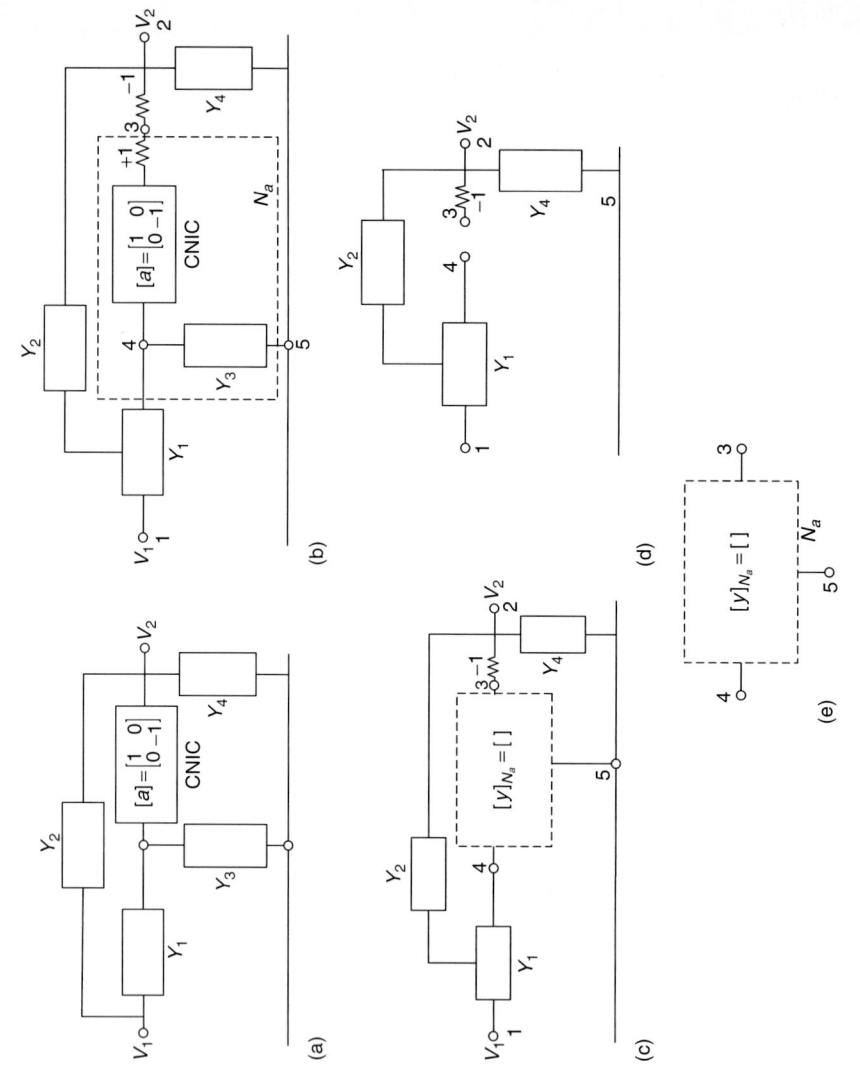

Figure 2.15 Derivation of the admittance matrix of a system consisting of one-port and two-port network elements: (a) A network containing a CNIC which does not have a [y] matrix description. (b) Addition of positive and negative resistances in series with one of the terminals of the CNIC. (c) A subsystem N_a containing Y_3, the CNIC, and the $-1\,\Omega$ resistor. (d) The subsystem excluding N_a. (e) The subsystem N_a as a three-terminal network.

IAM cannot be found directly for this element. For this purpose, we introduce two resistors of values +1 and −1 Ω in series, as shown in Figure 2.15b. Now we can consider the CNIC and the +1 Ω resistor in cascade, and find the overall chain matrix. Using this chain matrix, we can determine the associated [y] matrix of this subnetwork N_a as shown with dotted line boundary in Figure 2.15b. This matrix shown as $[y]_{Na}$ in Figure 2.15c can be derived as

$$[a]_{Na} = [a]_{\text{CNIC}}\,[a]_{1\,\Omega} = \begin{bmatrix} 1 & 0 \\ 0 & -1 \end{bmatrix}\begin{bmatrix} 1 & 1 \\ 0 & 1 \end{bmatrix} = \begin{bmatrix} 1 & 1 \\ 0 & -1 \end{bmatrix}$$

Hence,

$$[y]_{Na} = \begin{array}{c} \begin{matrix} 4 & \;\;3 \end{matrix} \\ \begin{bmatrix} -1 & 1 \\ -1 & 1 \end{bmatrix} \end{array} \begin{matrix} 4 \\ 3 \end{matrix}$$

The network of Figure 2.15c can be decomposed into two subnetworks as shown in Figure 2.15d,e, where the former corresponds to the one-ports of Figure 2.15c, and the latter to the two-port N_a.

We may write the IAM of Figure 2.15d as

$$[Y]_R = \begin{array}{c} \begin{matrix} 1 & \quad 2 & \quad 3 & \quad 4 & \quad 5 \end{matrix} \\ \begin{bmatrix} Y_1 + Y_2 & -Y_2 & 0 & -Y_1 & 0 \\ -Y_2 & Y_2 + Y_4 - 1 & 1 & 0 & -Y_4 \\ 0 & 1 & -1 & 0 & 0 \\ -Y_1 & 0 & 0 & Y_1 + Y_3 & -Y_3 \\ 0 & -Y_4 & 0 & -Y_3 & Y_3 + Y_4 \end{bmatrix} \end{array} \tag{2.22}$$

Also,

$$[Y]_{Na} = \begin{array}{c} \begin{matrix} 3 & \;\;4 & \;\;5 \end{matrix} \\ \begin{bmatrix} 1 & -1 & 0 \\ 1 & -1 & 0 \\ -2 & 2 & 0 \end{bmatrix} \end{array} \begin{matrix} 3 \\ 4 \\ 5 \end{matrix} \tag{2.23}$$

Adding rows and columns of zeros corresponding to nodes 1 and 2 in $[Y]_{Na}$, and adding it to $[Y]_R$, we get the overall IAM for Figure 2.15e as

$$[Y]_N = \begin{array}{c} \begin{matrix} 1 & \quad 2 & \quad 3 & \quad 4 & \quad 5 \end{matrix} \\ \begin{bmatrix} Y_1 + Y_2 & -Y_2 & 0 & -Y_1 & 0 \\ -Y_2 & Y_2 + Y_4 - 1 & 1 & 0 & -Y_4 \\ 0 & 1 & 0 & -1 & 0 \\ -Y_1 & 0 & 1 & Y_1 + Y_3 - 1 & -Y_3 \\ 0 & -Y_4 & -2 & 2 - Y_3 & Y_3 + Y_4 \end{bmatrix} \end{array}$$

Considering node 5 as the reference node, the NAM of the network N is given by (deleting column 5 and row 5 from $[Y]_N$)

$$[y]_N = \begin{array}{c} \\ \\ \\ \\ \end{array} \begin{array}{cccc} 1 & 2 & 3 & 4 \\ \left[\begin{array}{cccc} Y_1 + Y_2 & -Y_2 & 0 & -Y_1 \\ -Y_2 & Y_2 + Y_4 - 1 & 1 & 0 \\ 0 & 1 & 0 & -1 \\ -Y_1 & 0 & 1 & Y_1 + Y_3 - 1 \end{array} \right] \end{array}$$

Considering now node 4 as an internal node, that is suppressing node 4 using Eq. (2.16), the admittance matrix reduces to

$$[y]_N = \begin{array}{c} \\ \\ \\ \\ \end{array} \begin{array}{ccc} 1 & 2 & 3 \\ \left[\begin{array}{ccc} \dfrac{(Y_1 + Y_2)(Y_1 + Y_3 - 1) - Y_1^2}{Y_1 + Y_3 - 1} & -Y_2 & \dfrac{Y_1}{Y_1 + Y_3 - 1} \\ -Y_2 & Y_2 + Y_4 - 1 & 1 \\ -\dfrac{Y_1}{Y_1 + Y_3 - 1} & 1 & \dfrac{1}{Y_1 + Y_3 - 1} \end{array} \right] \end{array}$$

Suppressing the internal node 3 now, we get the two-port admittance matrix

$$[y]_N = \begin{array}{c} \\ \\ \end{array} \begin{array}{cc} 1 & 2 \\ \left[\begin{array}{cc} (Y_1 + Y_2)(Y_1 + Y_3 - 1) & -(Y_1 + Y_2) \\ Y_1 - Y_2 & (Y_2 - Y_1) - (Y_3 - Y_4) \end{array} \right] \end{array}$$

Hence, the VTF of the network is given by

$$\frac{V_2}{V_1} = -\frac{y_{21}}{y_{22}} = \frac{Y_2 - Y_1}{(Y_2 - Y_1) - (Y_3 - Y_4)}$$

2.7
Analysis of Constrained Networks

The IAM technique needs special modifications to be applied to controlled sources such as a VCVS, since the $[y]$ matrix does not exist, and we need to add positive- and negative-valued elements at an appropriate location in the circuit, as done in the previous example. We shall now introduce the method of constrained network that does not need such a special arrangement.

The case of constrained network implies existence of a specific relationship between the voltage and current variables at two distinct nodes in a network. If this relationship is known, a certain technique can be used in the matrix $I-V$ equation set to simplify the mathematical analysis. When a VCVS, such as an OA, exists in a network, the computation of the admittance matrix can be simplified considerably, by using the constraint equation. Consider that there is an OA connected between nodes i, j, and k in a system with n nodes, as shown in Figure 2.16. The $I-V$

2.7 Analysis of Constrained Networks

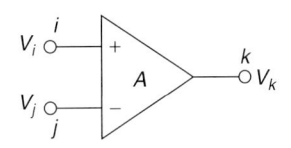

Figure 2.16 A differential-input voltage amplifier such as an OA.

equations at the different nodes can be described in the following form using admittance matrix elements:

$$I_1 = y_{11} V_1 + y_{12} V_2 + \cdots + y_{1i} V_i + y_{1j} V_j + y_{1k} V_k + \cdots + y_{1n} V_n$$
$$I_2 = y_{21} V_1 + y_{22} V_2 + \cdots + y_{2i} V_i + y_{2j} V_j + y_{2k} V_k + \cdots + y_{2n} V_n$$
$$\vdots$$
$$I_i = y_{i1} V_1 + y_{i2} V_2 + \cdots + y_{ii} V_i + y_{ij} V_j + y_{ik} V_k + \cdots + y_{in} V_n$$
$$I_j = y_{j1} V_1 + y_{j2} V_2 + \cdots + y_{ji} V_i + y_{jj} V_j + y_{jk} V_k + \cdots + y_{jn} V_n$$
$$I_k = y_{k1} V_1 + y_{k2} V_2 + \cdots + y_{ki} V_i + y_{kj} V_j + y_{kk} V_k + \cdots + y_{kn} V_n$$
$$\vdots$$
$$I_n = y_{n1} V_1 + y_{n2} V_2 + \cdots + y_{ni} V_i + y_{nj} V_j + y_{nk} V_k + \cdots + y_{nn} V_n \quad (2.24)$$

The constraint equation is

$$V_k = (V_i - V_j) A$$

or

$$V_i = V_j + \frac{V_k}{A}$$

On substituting the above constraint relation in the set of $I-V$ equations given by Eq. (2.24), one gets

$$I_1 = y_{11} V_1 + y_{12} V_2 + \cdots + (y_{1i} + y_{1j}) V_j + (y_{1k} + \frac{y_{1i}}{A}) V_k + \cdots + y_{1n} V_n$$
$$I_2 = y_{21} V_1 + y_{22} V_2 + \cdots + (y_{2i} + y_{2j}) V_j + (y_{2k} + \frac{y_{2i}}{A}) V_k + \cdots + y_{2n} V_n$$
$$\vdots$$
$$I_i = y_{i1} V_1 + y_{i2} V_2 + \cdots + (y_{ii} + y_{ij}) V_j + (y_{ik} + \frac{y_{ii}}{A}) V_k + \cdots + y_{in} V_n$$
$$I_j = y_{j1} V_1 + y_{j2} V_2 + \cdots + (y_{ji} + y_{jj}) V_j + (y_{jk} + \frac{y_{ji}}{A}) V_k + \cdots + y_{jn} V_n$$
$$I_k = y_{k1} V_1 + y_{k2} V_2 + \cdots + (y_{ki} + y_{kj}) V_j + (y_{kk} + \frac{y_{ki}}{A}) V_k + \cdots + y_{kn} V_n$$
$$\vdots$$
$$I_n = y_{n1} V_1 + y_{n2} V_2 + \cdots + (y_{ni} + y_{nj}) V_j + (y_{nk} + \frac{y_{ni}}{A}) V_k + \cdots + y_{nn} V_n$$

(2.25)

On examining the above, one can find that the node voltage V_i has got substituted by other node voltages. In matrix operation, this means that the column i corresponding to the voltage V_i has got discarded. Since node k is the output node of a voltage amplifier (VCVS), the current I_k is dependent solely on the networks connected to the output. Hence, this is not an independent variable and can be discarded in further manipulation of the matrix equation. Thus, application of the method of constraint involves the following steps.

Consider the admittance matrix description of the unconstrained part of the network (i.e., without considering the OA element); denote this to be $[y]_{UC}$.

1) Add to the element of column j, the element in column i, where i and j are the input nodes of the OA.
2) Add to the element of column k, $(1/A)$ times the element in column i, k being the output node of the OA.
3) Discard column i (or column j).
4) Discard row k.

The resulting matrix is the constrained $[y]$, the $[y]$ matrix of the constrained network.

Let us now consider some important cases.

Case 1: Infinite-gain differential-input OA

In this case, the gain $A \to \infty$, and hence we may obtain the $[y]$ of the constrained network by adding the columns corresponding to i and j (i.e., the input nodes of the OA), and deleting the row k (corresponding to the output node of the OA) in the unconstrained matrix $[y]_{UC}$.

Case 2: Single-input infinite-gain OA

In many instances node i is grounded, in which case $V_i = 0$, and $A \to \infty$. Hence, we may obtain the $[y]$ of the constrained network by discarding row k, and column j in $[y]_{UC}$. It should be noted that there is no row or column corresponding to the node i.

Case 3: Single-ended (input) finite-gain OA

Let the OA be of finite gain K and let V_i be the single-input node to the OA. Then $V_k = KV_i$. Also, there is no row or column corresponding to the node j, which is now grounded. Hence, we may obtain the $[y]$ matrix of the constrained network by simply adding K times the column k to the column i, and then deleting the row and column corresponding to node k.

Example 2.5. Consider the OA-based network shown in Figure 2.17. Find its VTF V_o/V_i.

$$[y]_{UC} = \begin{matrix} & 1 & 2 & 3 & 4 & \\ & \begin{bmatrix} y_a + y_c & -y_a & -y_c & 0 \\ -y_a & y_a + y_b & 0 & -y_b \\ -y_c & 0 & y_c + y_d & 0 \\ 0 & -y_b & 0 & y_b \end{bmatrix} & \begin{matrix} 1 \\ 2 \\ 3 \\ 4 \end{matrix} \end{matrix} \quad (2.26)$$

2.7 Analysis of Constrained Networks

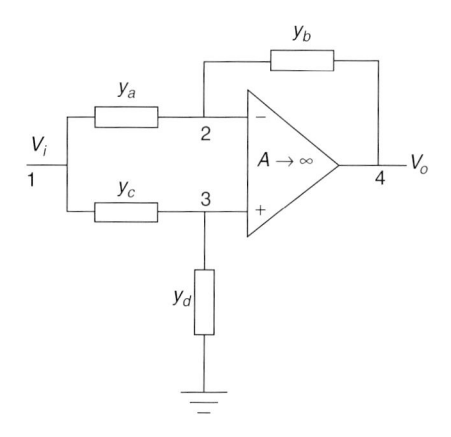

Figure 2.17 An OA used as an infinite-gain differential voltage amplifier.

Hence,

$$[y] = \begin{bmatrix} y_a + y_c & -(y_a + y_c) & 0 \\ -y_a & y_a + y_b & -y_b \\ -y_c & y_c + y_d & 0 \end{bmatrix} \begin{matrix} 1 \\ 2 \\ 3 \end{matrix} \quad \begin{matrix} 1 & 2 & 4 \end{matrix} \quad (2.27)$$

Then,

$$\frac{V_4}{V_1} = \frac{V_o}{V_i} = \frac{\begin{vmatrix} -y_a & y_a + y_b \\ -y_c & y_c + y_d \end{vmatrix}}{\begin{vmatrix} y_a + y_b & -y_b \\ y_c + y_d & 0 \end{vmatrix}} = \frac{y_b y_c - y_a y_d}{y_b(y_c + y_d)} \quad (2.28)$$

Example 2.6. For the network of Figure 2.18, find the VTF V_o/V_i.
Letting $G_i = 1/R_i$, $i = 1, 2$,

$$[y]_{UC} = \begin{bmatrix} G_1 & -G_1 & 0 & 0 \\ -G_1 & G_1 + G_2 + sC_3 + sC_5 & -sC_3 & -G_2 \\ 0 & -sC_3 & G_4 + sC_3 & 0 \\ 0 & -G_2 & 0 & G_2 \end{bmatrix} \quad \begin{matrix} 1 & 2 & 3 & 4 \end{matrix} \quad (2.29)$$

Since $V_4 = KV_3$,

$$[y] = \begin{bmatrix} G_1 & -G_1 & 0 \\ -G_1 & G_1 + G_2 + s(C_3 + C_5) & -(sC_3 + KG_2) \\ 0 & -sC_3 & G_4 + sC_3 \end{bmatrix} \quad \begin{matrix} 1 & 2 & 3 \end{matrix} \quad (2.30)$$

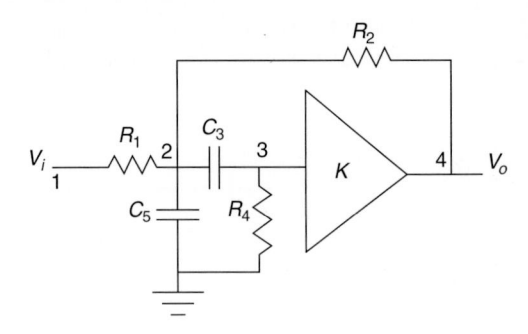

Figure 2.18 An OA used as a finite-gain single-input voltage amplifier.

Then,

$$\frac{V_3}{V_1} = \frac{\begin{vmatrix} -G_1 & G_1 + G_2 + s(C_3 + C_5) \\ 0 & -sC_3 \end{vmatrix}}{\begin{vmatrix} G_1 + G_2 + s(C_3 + C_5) & -(sC_3 + KG_2) \\ -sC_3 & G_4 + sC_3 \end{vmatrix}}$$

$$= \frac{sG_1 C_3}{s^2 C_3 C_5 + sC_3(G_1 + G_2 + G_4 - KG_2) + sG_4 C_3 + G_4(G_1 + G_2)} \quad (2.31)$$

Since $V_o = V_4 = KV_3$ and $V_1 = V_i$, we get

$$\frac{V_o}{V_i} = \frac{V_4}{V_1} = \frac{sKG_1 C_3}{s^2 C_3 C_5 + sC_3(G_1 + G_2 + G_4 - KG_2) + sG_4 C_3 + G_4(G_1 + G_2)}$$
$$(2.32)$$

2.8
Active Building Blocks for Implementing Analog Filters

Several semiconductor building blocks can be used as the active element for implementing an analog filter. Some examples are (i) OA, (ii) OTA, (iii) CC, and (iv) COA (current operational amplifier) . The OA behaves basically as a VCVS, the OTA as a VCCS, while the CC and COA behave as a CCCS. It is possible to use any of these active devices and several passive-RC components to realize active-RC filters. If the function of R is simulated by additional active device(s), the entire filter becomes implementable in a monolithic IC technology.

2.8.1
Operational Amplifier

An OA is perhaps the most widely used active device that has been employed to implement active-RC filter networks. In the ideal form, an OA has infinite input resistance, zero output resistance, and an infinite voltage gain. The device basically works as a VCVS. In practice, these ideal values are never achieved. But practical devices are quite often good enough to produce reasonably accurate transfer functions. Since a VCVS is a linear network element, the principles of linear

2.8 Active Building Blocks for Implementing Analog Filters

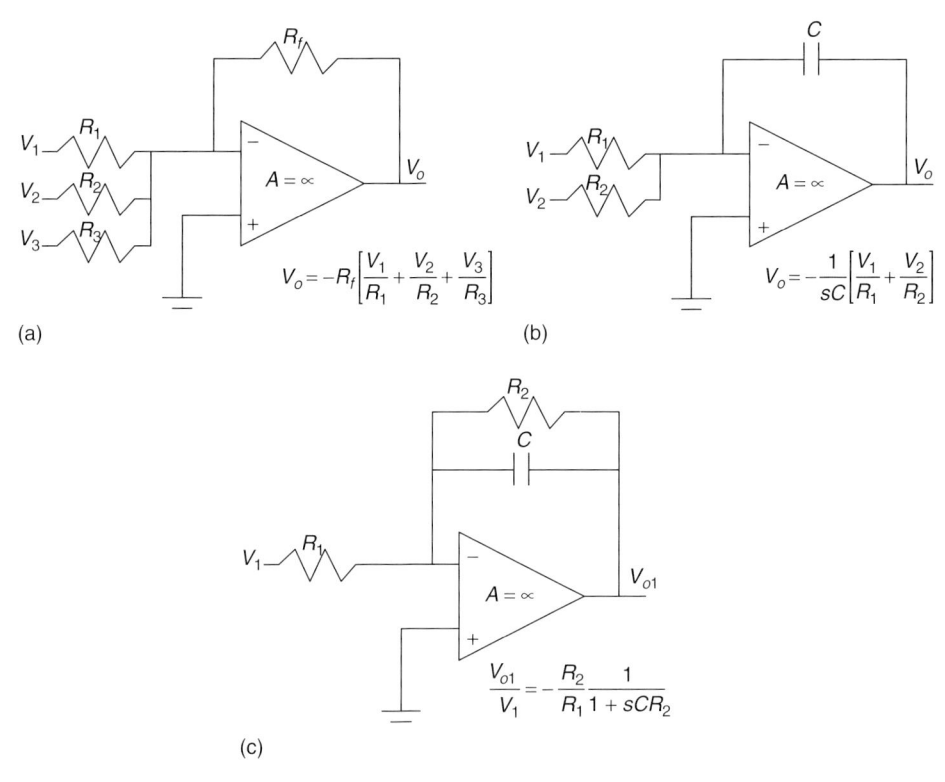

Figure 2.19 An OA used as (a) a summing inverting voltage amplifier, (b) a summing inverting integrating voltage amplifier, and (c) a lossy inverting integrating voltage amplifier.

(a) $V_o = -R_f \left[\dfrac{V_1}{R_1} + \dfrac{V_2}{R_2} + \dfrac{V_3}{R_3} \right]$

(b) $V_o = -\dfrac{1}{sC}\left[\dfrac{V_1}{R_1} + \dfrac{V_2}{R_2}\right]$

(c) $\dfrac{V_{o1}}{V_1} = -\dfrac{R_2}{R_1}\dfrac{1}{1+sCR_2}$

network analysis apply. Using superposition principle one can realize summing amplifiers and summing integrators with an OA and several RC components. Consider Figures 2.19a and 2.19b, which depict a summing amplifier and summing integrator respectively, realized using ideal OAs.

In the above, the voltages $V_1, V_2, \ldots$ are assumed to be available from voltage source nodes. Since an OA behaves as a VCVS, the output node of an OA appears like a voltage source. Thus, any one of the V_1, V_2 voltages could be derived from the output of the OA. When this is done in practice, we say that a feedback (negative) is being applied. Thus, in Figure 2.19b, if we label $V_o = V_{o1}$ and make $V_2 = V_{o1}$, we get $V_{o1} = -\dfrac{1}{sCR_1}V_1 - \dfrac{1}{sCR_2}V_{o1}$, which leads to the transfer function $\dfrac{V_{o1}}{V_1} = -\dfrac{R_2}{R_1}\dfrac{1}{1+sCR_2}$. This is the transfer function of a lossy integrator circuit as shown in Figure 2.19c.

It may be noted that by providing a feedback, a rational function of s has been generated. In principle, by combining ideal integrators, lossy integrators, and summing amplifiers, a frequency-selective transfer function of any high order (>1)

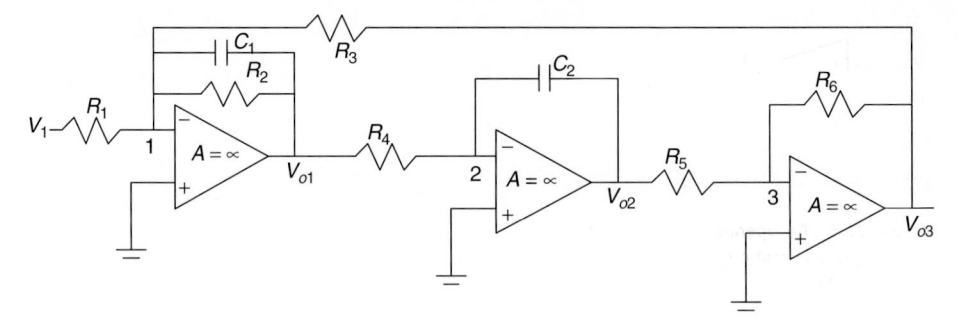

Figure 2.20 Use of an OA-based (inverting) integrator and (inverting) amplifier in realizing a second-order active-RC filter.

can be generated. Consider the case of realizing a second-order filter function as shown in Figure 2.20.

To analyze the system, one can proceed as follows: using nodal equations at nodes 1, 2, and 3, respectively, we get

$$V_{o1} = -\frac{V_1}{sC_1R_1} - \frac{V_{o1}}{sC_1R_2} - \frac{V_{o3}}{sC_1R_3}, \quad V_{o3} = -\frac{R_6}{R_5}V_{o2}, \quad \text{and} \quad V_{o2} = -\frac{V_{o1}}{sC_2R_4}$$

On substitution for V_{o3} and V_{o2}, one finally gets

$$\frac{V_{o1}}{V_1} = -\frac{N(s)}{D(s)} \qquad (2.33)$$

where

$$N(s) = \frac{R_6}{R_5} \frac{s}{R_1R_4C_1C_2}, \quad D(s) = s^2 + \frac{s}{C_1R_2} + \frac{R_6}{R_5} \frac{1}{R_3R_4C_1C_2}$$

The above represents a VTF $T(s) = V_{o1}/V_1$, of the form $As/(s^2 + Bs + C)$. If we let $s = j\omega$ and use suitable values for the RC elements, the plot of the magnitude response in terms of frequency ω rad s^{-1} will appear like a band-pass filter function. This serves as an example of implementing a band-pass filter using an OA as the active device.

2.8.2
Operational Transconductance Amplifier

The OTA functions as a VCCS, so that the significant transfer characteristic is the output short circuit small signal current to input small signal voltage (Schaumann, Ghausi, and Laker, 1990). If the output is terminated in a finite resistance, a VTF can be easily realized. OTAs can be easily configured to function as a driving point resistance and thus OTA-based filters may dispense with the use of any lumped resistances. This opens up the potential for implementing fully monolithic analog filters. The usual symbol and an associated AC equivalent network for an OTA are shown in Figures 2.21a and 2.21b.

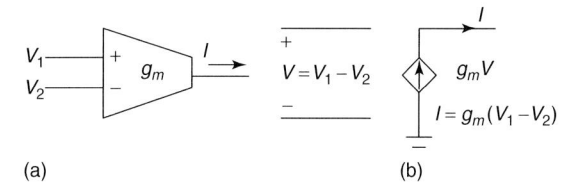

2.8 Active Building Blocks for Implementing Analog Filters

Figure 2.21 An operational transconductance amplifier (OTA): (a) schematic symbol; (b) AC equivalent circuit of an ideal OTA.

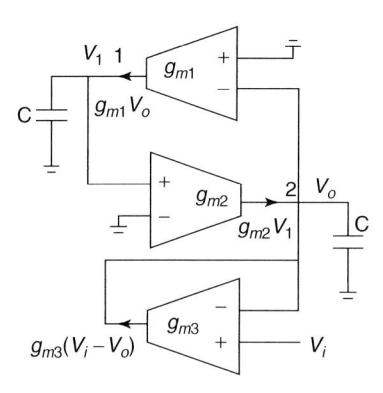

Figure 2.22 An OTA-based second-order filter.

Consider Figure 2.22 which presents the schematic of an OTA-based second-order filter.

At node 1, $sCV_1 = -g_{m1}V_o$, and at node 2, $g_{m2}V_1 + g_{m3}(V_i - V_o) = sCV_o$. From these two relations, we get

$$\frac{V_o}{V_i} = \frac{sg_{m2}C}{s^2C^2 + sg_{m3}C + g_{m1}g_{m2}} \tag{2.34}$$

$$\frac{V_1}{V_i} = \frac{-g_{m1}g_{m2}}{s^2C^2 + sg_{m3}C + g_{m1}g_{m2}} \tag{2.35}$$

V_o/V_i represents the VTF of a second-order band-pass filter and V_1/V_i that of a low-pass filter. This shows the potential of implementing a filtering function using OTAs.

2.8.3
Current Conveyor

A CC functions as a CCCS and in combination with RC elements can produce a frequency-dependent transfer function (Sedra and Smith, 1970). The schematic symbol of a CC is shown in Figure 2.23, and its input–output relations are given

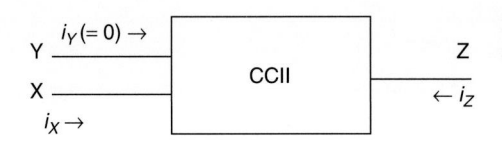

Figure 2.23 The schematic symbol of a type 2 current conveyor (CCII).

by

$$\begin{bmatrix} i_Y \\ v_X \\ i_Z \end{bmatrix} = \begin{bmatrix} 0 & 0 & 0 \\ 1 & 0 & 0 \\ 0 & \pm 1 & 0 \end{bmatrix} \begin{bmatrix} v_Y \\ i_X \\ v_Z \end{bmatrix} \quad (2.36)$$

The matrix relations imply that the X,Y terminals act like a virtual short circuit (as is in an OA), while the X, Z terminals function like a current mirror. It is possible to arrange $I_z = \pm k I_X$, where k is a constant. The CC with positive current mirroring is named as positive current conveyor type 2 (CCII+) and the CC with negative current mirroring is termed as a negative current conveyor type 2 (CCII−). The symbol II represents CC type 2, which has been established as a more popular and practical device.

If one considers the network with a CCII− of current gain factor k (Figure 2.24), it will be possible to establish by analysis, that a low-pass CTF is realized.

Note that $V_X = V_Y = 0$. Then KCL at the input node gives

$$I_X = I_S + sCV_1$$

At node Z,

$$I_Z = sCV_2 + G(V_2 - V_1) = kI_X$$

where $G = (1/R)$. Substituting for I_X and rearranging, we get

$$(sC + G)V_2 - (G + ksC)V_1 = kI_S \quad (2.37)$$

At the output node

$$(V_1 - V_2)G + V_1 G + V_1 sC = 0 \quad (2.38)$$

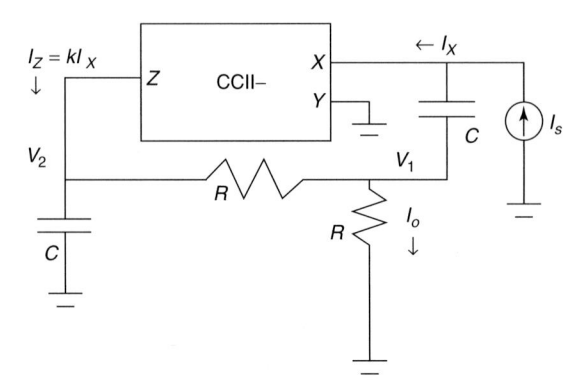

Figure 2.24 A negative CCII-based low-pass current transfer function filter.

Equation (2.38) gives

$$V_2 = V_1(sC + 2G)/G \qquad (2.39)$$

Substituting for V_2 from Eq. (2.39) into Eq. (2.36), we have

$$V_1 \frac{sC + 2G}{G}(sC + G) - V_1(G + ksC) = kI_s \qquad (2.40)$$

Finally, noting that $I_o = V_1 G$, one arrives at the current-mode transfer function

$$\frac{I_o}{I_s} = \frac{kG^2/C^2}{s^2 + (3-k)(G/C)s + G^2/C^2} \qquad (2.41)$$

Equation (2.41) represents a second-order low-pass filter transfer function in terms of current signals. The above derivation establishes that a filtering function can be implemented using a CC device.

Practice Problems

2.1 Consider the ladder network with component values as shown in Figure P2.1. Derive the transfer function for the voltage ratio V_L/V_s.
2.2 Find the VTF of the ladder filter shown in Figure P2.2.
2.3 Consider the network shown in Figure P2.3 that uses ideal OAs. (a) Find the chain matrix of the network and show that it corresponds to that of an impedance converter. (b) If a load Z_L is connected at port 2, what will be the input impedance at port 1? (c) If $Y_1 = Y_2 = Y_3 = G$, $Y_4 = sC$, and $Z_L = R_L$ what is the input impedance at port 1?

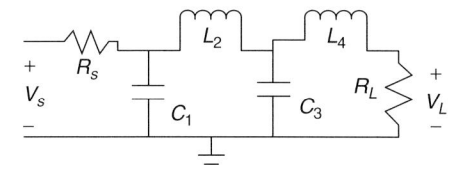

$R_s = 1$ $C_1 = 0.0114$ F $L_2 = 217$ H $C_3 = 0.0134$ F
$L_4 = 136$ H $R_L = 1$

Figure P2.1

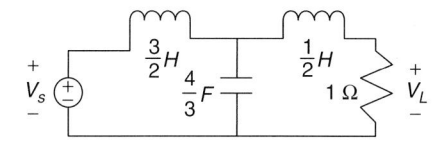

Figure P2.2

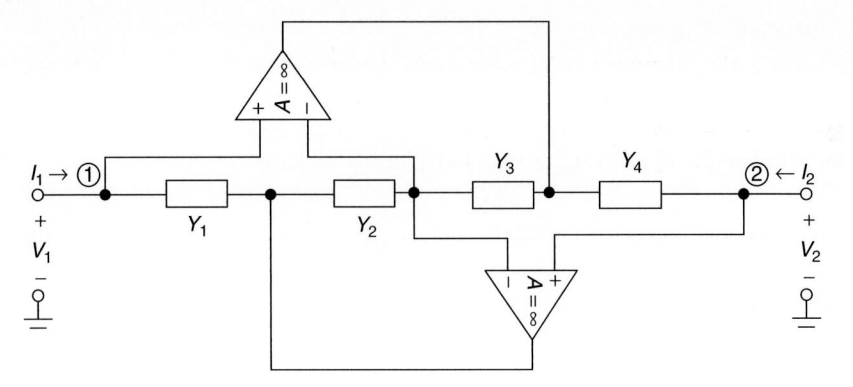

Figure P2.3

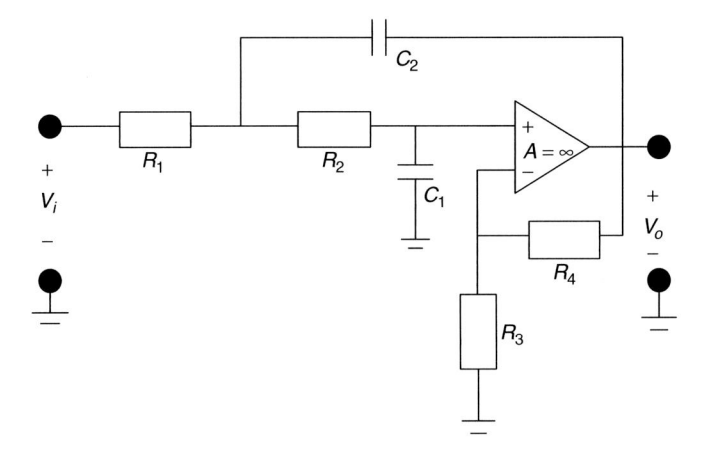

Figure P2.4

2.4 Assuming that the OA is ideal, show that the VTF of the active-RC network of Figure P2.4 is given by

$$\frac{V_o(s)}{V_i(s)} = \frac{K/(C_1 C_2 R_1 R_2)}{s^2 + s\left[\dfrac{1}{C_2 R_1} + \dfrac{1}{C_2 R_2} + \dfrac{1}{C_1 R_2} - \dfrac{K}{C_1 R_2}\right] + \dfrac{1}{C_1 C_2 R_1 R_2}}$$

where

$$K = 1 + \left(\frac{R_4}{R_3}\right)$$

2.5 For the network, shown in Figure P2.5, with a VCVS of gain K, find an expression for the VTF, $V_2(s)/V_1(s)$.

2.6 For the network of Figure P2.6, show that the VTF is given by

$$\frac{V_2(s)}{V_1(s)} = \frac{Ks^2}{s^2 + s\left[\dfrac{1}{C_1 R_2} + \dfrac{1}{C_3 R_4} + \dfrac{1}{C_1 R_4} - \dfrac{K}{C_1 R_2}\right] + \dfrac{1}{C_1 C_3 R_2 R_4}}$$

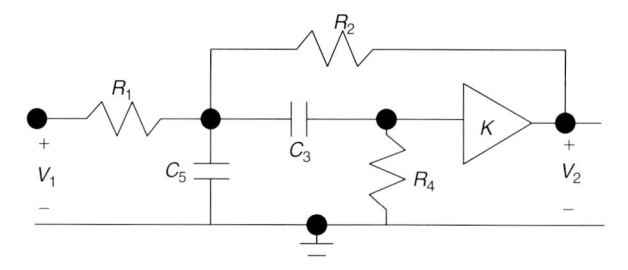

Figure P2.5

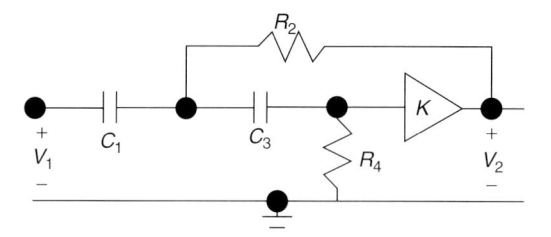

Figure P2.6

If $R_2 = R_4 = 1\,\Omega$, $C_1 = C_3 = 1F$, what value of K will make the coefficient of the "s" term in the denominator equal to 1.414?

2.7 The network of Figure P2.7 represents a low-pass filter. Derive the expression for the VTF of the network using the method of constrained networks. The OA is ideal.

2.8 Assuming the OA to be ideal, for the band-pass network shown in Figure P2.8, verify that the VTF is given by

$$\frac{V_2(s)}{V_1(s)} = \frac{sG_1 C_3}{s^2 C_3 C_4 + sG_5(C_3 + C_4) + G_5(G_1 + G_2)}$$

Note that G_1, G_2, and G_5 are conductances.

2.9 Consider the second-order filter network (Ackerberg and Mossberg, 1974) implemented using three OAs, and shown in Figure P2.9. Assuming that the OAs are identical and that each has a finite gain A, show that the VTF,

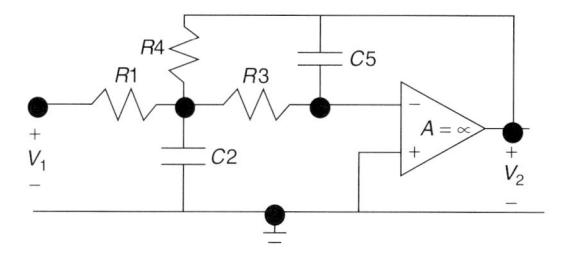

Figure P2.7

2 A Review of Network Analysis Techniques

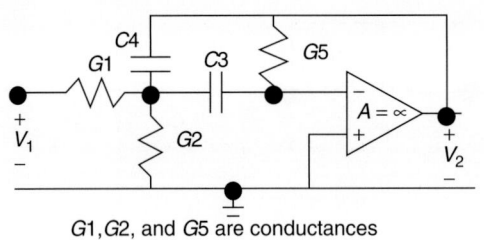

Figure P2.8

G1, G2, and G5 are conductances

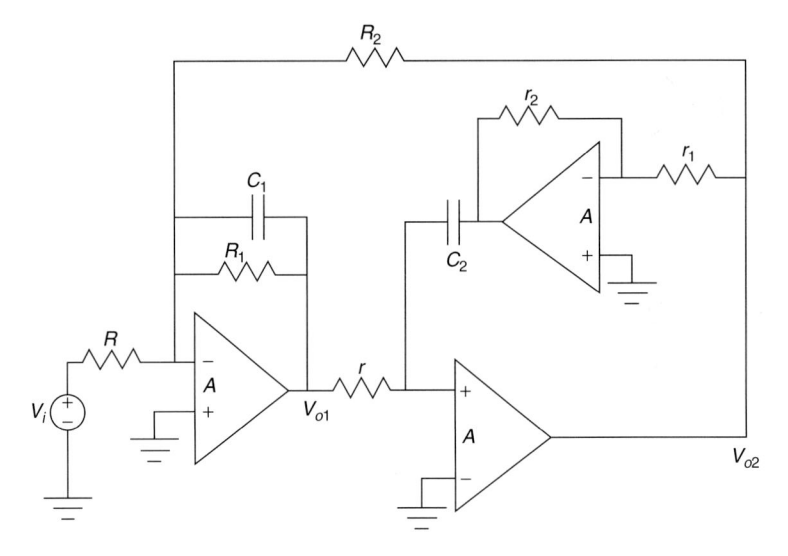

Figure P2.9

$V_{o2}(s)/V_i(s)$ is given by

$$\frac{V_{o2}}{V_i} = \frac{-\dfrac{1}{rR}}{\dfrac{1}{rR_2} + \left(\dfrac{\dfrac{1}{R}+\dfrac{1}{R_2}}{A} + \left(1+\dfrac{1}{A}\right)\left(\dfrac{1}{R_1}+sC_1\right)\right)\left(\dfrac{\dfrac{sC_2}{r_1}}{\dfrac{1}{r_2}+\dfrac{1}{r_1}} + \dfrac{\dfrac{1}{r}+sC_2}{A}\right)}$$

2.10 Consider the circuit of Figure P2.10, which is fed from a current source. Each OA has a finite gain A. Show that the current transfer function (CTF), I_o/I_{i2}, is given by

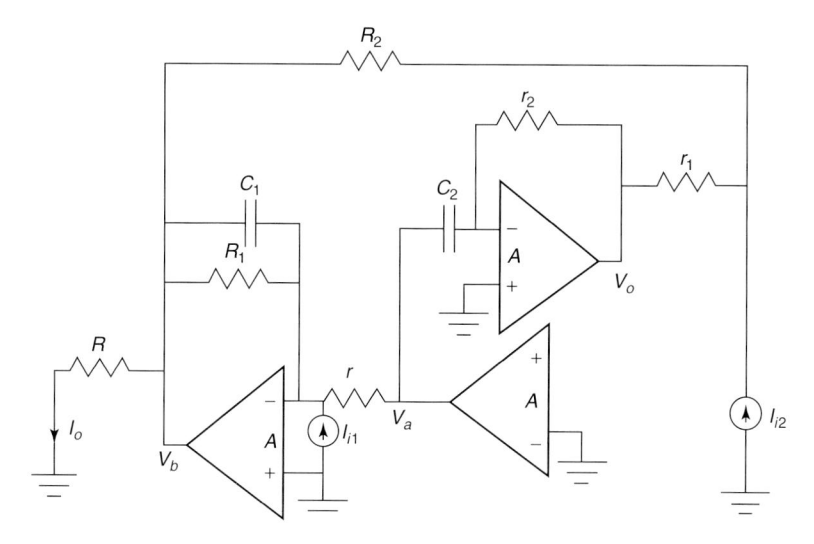

Figure P2.10

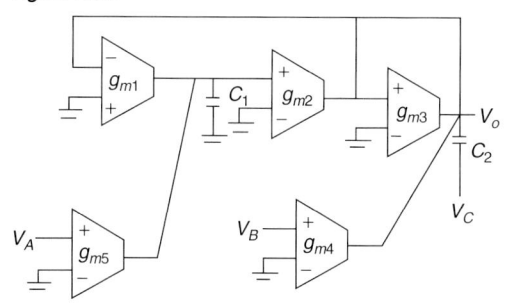

Figure P2.11

$$\frac{I_o}{I_{i2}} = \frac{-\dfrac{1}{rR}}{\dfrac{1}{rR_2} + \left(\dfrac{1}{Ar} + \left(1 + \dfrac{1}{A}\right)\left(\dfrac{1}{R_1} + sC_1\right)\right)\left(\dfrac{\dfrac{sC_2}{r_1} + \dfrac{1}{r_2} + sC_2}{\dfrac{1}{r_2} + \dfrac{1}{A}} + \dfrac{\dfrac{1}{r_1} + \dfrac{1}{r_2}}{A}\right)}$$

2.11 Using the ideal small signal model for each OTA in Figure P2.11, derive that

$$V_o = \frac{g_{m2}g_{m5}V_A - sg_{m4}C_1V_B + s^2C_1C_2V_C}{s^2C_1C_2 + sg_{m3}C_1 + g_{m1}g_{m2}}$$

2.12 Figure P2.12 shows a twin-Tee network frequently used in connection with design of equalizers. Write a subset of two-port parameters which could be used very easily to derive the overall two-port parameters of the system.

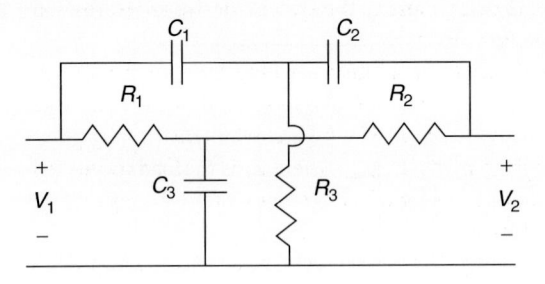

Figure P2.12

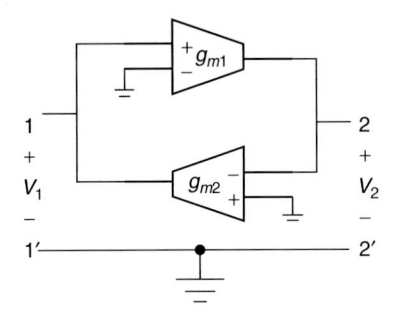

Figure P2.14

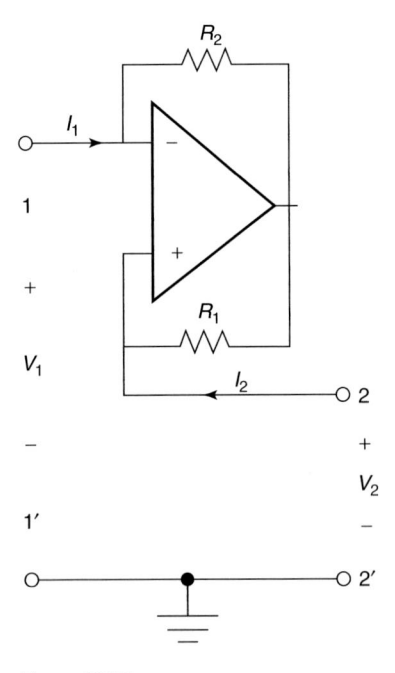

Figure P2.16

Using the overall two-port parameters of the system, derive an expression for the voltage transfer function V_2/V_1.

2.13 Solve Problem 2.12 using the method of indefinite admittance matrix (IAM).

2.14 (a) Derive the chain matrix [a] for the circuit of Figure P2.14, which employs two OTAs. Show that it corresponds to a PII. (b) Show that the driving point impedance $Z_1 = V_1/I_1$ is given by $Y_L/g_{m1}g_{m2}$, where Z_L is the load connected at port 2. (c) If Z_L corresponds to that of a capacitor C, find the equivalent element seen at port 1 and its value.

2.15 Suggest a circuit for an NII using OTAs. Use the knowledge you have gained from solving Problem 2.14.

2.16 Show that the circuit of Figure P2.16 realizes a CNIC. What is the input impedance at port 1, if a load Z_L is connected at port 2?

2.17 Show that the forward VTF $V_{o2}(s)/V_i(s)$ of Problem 2.9 is the same as the reverse CTF I_o/I_{i2} of Problem 2.10, when the OAs are ideal. Observe that the two circuits are the same except that the input and output terminals of each of the OAs are interchanged in the two circuits.

3
Network Theorems and Approximation of Filter Functions

In the previous chapter, we considered different methods of analyzing electrical circuits containing passive and active elements, and discussed various ways of characterizing a two-port network, which is extremely useful in understanding filter design. In this chapter, we introduce some general network theorems that are useful in filter design. This is followed by magnitude and phase approximations for the synthesis of the filter function. The ideal frequency response characteristic of a filter assumes brick-wall characteristics with abrupt jumps at the transition between the passbands (PBs) and stopbands (SBs). In reality, such a response is not achievable. Instead, filter designers contend themselves by attempting to approximate the idealized characteristics with some special mathematical functions, which very closely match the ideal brick-wall characteristics. By doing so, some of the idealness is sacrificed, but the approximation functions lead to the realization of filters using real components. This part of the filter design is known as the *task of approximation*. For magnitude approximation, the cases of maximally flat, Chebyshev (CHEB), and elliptic functions are discussed. For phase approximation, Bessel–Thomson (BT) approximation is introduced.

3.1
Impedance Scaling

Consider a linear time-invariant (LTI) n-port network N. Let N contain one-port elements as well as two-port elements. This has been assumed for the sake of simplicity, but the results to be established are true even if the subnetworks are n-port networks. From Chapter 2, we know that we can write the IAM of the n-port as

$$[I] = [Y]_N[V] \qquad (3.1)$$

Let us multiply all the one-port impedances and the impedance matrices of the internal two-ports by a constant positive value k. It is clear that as a consequence, all the one-port admittances as well as the $[y]$ matrices of the internal two-ports are multiplied by $(1/k)$. Hence, Eq. (3.1) becomes

$$[I] = (1/k)[Y]_N[V] \qquad (3.2)$$

Modern Analog Filter Analysis and Design: A Practical Approach. Rabin Raut and M. N. S. Swamy
Copyright © 2010 WILEY-VCH Verlag GmbH & Co. KGaA, Weinheim
ISBN: 978-3-527-40766-8

It is seen from Eqs. (2.20) and (2.21) that the DPI between the nodes q and l also get multiplied by the same factor k, whereas the VTF between the node pairs (i, j) and (q, l) is unaltered. This process is known as *impedance scaling*. It is very clear that impedance scaling by a factor of k is equivalent to changing a resistor of value R to kR, an inductor of value L to another inductor of value kL, and a capacitor of value C to another capacitor of value C/k. We will see what happens to various types of internal two-ports a little later.

3.2
Impedance Transformation

By applying the same arguments as above, it is obvious that if we multiply all the internal one-port impedances as well as the two-port $[z]$ matrices by a function $f(s)$, then the DPIs between any two nodes k and l will also be multiplied by $f(s)$, while the VTF between the pairs (i, j) and (k, l) will be unaffected. We shall call the operation of multiplying the internal impedances or impedance matrices by $f(s)$ as *impedance transformation* by $f(s)$ (Swamy, 1975). Since most of the time we will be dealing with two-port networks, let us assume that we are dealing with a two-port, which consists of one-ports and two-ports as subnetworks. Let its chain matrix be denoted by

$$\begin{bmatrix} V_1 \\ I_1 \end{bmatrix} = \begin{bmatrix} A & B \\ C & D \end{bmatrix} \begin{bmatrix} V_2 \\ -I_2 \end{bmatrix} \tag{3.3}$$

It is clear that as a consequence of the impedance transformation, an internal resistor or inductor gets multiplied by $f(s)$, while the capacitor gets divided by $f(s)$. Let us see what happens to an internal two-port, which is represented by its chain matrix as

$$[a_j] = \begin{bmatrix} A_j & B_j \\ C_j & D_j \end{bmatrix} \tag{3.4}$$

where A_j, B_j, C_j, and D_j are, in general, functions of s. In view of the interrelationships between the chain and $[z]$ matrices of two-ports, the consequence of the impedance transformation is to render the chain matrix of the transformed network to be

$$[a'] = \begin{bmatrix} A & Bf(s) \\ C/f(s) & D \end{bmatrix} \tag{3.5}$$

while the chain matrix of the internal two-port becomes

$$[a'_j] = \begin{bmatrix} A_j & B_j f(s) \\ C_j/f(s) & D_j \end{bmatrix} \tag{3.6}$$

Thus, if the network consists of VCVS or CCCS, these elements will remain as VCVS or CCCS with the same voltage or current amplification factor under impedance transformation. Using Eq. (3.6), we may similarly find the transformed elements for the other two-port elements listed in Table 2.1.

3.2 Impedance Transformation

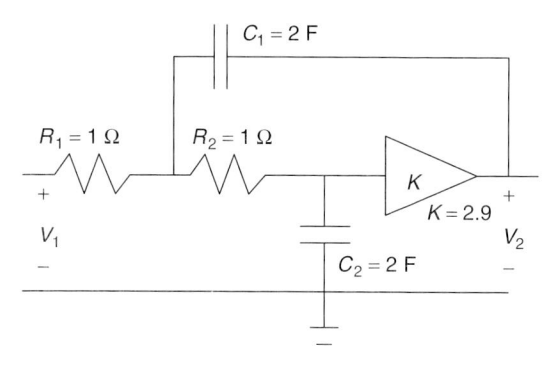

Figure 3.1 An active-RC filter using one OA.

Let us see what has been the effect on the overall two-port of impedance transformation by $f(s)$ (or scaling by a factor k). We can conclude the following using Eq. (3.5):

1) Driving point and transfer impedances are multiplied by $f(s)$.
2) Driving point and transfer admittances are multiplied by $1/f(s)$.
3) The open circuit VTF ($1/A$) is unaltered.
4) The short circuit CTF ($1/D$) is unaltered.

Example 3.1. Figure 3.1 shows an active-RC filter using one OA. Determine the transfer function (TF) V_2/V_1 and hence the nature of the filter. Determine the value of the element if R_1 is chosen as (scaled to) $R_1 = 1$ kΩ. Also find the new TF.

It can readily be seen that

$$T_v = \frac{V_2}{V_1} = \frac{K/C_1 C_2 R_1 R_2}{s^2 + s\left[\frac{1}{C_1}\left(\frac{1}{R_1} + \frac{1}{R_2}\right) + (1-K)/C_2 R_2\right] + \frac{1}{C_1 C_2 R_1 R_2}}$$

Since $C_1 = C_2 = 2$ F, $R_1 = R_2 = 1$ Ω, and $K = 2.9$

$$\frac{V_2}{V_1} = T_v = \frac{2.9/4}{s^2 + 0.05s + 0.25}$$

Hence, the given network is a low-pass (LP) filter. If now R_1 is scaled to 1 kΩ, the impedances are all scaled by 1000, but not K. Hence, the capacitive impedances are multiplied by 1000. In other words, the capacitor values are divided by 1000. Hence, the new element values are

$$R_1 = R_2 = 1 \text{ k}\Omega, C_1 = C_2 = 2 \text{ MF}, K = 2.9$$

Also since the voltage TF is unaltered by impedance scaling, the value of the new TF is still the same as before.

Example 3.2. Figure 3.2 shows a normalized filter working into a 1 Ω load resistance. Obtain the system function V_2/I_1. Find the filter corresponding to a 100 Ω load resistance. Find the new system function.

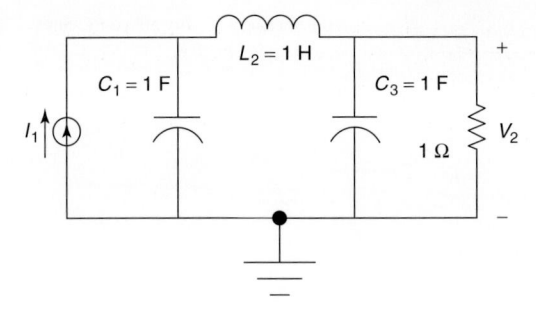

Figure 3.2 An LC filter with a normalized load resistance of 1 Ω.

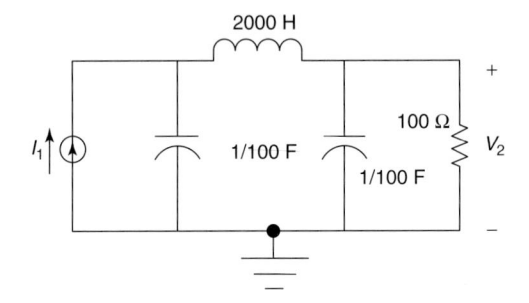

Figure 3.3 The LC filter corresponding to that of Figure 3.2 for a load of 100 Ω.

The system function given by $V_2/I_1 = Z_{21}$ is

$$Z_{21} = \frac{1}{s^3 C_1 L_2 C_3 + s^2 L_2 C_1 + s(C_1 + C_3) + 1} = \frac{1}{2s^3 + 2s^2 + 2s + 1}$$

To obtain the filter corresponding to a 100 Ω resistance at the load we perform the following scaling:

$$R_i^* = 100 R_i = 100 \ \Omega, \ L_2^* = 100 L_2 = 200 \ H, \ C_1^* = C_2^* = \frac{1}{100} \ F$$

Hence,

$$Z_{21}^* = 100 Z_{21} = \frac{100}{2s^3 + 2s^2 + 2s + 1}$$

The scaled circuit is shown in Figure 3.3.

3.3
Dual and Inverse Networks

3.3.1
Dual and Inverse One-Port Networks

It is well known that if two one-port RLC networks N_1 and N_2 have their impedances z_1 and z_2 related by

$$z_1 z_2 = K \tag{3.7}$$

then N_1 and N_2 are said to be inverses of each other with respect to K. If the network N_1 is a planar one-port network, it is also known that we can obtain N_2 by obtaining the topological dual of N_1 by the "dot window" technique (Swamy and Thulasiraman, 1981) and associating an impedance of z_{iD} to the ith branch of N_2 such that

$$z_i z_{iD} = K \tag{3.8a}$$

or

$$z_{iD} = K y_i \tag{3.8b}$$

where z_i is the impedance of the corresponding ith branch in N_1. It may be noted that the unit of K is ohms square. When N_2 is obtained from N_1 in this way, N_2 is known as the *dual of N_1* w.r.t. K. (If $K = 1$, the two networks are simply known as *duals of each other*). Hence, a resistor R in N_1 would become a resistor of value K/R_1 in N_2, an inductor of value L would become a capacitor of value L/K in N_2 and a capacitor C in N_1 would become an inductor of value KC in N_2.

If $K = f(s)$, a function of s, then we call N_2 the generalized dual of N_1 w.r.t. $f(s)$ (Mitra, 1969). N_1 is not restricted to be an RLC network. Of course, the individual elements in the two networks are also correspondingly generalized duals of each other w.r.t. $f(s)$. When $f(s) = K/s$, the network N_2 is called the *capacitive dual of N_1* and is useful in RC-active synthesis where we would like to obtain one RC-active network from another (an ordinary resistive dual, $f(s) = K$, would convert an RC network to an RL network). In this book, we do not make a distinction between duals and inverses, and simply call them as duals, even if we do not obtain the duals topologically.

3.3.2
Dual Two-Port Networks

It is possible to extend the generalized dual concept for a planar two-port network consisting of one-ports and three-terminal two-ports as internal elements, wherein the dual is topologically obtained from the original two-port (Swamy, Bhusan, and Bhattacharyya, 1974). However, it is not done so here, but the generalized dual or generalized inverse is defined through the two-port chain matrix (Swamy, 1975). Let N be an LTI 2-port network with a chain matrix

$$[a] = \begin{bmatrix} A & B \\ C & D \end{bmatrix} \tag{3.9}$$

We define N_D to be the generalized dual (or generalized inverse) of N w.r.t. $f(s)$ if the chain matrix $[a]_D$ of N_D is related to $[a]$ as

$$[a]_D = \begin{bmatrix} D & Cf(s) \\ B/f(s) & A \end{bmatrix} \tag{3.10}$$

As in the case of one-ports, if $f(s) = 1/s$, we call the dual as the capacitive dual. It is obvious that the dual of the dual is the original network. Now, using Table 2.1, we can show that

$$[z]_D = f(s) \begin{bmatrix} y_{11} & -y_{12} \\ -y_{21} & y_{22} \end{bmatrix} \qquad (3.11)$$

Hence, the VTF and CTF are given by

$$T_v(s) = \frac{V_2}{V_1} = -\frac{y_{21}}{y_{22}} = \frac{(z_{21})_D/f(s)}{(z_{22})_D/f(s)} = \frac{(z_{21})_D}{(z_{21})_D} = (T_I)_D \qquad (3.12)$$

$$T_I(s) = \frac{-I_2}{I_1} = -\frac{y_{21}}{y_{11}} = \frac{(z_{21})_D/f(s)}{(z_{11})_D/f(s)} = \frac{(z_{21})_D}{(z_{11})_D} = (T_V)_D \qquad (3.13)$$

Thus, we have the following properties:

1) The open circuit voltage TF T_V of N is the same as the short circuit current TF $(T_I)_D$ of N_D and vice versa, that is,

$$T_V = (T_I)_D, (T_V)_D = T_I \qquad (3.14)$$

2) The DPIs z_{11} and z_{22} of N are $f(s)$ times the driving point admittances y_{11} and y_{22}, respectively, of N_D and vice versa, that is,

$$(z_{11})_D = f(s)y_{11}, (z_{22})_D = f(s)y_{22}; z_{11} = f(s)(y_{11})_D,$$
$$z_{22} = f(s)(y_{22})_D \qquad (3.15)$$

3) The transfer impedances z_{12} and z_{21} of N are $-f(s)$ times the transfer admittances y_{12} and y_{21}, respectively, of N_D and vice versa, that is,

$$(z_{12})_D = -f(s)y_{12}, (z_{21})_D = -f(s)y_{21}; z_{12} = -f(s)(y_{12})_D,$$
$$z_{21} = -f(s)(y_{21})_D \qquad (3.16)$$

The property (1) above is particularly useful in obtaining a CTF realization from a known VTF realization. Using the definition of the generalized dual, the following results can easily be established.

Theorem 3.1. *The generalized dual of a cascade of networks w.r.t. $f(s)$ is the cascade of the generalized duals w.r.t. $f(s)$, that is,*

$$\begin{bmatrix} N_1 & N_2 & \cdots & N_m \end{bmatrix}_D = [N_1]_D \ [N_2]_D \ \cdots \ [N_m]_D \qquad (3.17)$$

Theorem 3.2. *The generalized dual Z_D of a series (shunt) element of impedance Z is a shunt (series) element of impedance $f(s)/Z$. Thus, $Z Z_D = f(s)$, that is, the one-ports Z and Z_D are themselves generalized duals of each other. Figure 3.4 illustrates these operations.*

Example 3.3. Obtain the capacitive dual of the RC ladder structure shown in Figure 3.5a.

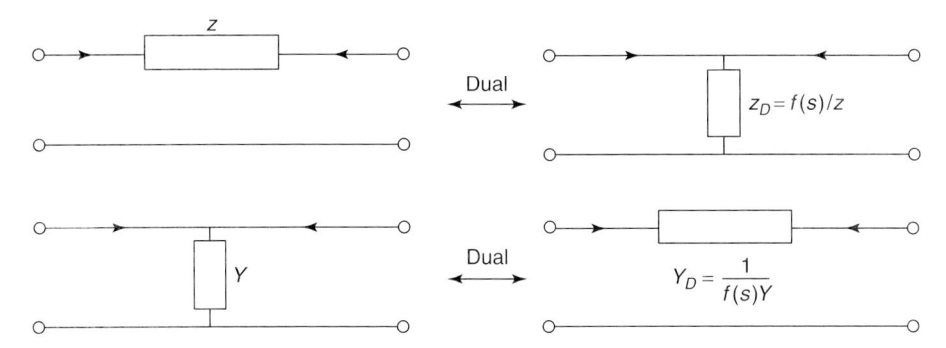

Figure 3.4 Series and shunt elements and their generalized duals.

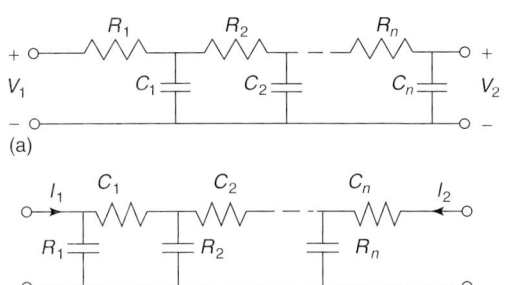

Figure 3.5 (a) An RC ladder network and (b) its capacitive dual.

Using the results of Theorems 3.1 and 3.2, it is easily seen that the capacitive dual is another RC ladder as shown in Figure 3.5b. It may be verified that the open circuit VTF of the RC ladder (Figure 3.5a) is the same as the short circuit CTF of the RC ladder of Figure 3.5b.

3.4 Reversed Networks

Consider a two-port network N whose $[z]$ matrix is given by

$$[z] = \begin{bmatrix} z_{11} & z_{12} \\ z_{21} & z_{22} \end{bmatrix} \quad (3.18)$$

The reversed network N_R is obtained by reversing the roles of the input and outputs in N. Hence, the z matrix of the reversed network is

$$[z_R] = \begin{bmatrix} z_{22} & z_{21} \\ z_{12} & z_{11} \end{bmatrix}$$

Hence, from Table 2.1, we have that the chain matrix of the reversed network is

$$[a_R] = \frac{1}{AD - BC} \begin{bmatrix} D & B \\ C & A \end{bmatrix} \quad (3.19)$$

If the network is reciprocal, $AD - BC = 1$.

3.5
Transposed Network

Consider a two-port network N with its impedance matrix given by $[z]$. Bhattacharyya and Swamy (1971) defined the two-port network having an impedance matrix that is the transpose of $[z]$ as the transposed network (Bhattacharyya and Swamy, 1971), or simply the transpose of N, and denoted it by N^T. Denoting the impedance and admittance matrices of N^T by $[z^T]$ and $[y^T]$, we thus have $[z^T] = [z]^T$ and $[y^T] = [y]^T$. Correspondingly, the chain matrix $[a^T]$ of N^T is related to the chain matrix $[a]$ of N by

$$[a^T] = \frac{1}{AD - BC} [a] \qquad (3.20)$$

This definition could be of course extended to n-ports (Bhattacharyya and Swamy, 1971). It is clear that for a reciprocal network, since $z_{12} = z_{21}$, the transpose is itself. It can be shown that if the network N consists of one- and two-port elements, the transposed network N^T can be obtained from N by replacing the internal nonreciprocal two-ports by their transposes and leaving the one-ports and reciprocal two-ports unaltered (Bhattacharyya and Swamy, 1971).

If we now denote the reversed transpose of N by $(N^T)_R$, then the chain matrix of $(N^T)_R$ is given by

$$[a_R^T] = \begin{bmatrix} D & B \\ C & A \end{bmatrix} \qquad (3.21)$$

We can use the above relation to determine the transposes of the various nonreciprocal two-ports. The reversed transposes of the various nonreciprocal two-ports (including the controlled sources) are listed in Table 3.1.

From Table 3.1, we observe that the transpose of a VCVS is a CCCS with its input and output ports reversed and with a current gain equal to the voltage gain of the VCVS, and vice versa. Also, the transpose of an impedance inverter, a VCCS or a CCVS, is itself with input and output ports reversed. Thus, we can very easily construct the transpose of a given network with one-ports and two-ports as subnetworks using Table 3.1.

From the above analysis, the following important conclusions can be drawn:

1) The driving point functions of N^T are the same as those of N.
2) The forward open circuit VTF of a network N is the same as the reverse short circuit CTF of its transpose N^T, and vice versa.
3) The forward transfer impedance (admittance) of N is the same as the reverse transfer impedance (admittance) of N^T, and vice versa.

It should be noted again that if N is reciprocal, N_R^T is nothing but N_R, the reversed network of N. Thus given a network N, we can obtain three other networks, namely, the dual N_D, the reversed transpose N_R^T, and $\overline{N} = (N_R^T)_D = (N_D)_R^T$. The interrelationships between these four networks are pictorially shown in Figure 3.6 along with their chain matrices (Swamy, Bhusan, and Bhattacharyya, 1976).

3.5 Transposed Network

Table 3.1 Reversed transposes of some of the common two-port elements.

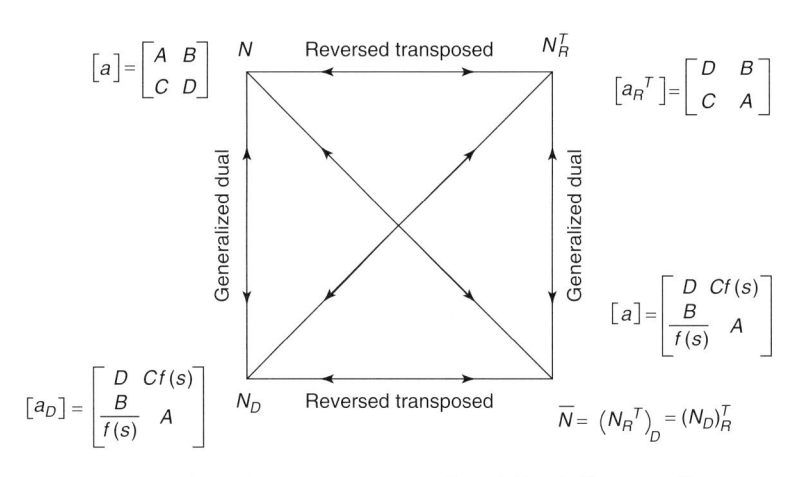

Two-port element N	[a] Chain matrix of N	$[a_R^T]$ Chain matrix of N_R^T	Reversed transpose of element N
VCVS	$\begin{bmatrix} \frac{1}{\mu} & 0 \\ 0 & 0 \end{bmatrix}$ μ = voltage gain	$\begin{bmatrix} 0 & 0 \\ 0 & \frac{1}{\mu} \end{bmatrix}$	CCCS of gain μ
CCCS	$\begin{bmatrix} 0 & 0 \\ 0 & \frac{1}{\alpha} \end{bmatrix}$ α = current gain	$\begin{bmatrix} \frac{1}{\alpha} & 0 \\ 0 & 0 \end{bmatrix}$	VCVS of gain α
VCCS	$\begin{bmatrix} 0 & \frac{1}{g} \\ 0 & 0 \end{bmatrix}$ g = transconductance	$\begin{bmatrix} 0 & \frac{1}{g} \\ 0 & 0 \end{bmatrix}$	VCCS of transconductance g
CCVS	$\begin{bmatrix} 0 & 0 \\ \frac{1}{r} & 0 \end{bmatrix}$ r = transresistance	$\begin{bmatrix} 0 & 0 \\ \frac{1}{r} & 0 \end{bmatrix}$	CCVS of transresistance r
Impedance converter	$\begin{bmatrix} K_1 & 0 \\ 0 & K_2 \end{bmatrix}$	$\begin{bmatrix} K_2 & 0 \\ 0 & K_1 \end{bmatrix}$	Another impedance converter
Impedance inverter (II)	$\begin{bmatrix} 0 & \frac{1}{G_1} \\ G_2 & 0 \end{bmatrix}$	$\begin{bmatrix} 0 & \frac{1}{G_1} \\ G_2 & 0 \end{bmatrix}$	Same impedance inverter as N

$$[a] = \begin{bmatrix} A & B \\ C & D \end{bmatrix} \quad N \xrightarrow{\text{Reversed transposed}} N_R^T \quad [a_R^T] = \begin{bmatrix} D & B \\ C & A \end{bmatrix}$$

(Generalized dual ↕ Generalized dual)

$$[a_D] = \begin{bmatrix} D & Cf(s) \\ \frac{B}{f(s)} & A \end{bmatrix} \quad N_D \xrightarrow{\text{Reversed transposed}} \quad \bar{N} = (N_R^T)_D = (N_D)_R^T$$

$$[a] = \begin{bmatrix} D & Cf(s) \\ \frac{B}{f(s)} & A \end{bmatrix}$$

Figure 3.6 Interrelationships among N, N_D, N_R^T, and $\bar{N} = (N_R^T)_D = (N_D)_R^T$.

As a consequence, we see that

1) The open circuit VTFs (short circuit CTFs) of N and $\bar{N}$ are identical. Hence, N and $\bar{N}$ are alternate structures for a VTF or a CTF.
2) The open circuit VTFs (short circuit CTFs) of N_D and N_R^T are identical.
3) The open circuit VTF (CTF) of N and $\bar{N}$ is the same as the CTF (VTF) of N_D and N_R^T.

4) The DPIs of N and N^T are the same.
5) The DPIs of N_D and N_R^T are the same.

3.6
Applications to Terminated Networks

Lossless networks terminated by resistors, either at one end or at both ends, are used in communication systems to couple an energy source to a load. The design of lossless ladder networks is considered in detail in any text book on passive network synthesis (see, for example, Van Valkenburg, 1960; Weinberg, 1962). It has been shown (Temes and Mitra, 1973) that when the ladder is designed properly, the first-order sensitivities of the magnitude of the TF to each of the inductors and capacitors are zero at the frequencies of maximum power transfer, and further that they remain low in the intermediate frequencies throughout the PB. In this section, we consider the applications of dual, transposed, and dual-transposed networks in doubly terminated networks. Similar results can be obtained for singly terminated networks.

A general coupling network may be excited by a voltage source or a current source, which may be nonideal. Further, the network may be working into an arbitrary load. Such a situation can be represented by the network of Figure 3.7a or b, and is known as a *doubly terminated network*.

For the network of Figure 3.7a, the suitable specifications are

$$T_V(s) = \frac{V_2}{V_s} = \frac{R_L}{AR_L + B + CR_s R_L + DR_s} \tag{3.22}$$

and

$$Y_{21}(s) = \frac{-I_2}{V_s} = \frac{1}{AR_L + B + CR_s R_L + DR_s} \tag{3.23}$$

while for the network of Figure 3.7b, the suitable specifications are

$$T_I(s) = -\frac{I_2}{I_s} = \frac{R_s}{AR_L + B + CR_s R_L + DR_s} \tag{3.24}$$

and

$$Z_{21}(s) = \frac{V_2}{I_s} = \frac{R_s R_L}{AR_L + B + CR_s R_L + DR_s} \tag{3.25}$$

Thus, if we know how to realize a rational function $H(s)$, say, as the $T_V(s)$ of a doubly terminated network with a load resistance R_L and source resistance R_S, then

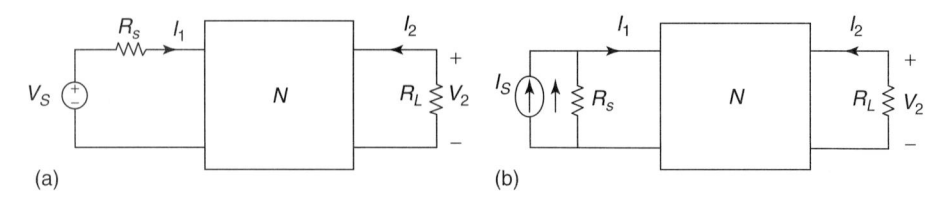

Figure 3.7 Doubly terminated networks fed by (a) a voltage source and (b) a current source.

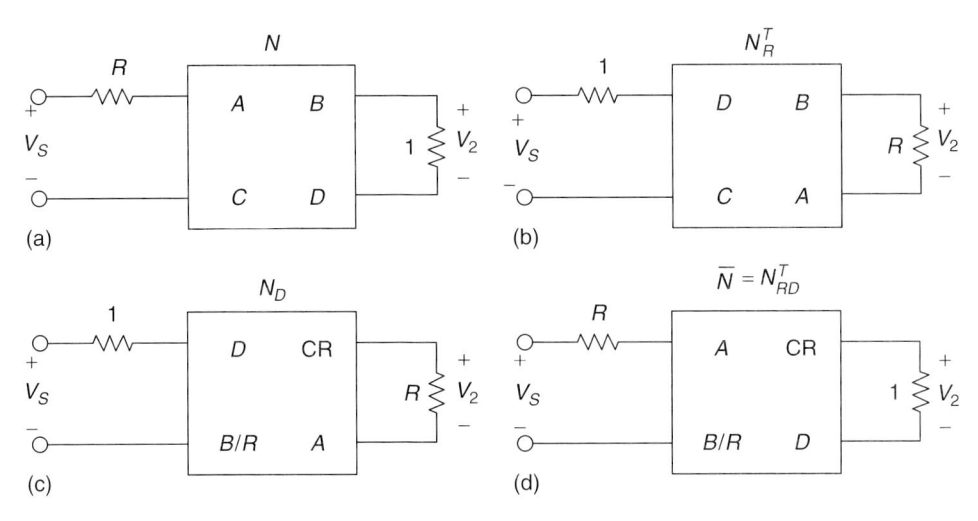

Figure 3.8 Doubly terminated networks (a) N, (b) N_D, (c) N_R^T, and (d) $\overline{N} = (N_D)_R^T$.

the designs for $T_I(s)$, $Y_{21}(s)$, and $Z_{21}(s)$ are directly obtained. Equations (3.22–3.25) may also be used in the case of singly terminated networks.

Let us denote $R_S R_L$ by R and assume without loss of generality that $R_L = 1$, since we can always scale the impedances for any other load resistance. Consider the networks of Figures 3.8a–3.8d, where N_D is the dual of N w.r.t. $f(s) = R$, N_R^T is the reversed transpose of N, and $\overline{N} = (N_D)_R^T$. The TF of these structures may be obtained using Eqs. (3.22–3.25) as

$$T_V^{(a)}(s) = \frac{1}{R}T_V^{(b)}(s) = \frac{1}{R}T_V^{(c)}(s) = T_V^{(d)}(s) = H(s) \tag{3.26}$$

where

$$H(s) = \frac{1}{(A+B) + R(C+D)} \tag{3.27}$$

Thus, if we know how to synthesize a given $H(s)$ as the $T_V(s)$ of the structure of Figure 3.8a and obtain the two-port N, then three other realizations may be found for $H(s)$ using N_D, N_R^T, and $\overline{N}$. It should be noted, however, that the source and load conditions are the same for Figures 3.8a and 3.8d, and again for Figures 3.8b and 3.8c. Further, source and load resistances for the former networks are the load and source resistances for the latter. Hence (Swamy, 1975),

1) Figures 3.8a and 3.8d show alternate structures realizing $H(s)$ as a voltage TF with a load of 1 Ω and a source of RΩ, and one may be obtained from the other.
2) Figures 3.8b and 3.8c show alternate structures, but have a load of RΩ and a source of 1 Ω; further these structures may be directly obtained from that of Figure 3.8a.

3.7 Frequency Scaling

Consider a general LTI network N. If we change the complex frequency s to s/b, b being a constant, it is clear that in any of the system functions, whether it be the TIF $Z_{21}(s)$, the TAF $Y_{21}(s)$, the VTF $T_V(s)$, or the CTF $T_I(s)$, the frequency s is changed to s/b. Hence, whatever was the magnitude or phase of a system function $F(s)$ at a frequency ω, the new system function $F(s/b)$ would have the same magnitude and phase at the frequency ω/b, that is, the frequency has been scaled by a factor $1/b$. Such an operation is called *frequency scaling* by the factor $1/b$. As a consequence of the frequency scaling, it is clear that an inductance of value L and a capacitor of value C would now have reactances of sL/b and $1/(sC/b)$, respectively. Thus, these will become, respectively, an inductor of value L/b and a capacitor of value C/b, while a resistor will remain unchanged. It is also seen that the ideal controlled sources also remain unaltered. Frequency scaling is used to obtain the response of a given filter in a given frequency band scaled up or down to a different frequency band.

Example 3.4. The Sallen and Key structure of Figure 3.1 realizes the LP TF

$$\frac{V_2}{V_1} = T_V(s) = \frac{2.9/4}{s^2 + 0.05s + 0.25}$$

According to the standard notations for a second-order filter, namely,

$$T_V(s)|_{LP} = \frac{H_{LP}\,\omega_p^2}{s^2 + (\omega_p/Q_p)s + \omega_p^2}$$

the above filter has a pole frequency $\omega_p = \sqrt{0.25} = 0.5$.

(a) Find the TF if ω_p is to be 1000 rad s^{-1}.

(b) If in addition, we want to use 1 K resistors in the circuit, how is the TF altered. Also, how do the values of the components change in the LP structure?

Solution: (a) The frequency scaling factor is $b = 1000/0.5 = 2000$. Hence, s is changed to $s/2000$ in the expression for $T_V(s)$, giving the new TF to be

$$T'_V(s) = \frac{(2.9/4)(2000)^2}{s^2 + 0.05(2000)s + 0.25(2000)^2}$$

This does not alter the values of the resistors, but the capacitance values become 0.001.

(b) If all the resistors have to be 1 K, we have to scale all the impedances by the factor $a = 1000$. Since impedance scaling does not alter a VTF, the TF remains $T'_V(s)$. However, the new values of the components are $R_1 = R_2 = 1000\ \Omega$ and $C_1 = C_2 = 1$ pF, with the gain $K = 2.9$ of the voltage amplifier remaining unchanged.

3.8 Types of Filters

Filters are categorized depending on the type of filtering function that they perform. If the primary consideration is the magnitude or attenuation characteristic, then we classify them as LP, high-pass (HP), band-pass (BP), and bandstop (BS) or band-reject (BR). However, if our main concern is the phase or delay specification, then the type of filters are all-pass (AP) and delay equalizers. In all these cases, the TF to be realized is of the form

$$H(s) = \frac{N(s)}{D(s)} = k\frac{s^m + \cdots + a_1 s + a_0}{s^n + \cdots + b_1 s + b_0}, \quad m \leq n \tag{3.28}$$

An LP filter is to pass low frequencies from DC to a desired frequency ω_c, called the *cutoff frequency*, and to attenuate frequencies beyond the cutoff. An ideal LP filter would have a magnitude for $H(j\omega)$ that is constant from zero to ω_c and zero beyond ω_c. The frequency band from DC to ω_c is called the *passband*. However, such a brick-wall characteristic is impossible to realize in practice, and the LP filter specifications are always given in terms of its cutoff frequency ω_c, an SB frequency ω_s, maximum loss (A_p) allowed in the PB, and minimum SB attenuation (A_s). The

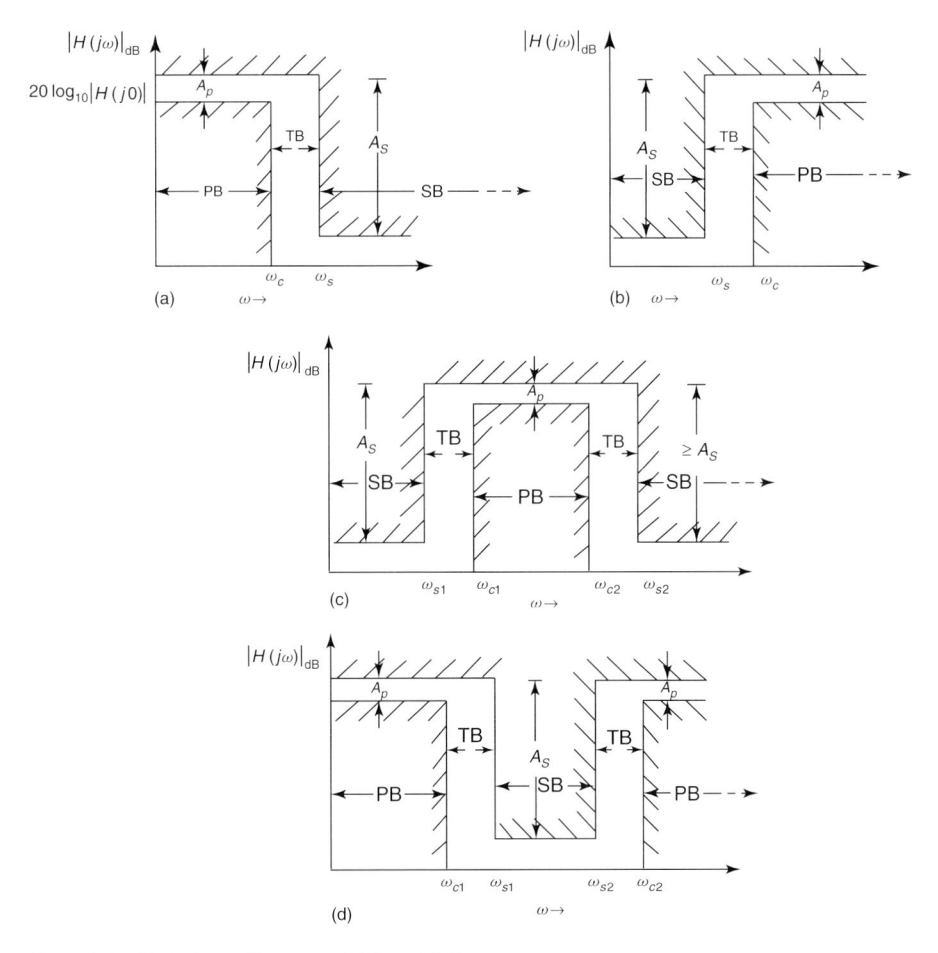

Figure 3.9 Typical specifications of (a) an LP filter, (b) an HP filter, (c) a BP filter, and (d) a BR filter.

frequency range from ω_c to ω_s is termed the *transition band* (TB) (see Figure 3.9a). An ideal HP filter passes all frequencies beyond the cutoff frequency ω_c and attenuates all frequencies below it. However, a practical HP filter is characterized in a way similar to the LP filter in terms of an SB frequency and a maximum PB loss (see Figure 3.9b). A BP filter passes a finite band of frequencies bounded by two cutoff frequencies ω_{c1} and ω_{c2}, and attenuates all frequencies below ω_{c1} and those above ω_{c2}. The attenuations in the two SBs may be different, as well as the widths of the two TBs (see Figure 3.9c). A BR filter attenuates a finite band of frequencies bounded by two cutoff frequencies ω_{c1} and ω_{c2}, and passes all frequencies below ω_{c1} and those above ω_{c2}. Thus, a BR filter has two PBs and the attenuations in the two PBs may be different; also the widths of the two TBs may be different (see Figure 3.9d).

Sometimes we need a filter which has a null at a particular frequency; such filters are called *null or notch filters*. Apart from these filters, we also have gain equalizers that are used to shape the gain versus frequency spectrum of a given signal. For audio signal processing, phase is not as important as the magnitude of the TF. However, it is very important in the case of video signal processing. The phase distortion causes a variable delay. In such a case, an AP filter is used for phase correction or to provide delay equalization. An AP is one for which the magnitude of the TF is constant, whereas its phase is a function of frequency.

3.9
Magnitude Approximation

The specifications for a filter, whether it be a LP, HP, BP, or BR, are given in terms of the PB, TB, and SB regions as well as in terms of the loss requirements in the PB and the SB. The magnitude approximation problem consists of finding a suitable magnitude function, whose magnitude characteristics satisfy the given specifications, and which is realizable in practice. We first consider the approximation problem for a LP filter, and then introduce frequency transformations so that we can design HP, symmetrical BP, or BR filters for a given set of specifications. For an LP filter, these specifications are usually given in terms of (i) the *maximum loss* in the PB, A_p, in decibels (dB), (ii) the PB edge (i.e., cutoff) frequency, ω_c, in radians per second (rad s^{-1}), (iii) the loss in the SB, A_s, in decibels, and (iv) the SB edge frequency, ω_s, where the loss is *at least* A_s decibels, as shown in Figure 3.9. Without loss of generality, we assume that the cutoff frequency for the LP filter is 1 rad s^{-1}, since we can always employ frequency scaling to convert it to an LP filter of any given cutoff frequency. Such an LP filter is called the *normalized LP filter*, and we shall denote the magnitude response of such a normalized filter by $|H_N(j\omega)|$. In the ideal case, $|H_N(j\omega)| = 1$ in the PB ($0 \leq \omega \leq 1$) and zero in the SB ($\omega > 1$). One can define the characteristic function $K(s)$ of the filter to be such that

$$|K(j\omega)|^2 = \frac{1}{|H_N(j\omega)|^2} - 1 \qquad (3.29)$$

Thus, $|K(j\omega)|^2 \to 0$ in the PB and $|K(j\omega)|^2 \to \infty$ in the SB. The magnitude approximation principle assumes that $|K(j\omega)|^2$ remain *less than or equal to* ε^2 in the PB, where ε is a small number. Similarly in the SB, $|K(j\omega)|^2$ approximates to a value *greater than or equal to* δ^2, where δ is a very large number. In the PB, the loss A_p in decibels is related to the number ε as follows:

$$A_p = 10 \log \frac{1}{|H_N(j\omega)|^2}$$

so that

$$\varepsilon^2 = |K(j\omega)|^2 = \frac{1}{|H_N(j\omega)|^2} - 1 = 10^{0.1 A_p} - 1$$

Hence,

$$\varepsilon = \sqrt{10^{0.1 A_p} - 1} \tag{3.30}$$

By similar reasoning, we arrive at

$$\delta = \sqrt{10^{0.1 A_s} - 1} \tag{3.31}$$

It must be noted that now onwards ω is the normalized angular frequency, normalized w.r.t. ω_c. Similarly, ω_s is the normalized SB edge frequency, i.e., $\omega_s = \omega_a/\omega_c$, where ω_a is the given SB edge frequency

3.9.1
Maximally Flat Magnitude (MFM) Approximation

Assuming the function $K(s)$ to be

$$K(s) = \varepsilon s^n \tag{3.32}$$

we get from Eq. (3.29)

$$|H_n(j\omega)| = \frac{1}{\sqrt{1 + \varepsilon^2 \omega^{2n}}} \tag{3.33}$$

Expanding the RHS of the above expression, we have

$$|H_n(j\omega)| = 1 - \varepsilon^2 \frac{1}{2} \omega^{2n} + \varepsilon^4 \frac{3}{8} \omega^{4n} - \varepsilon^6 \frac{5}{16} \omega^{6n} + \cdots \tag{3.34}$$

It is clear from Eq. (3.34) that the first $(n-1)$ derivatives of $|H_n(j\omega)|$ are zero at $\omega = 0$ (i.e., at DC), and further that $|H_n(0)| = 1$ from Eq. (3.33). Thus, the slope is as flat as can be made since $K(s)$ was chosen as the nth-order polynomial. This is why Eq. (3.33) is called *maximally flat magnitude* (MFM) approximation for the LP filter.

At the edge of the PB, the frequency is $\omega_c = 1$, and $\varepsilon^2 = (10^{0.1 A_p} - 1)$. Also, at the SB frequency ω_s, $(10^{0.1 A_s} - 1) = \varepsilon^2 \omega_s^{2n}$. Hence, the minimum order of the MFM filter necessary to meet the specifications is given by

$$n = \frac{\log \eta}{2 \log \omega_s} \tag{3.35a}$$

where

$$\eta = \frac{10^{0.1 A_s} - 1}{10^{0.1 A_p} - 1} \tag{3.35b}$$

If n is not an integer, the next higher integer value is chosen.

One of the important special cases of the MFM approximation is the *Butterworth approximation*, which is obtained by choosing $\varepsilon = 1$. It is seen in this case that at the PB edge, the magnitude of the Butterworth filter is always 0.707, that is, 3 dB down from its DC ($\omega = 0$) value, irrespective of the order of the filter.

3.9.1.1 MFM Filter Transfer Function

It is clear from Eq. (3.33) that the TF $H_N(s)$ will be an all-pole function. Assuming the denominator of $H_N(s)$ to be $D(s)$, we have

$$|H_N(s)|^2 = \frac{1}{D(s)D(-s)} \tag{3.36}$$

Since the poles of $H_N(s)$ have to be in the left half (LH) of the s-plane for stability, we associate the LH plane poles to $H_N(s)$ and the right-half ones to $H_N(-s)$. Now extending the MFM approximation given by Eq. (3.33) to the s-domain, we get

$$|H_N(s)|^2 = \frac{1}{1 + \varepsilon^2 (-s^2)^n} \tag{3.37}$$

Hence, we associate the LH plane roots of

$$1 + \varepsilon^2(-s^2)^n = 0 \quad \text{or} \quad \varepsilon^{2n} s^{2n} = (-1)^{n+1} \tag{3.38}$$

with the poles of $D(s)$. The roots of Eq. (3.38) are complex and are given by

$$s_k = \varepsilon^{\frac{-1}{n}} \exp\left(j \frac{2k+n-1}{2n} \pi\right) \quad k = 1, 2, \ldots, 2n \tag{3.39}$$

where we have used the fact that $\exp(j\pi) = -1$. These roots are all uniformly located on a circle of radius $\varepsilon^{\frac{-1}{n}}$; hence, for a Butterworth filter approximation, they are all uniformly located on the unit circle. Denoting the poles of the MFM filter by p_k, and associating the LH plane roots of Eq. (3.39) with p_k, we get the poles to be

$$p_k = \varepsilon^{\frac{-1}{n}} \exp\left(j \frac{2k+n-1}{2n} \pi\right) \quad k = 1, 2, \ldots, n \tag{3.40}$$

The denominator $D(s)$ of the normalized MFM filter is given by

$$D(s) = \prod_k (s - p_k) \tag{3.41}$$

For Butterworth filters, the polynomial $D(s)$ for various values of n is tabulated in Appendix A, and these polynomials are called *Butterworth polynomials*.

Example 3.5. Derive the TF for an MFM filter with a PB loss of 3 dB at 10 kHz and an attenuation of at least 100 dB at $f \geq 100$ kHz.

It is clear that $A_p = 3$ dB, $A_s = 100$ dB, $\omega_c = 2\pi \times 10^4$ rad s^{-1}, and $\omega_s = 2\pi \times 10^5$ rad s^{-1}. Using these values in Eq. (3.35) we obtain the order of the filter as $n = 5$. Since the PB loss is 3 dB, it is obvious that it corresponds to a Butterworth filter. Hence, we can use Appendix A to obtain the Butterworth polynomial of degree 5, which will be the denominator of the normalized Butterworth filter of order 5. Thus, the normalized TF is

$$H_N(s) = \frac{1}{(s+1)(s^2 + 0.618s + 1)(s^2 + 1.618s + 1)} \tag{3.42}$$

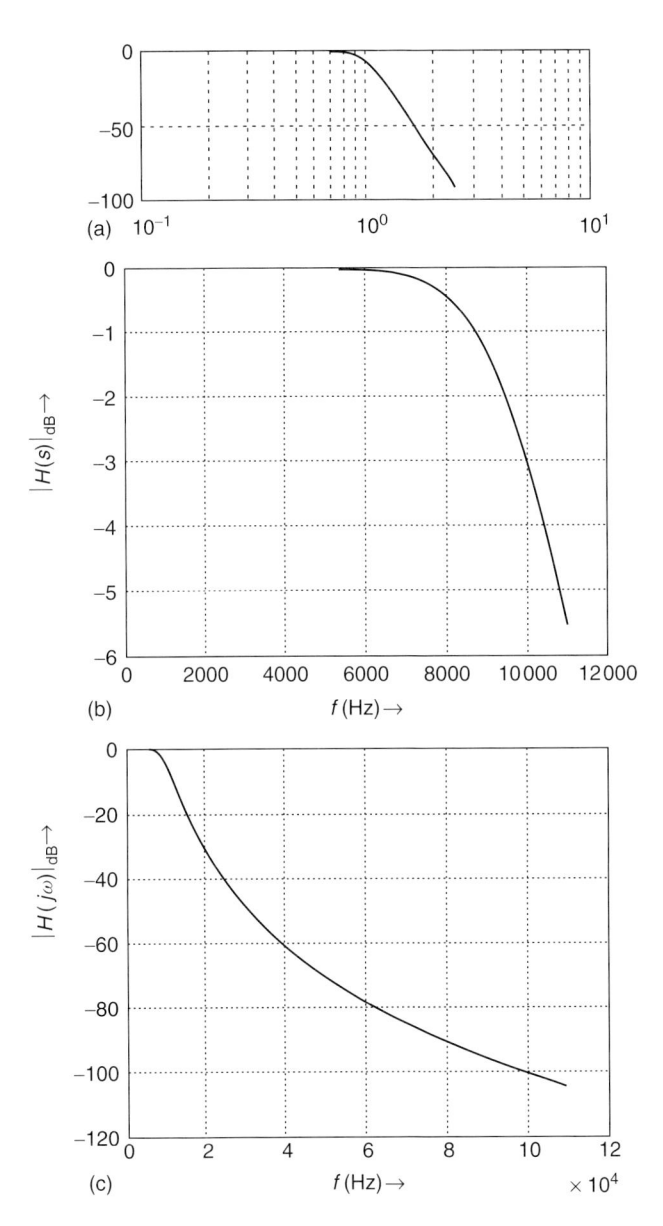

Figure 3.10 Magnitude response of (a) Butterworth fifth-order normalized LP filter. (b) Response of the desired filter around the PB edge. (c) The overall response of the desired filter.

The next task is to derive the frequency-denormalized filter TF $H(s)$ by frequency scaling:

$$H(s) = H_N \left(\frac{s}{\omega_c} \right)$$

$$= \frac{9.7967 \times 10^{23}}{[(s + 6.2842 \times 10^4)(s^2 + 3.8835 \times 10^4 s + 3.9488 \times 10^9) \times (s^2 + 1.0166 \times 10^5 s + 3.9488 \times 10^9)]}$$

The response of the normalized filter as well as that of the required Butterworth filter, obtained using MATLAB program, is depicted in Figures 3.10a–3.10c, and the MATLAB program for the magnitude response of the normalized filter is provided in Program 3.1, while that for the denormalized filter is provided in Program 3.2.

Program 3.1 MATLAB code for the normalized Butterworth filter of Example 3.5

```
%MATLAB program listing for H_N(s), 5th order Butterworth Filter
w=logspace(-1,1);
s=0+w*i;b=[0,0,0,0,0,1];a=[1,3.2361,5.2361,5.2361,3.2361,1];
h=freqs(b,a,w);amag=abs(h);
y=20*log(amag);subplot(2,1,1),semilogx(w,y);
end
```

Program 3.2 MATLAB code for the denormalized Butterworth filter of Example 3.5

```
%MATLAB program listing for |H(s)|, 5th order Butterworth
Filter
pi=3.14159;
fc=1E4;
fs=1E5;
n=2*pi*fc;% to be used for frequency scaling factor
b6=n^5;
a1=1.0;a2=3.2361*n;a3=5.2361*n^2;a4=5.2361*n^3;a5=3.2361*n^4;
a6=b6;
b=[0,0,0,0,0,b6];a=[a1,a2,a3,a4,a5,a6];
r1=1.1;
r2=1.1*(fs/fc);
np=100; % number of points in the graph
df=(r1*fc)/np;% this will show response up to 1.1 times the
passband
% edge frequency
df=(r2*fc)/np;% this will show re-
sponse up to 11.1 times the passband
% edge frequency, i.e., beyond the given stopband edge
frequency
for n=1:(np+1)
w=(n-1)*2*pi*df;
s=0+w*i;
anum=b6;
denm=a1*s^5+a2*s^4+a3*s^3+a4*s^2+a5*s+a6;
```

```
tfs=anum/denm;
amag=abs(tfs);
y(n)=20*log10(amag);
f=w/(2*pi);
x(n)=f;
end;
plot(x,y)
grid
end
```

Example 3.6. Repeat the previous problem, but for a PB loss of only 1 dB, with all the other specifications remaining the same.

Using Eq. (3.35) for n, one finds $n = 5.29$. Hence, we choose $n = 6$. Further for $A_p = 1$ dB, $\varepsilon = 0.50885$. Thus, $\varepsilon^{-1/n} = 1.11918$. An examination of Eq. (3.40) shows that the poles of the MFM filter for a given n for any value of ε are the same as those of the Butterworth filter of order n, except that they are multiplied by $\varepsilon^{-1/n}$. We shall first use the table of Butterworth functions for $n = 6$ to obtain the denominator polynomial as

$$D_B(s) = s^6 + 3.8637s^5 + 7.4641s^4 + 9.1416s^3 + 7.4641s^2 + 3.8637s + 1$$

(3.43)

In order to get the $D(s)$ for $A_p = 1$ dB, we substitute $\varepsilon^{1/n}s = 0.8935s$ in place of s in $D_B(s)$ given by Eq. (3.43). Hence, the denominator polynomial for the required MFM filter is given by

$$D(s) = D_B(0.8935s)$$

and the normalized MFM by $H_N(s) = 1/D(s)$. The MFM filter required to satisfy the given specifications is obtained by frequency scaling of $H_N(s)$, as was done before in the previous problem by scaling $s \to s/(2\pi \times 10^4)$. The final TF is given by

$$H(s) = \frac{1}{\begin{array}{l}0.82695 \times 10^{-29}s^6 + 0.22468 \times 10^{-23}s^5 + 0.30523 \times 10^{-18}s^4 \\ + 0.26288 \times 10^{-13}s^3 + 0.15094 \times 10^{-8}s^2 + 0.54943 \times 10^{-4}s + 1\end{array}}$$

The magnitude response plot, obtained from MATLAB code, is shown in Figures 3.11a and 3.11b, and the MATLAB code listing is given in Program 3.3.

Program 3.3 MATLAB code for the magnitude response of the LP filter of Example 3.6

```
%MATLAB program listing for |H(s)| of Example 3.6
pi=3.14159;
eps=0.50885;
mord=6;
an=[1,3.8637,7.4641, 9.1416, 7.4641, 3.8637, 1];
bn=[ 0,0,0,0,0,0,1];
scf=eps^(1/mord);
a1=an(1)*scf^mord;
a2=an(2)*scf^(mord-1);
```

```
a3=an(3)*scf^(mord-2);
a4=an(4)*scf^(mord-3);
a5=an(5)*scf^(mord-4);
a6=an(6)*scf^(mord-5);
a7=an(7)*scf^(mord-6);
fc=1E4;
fs=1E5;
r1=1.1;
r2=1.1*(fs/fc);
n=2*pi*fc;
b7=n^6;
a1=a1; a2=a2*n; a3=a3*n^2; a4=a4*n^3; a5=a5*n^4; a6=a6*n^5;
a7=b7;
b=[0,0,0,0,0,0,b7];a=[a1,a2,a3,a4,a5,a6,a7];
np=100; % number of points
df=(r1*fc)/np;% use this for response up to around the pass-
band edge frequency
% df=(r2*fc)/np; % use this for the overall response
for n=1:(np+1)
f=(n-1)*df;
w=(n-1)*2*pi*df;
f=w/(2*pi);
s=0+w*i;
anum=b7;
denm=a1*s^6+a2*s^5+a3*s^4+a4*s^3+a5*s^2+a6*s+a7;
tfs=anum/denm;
amag=abs(tfs);
y(n)=20*log10(amag);
x(n)=f;
end;
plot(x,y)
grid
end
```

3.9.2
Chebyshev (CHEB) Magnitude Approximation

The response of the MFM filter is very good around $\omega = 0$ and for large values of ω. However, in the neighborhood of the cutoff frequency, the selectivity is not very good. This is so because we chose $|K(j\omega)|^2$ to be just $\varepsilon^2 \omega^{2n}$, all of whose zeros are at $\omega = 0$. However, if we chose it to be a polynomial $\varepsilon^2 f(\omega^2)$ where $f(\omega^2)$ had a number of zeros spread out in the PB, then the response would have ripples in the PB, and the performance would improve. For this purpose, in this section we assume $|K(j\omega)| = \varepsilon C_n(\omega)$, where $C_n(\omega)$ is the nth-order CHEB polynomial, defined as

$$C_n(\omega) = \cos(n \cos^{-1} \omega) \tag{3.44}$$

It can be shown that these polynomials satisfy the recurrence relation (Van Valkenburg, 1960)

$$C_{n+2}(\omega) = 2\omega C_{n+1}(\omega) - C_n(\omega) \quad \text{with} \quad C_0(\omega) = 1, C_1(\omega) = \omega \tag{3.45}$$

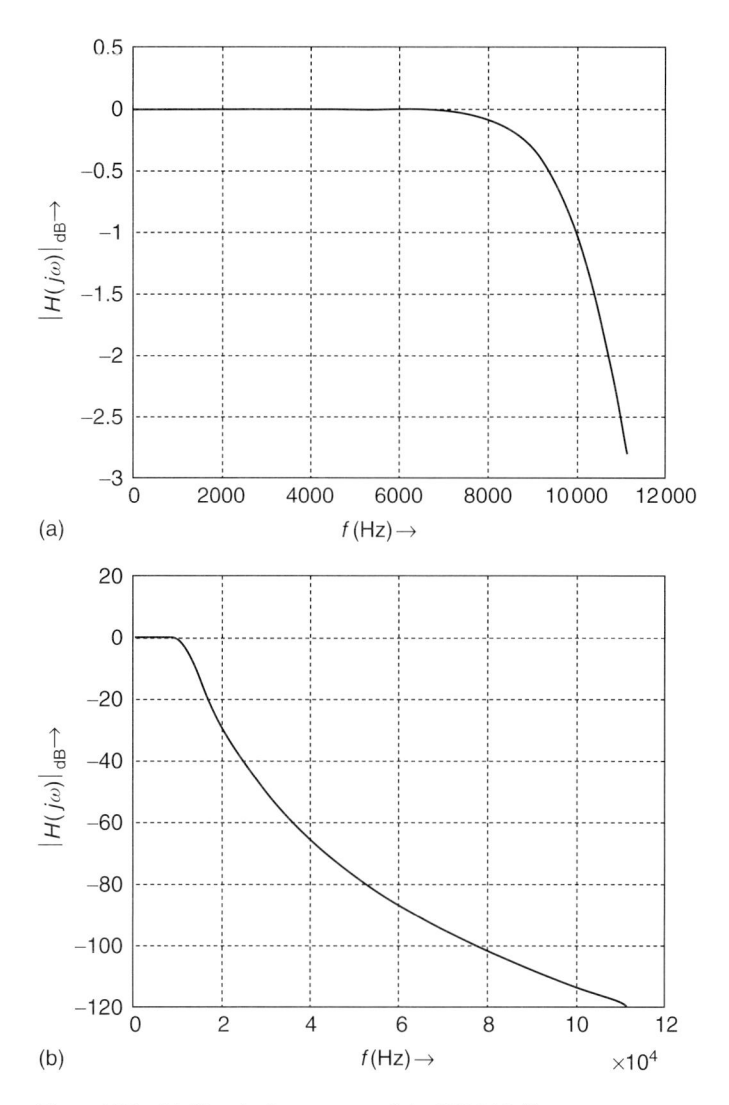

Figure 3.11 (a) Magnitude response of the MFM LP filter of Example 3.6 in the neighborhood of the PB edge. (b) The overall response.

We can easily prove the following properties of $C_n(\omega)$:

1) The zeros of $C_n(\omega)$ are given by

$$\omega_k = \cos\{(2k-1)\pi/2n\}, k = 1, 2, \ldots, n \qquad (3.46)$$

and thus, all the n zeros of the polynomial are real, distinct, and located in $-1 \leq \omega < 1$.

2) $C_n(\omega)$ is an even function for even n and an odd function for odd n.

3) $|C_n(j\omega)| \leq 1$ in $-1 \leq \omega \leq 1$.
4) $C_n(1) = 1$.
5) $|C_n(\omega)|$ is monotonically increasing for $|\omega| > 1$.
6) For $|\omega| > 1$, $C_n(\omega)$ can be written as

$$C_n(\omega) = \cosh(n \cosh^{-1} \Omega) = \frac{1}{2}\left[(\omega + \sqrt{\omega^2 - 1})^n + (\omega + \sqrt{\omega^2 - 1})^{-n}\right] \tag{3.47}$$

A few of the CHEB polynomials are given in Appendix A. Substituting $|K(j\omega)| = \varepsilon C_n(\omega)$ in Eq. (3.25), we get the magnitude of the TF for CHEB approximation to be

$$|H_N(j\omega)|^2 = \frac{1}{1 + \varepsilon^2 C_n^2(\omega)} \tag{3.48}$$

From the properties of $C_n(\omega)$ listed above, we can conclude the following properties regarding $|H_N(j\omega)|$:

1) $|H_N(0)| = 1$ if n is odd and is $1/\sqrt{1+\varepsilon^2}$ if n is even.
2) $|H_N(j1)| = 1/\sqrt{1+\varepsilon^2}$ for all n, that is, the magnitude of the CHEB function is always $1/\sqrt{1+\varepsilon^2}$ at the edge of the PB.
3) $|H_N(j\omega)|$ oscillates with equal-ripple throughout the PB between 1 and $1/\sqrt{1+\varepsilon^2}$.
4) The number of peaks and troughs of the response within the PB is equal to the order n.
5) $|H_N(j\omega)|$ monotonically decreases to zero for $|\omega| > 1$.

The responses for $n = 2$ and 3 with $\varepsilon = 0.5089$ are shown in Figure 3.12, and the MATLAB program listing for LP CHEB filter of orders 2 and 3 is given in Program 3.4.

Program 3.4 MATLAB program code for LP CHEB filter of orders 2 and 3

```
%CHEB filter response
ep=.5089;
for n=1:300
w(n)=0.01*n;
if n<101
a2(n)=cos(2*acos(w(n)));
a3(n)=cos(3*acos(w(n)));
else
a2(n)=cosh(2*acosh(w(n)));
a3(n)=cosh(3*acosh(w(n)));
y1(n)=1/(1+ep^2*a2(n)^2);
y2(n)=1/(1+ep^2*a3(n)^2);
end;
grid
plot(w,y1,'b.',w,y2,'r--
');% blue dots ,order=2, red dashes, order=3)
xlabel('Normalized frequency')
ylabel('..m=2,--m=3')
end
```

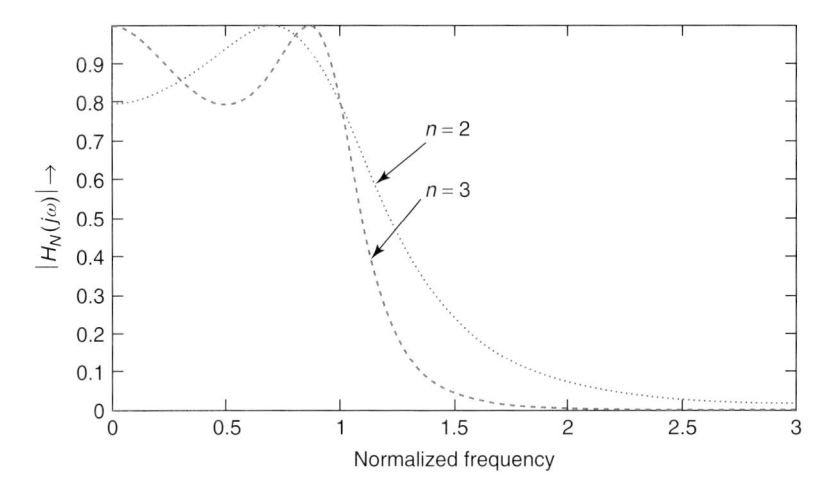

Figure 3.12 Magnitude response of the normalized CHEB filter for $n = 2$ and 3 with $\varepsilon = 0.5089$.

Given the parameter A_p in decibels, we can find the ripple factor ε using

$$\varepsilon = \sqrt{10^{0.1A_p} - 1} \tag{3.49}$$

Also, at the edge of the SB, that is ω_s, the loss in decibels is A_s. Hence,

$$-A_s = 10 \log 1/[1 + \varepsilon^2 C_n^2(\omega_s)]$$

Using Eqs. (3.47) and (3.49), we get the order of the filter, n, to be the lowest integer when

$$n \geq \frac{\cosh^{-1} \sqrt{\eta}}{\cosh^{-1} \omega_s} \tag{3.50a}$$

where

$$\eta = \frac{10^{0.1A_s} - 1}{10^{0.1A_p} - 1} \tag{3.50b}$$

An alternate approximate expression is (Schaumann, Ghausi, and Laker, 1990)

$$n \geq \frac{\ln \sqrt{4(10^{0.1A_s} - 1)/\varepsilon^2}}{\ln(\omega_s + \sqrt{\omega_s^2 - 1})} \tag{3.51}$$

3.9.2.1 CHEB Filter Transfer Function

Just as in the case of the MFM approximation, we start with

$$D(s)D(-s) = 1 + \varepsilon^2 C_n^2(\omega)$$

where $\omega = s/j$, and obtain the roots of $1 + \varepsilon^2 C_n^2(\omega) = 0$, and associate the LH plane roots with those of the poles of $D(s)$. The above equation can be rewritten as

$$\cos\{n \cos^{-1}(-js)\} = \pm j/\varepsilon \tag{3.52}$$

The roots of the above equation can be shown to be $s_k = \sigma_k \pm j\omega_k$, where

$$\sigma_k = -\sinh\alpha \sin\left(\frac{2k-1}{2n}\right)\pi, \quad \omega_k = \cosh\alpha \cos\left(\frac{2k-1}{2n}\right)\pi$$
$$k = 1, 2, \ldots, n \tag{3.53a}$$

and

$$\alpha = (1/n)\sinh^{-1}(1/\varepsilon) \tag{3.53b}$$

It is readily seen from Eq. (3.53) that the real and imaginary parts of s_k satisfy the relation

$$\frac{\sigma_k^2}{\sinh^2\alpha} + \frac{\omega_k^2}{\cosh^2\alpha} = 1 \tag{3.54}$$

Equation (3.54) shows that the roots of $D(s)$ $D(-s)$ all lie on an ellipse with semimajor axis of length $\cosh\alpha$ and semiminor axis of length $\sinh\alpha$. Since the poles of $H_N(s)$ are the left half plane (LHP) zeros of $D(s)$ $D(-s)$, we see that the poles of the CHEB TF are on the ellipse. Further, the TF $H_N(s)$ is given by

$$H_N(s) = \frac{1}{2^{n-1}\varepsilon} \frac{1}{\prod_k (s - p_k)} = \frac{1/(2^{n-1}\varepsilon)}{s^n + b_{n-1}s^{n-1} + \cdots + b_1 s + b_o} \tag{3.55}$$

where

$$p_k = -\sinh\alpha \sin\left(\frac{2k-1}{2n}\right)\pi \pm j\cosh\alpha \cos\left(\frac{2k-1}{2n}\right)\pi,$$
$$k = 1, 2, \ldots, n \tag{3.56}$$

and the factor $2^{n-1}\varepsilon$ appears in the denominator of $H_N(s)$, since the highest term ω^n in $C_n(\omega)$ has a coefficient of $2^{n-1}\varepsilon$. For different values of ε and n, there are tables available giving the denominator polynomial in Eq. (3.55) (Christian and Eisermann, 1977; Weinberg, 1962; Zverev, 1967). For some values of ε and n, these are listed in Appendix A.

Example 3.7. Obtain the TF for the CHEB LP filter, which has a PB loss ripple of 0.5 dB up to 3 kHz and a minimum attenuation of 60 dB at 30 kHz.

The specifications for the normalized CHEB filter are $A_p = 0.5$ dB, $A_s = 60$ dB, $\omega_c = 1$, and $\omega_s = 30/3 = 10$. Hence, from Eqs. (3.49) and (3.51), we have $\varepsilon = 0.3493$ and $n = 3$. Referring to the tables for CHEB filter for $\varepsilon = 0.3493$ and $n = 3$, we get the denominator in Eq. (3.55) to be

$$(s + 0.6264)(s^2 + 0.626s + 1.142) = s^3 + 1.253s^2 + 1.535s + 0.716$$

Hence, the normalized CHEB LP TF according to Eq. (3.55) is

$$H_N(s) = \frac{1/(2^{3-1})(0.3493)}{s^3 + 1.253s^2 + 1.535s + 0.716}$$

Now, we apply frequency scaling as $s \to s/(2\pi)(3)(10^3)$ to obtain the TF $H(s)$ realizing the specifications of the problem. Thus (using a MATLAB program), we get

$$H(s) = \frac{7.1572 \times 10^{27}}{(0.1493 \times 10^{16} s^3 + 3.5265 \times 10^{19} s^2 + 8.1434 \times 10^{23} s + 7.1599 \times 10^{27})}$$

The associated MATLAB program is given in Program 3.5.

Program 3.5 MATLAB code for H(s) of Example 3.7

```
%normalized to denormalized TF derivation
m=3;
ep=.3493;
den='(s^3+1.253*s^2+1.535*s+.716)';
pi=3.14159;
snew='s/(2*pi*3e3)';
ys=subs(den,snew);
ys2=inverse(ys);
ys3=simplify(ys2);
x1=1/(2^(m-1)*ep);
ys4=symmul(ys3,x1)
end
```

Figures 3.13a and 3.13b illustrate the magnitude responses of $H_N(j\omega)$ and $H(j\omega)$, respectively, derived using MATLAB program.

3.9.3
Elliptic (ELLIP) Magnitude Approximation

In Sections 3.9.1 and 3.9.2, we have dealt with all-pole LP filters, whose magnitude square functions were of the form $|H_N(j\omega)|^2 = \frac{H^2}{1+|K(j\omega)|^2}$, where $|K(j\omega)|^2$ was either $\varepsilon^2 \omega^{2n}$ as in the case of MFM, or $\varepsilon^2 C_n^2(\omega)$ as in the case of the CHEB approximation. The all-pole functions do not have zeros at finite frequencies, but have all their zeros at infinity. If we substitute a rational function, $\varepsilon^2 R_n^2(\omega)$ for $|K(j\omega)|^2$, then the TF $H_N(s)$ will have zeros at finite frequencies in addition to poles at finite

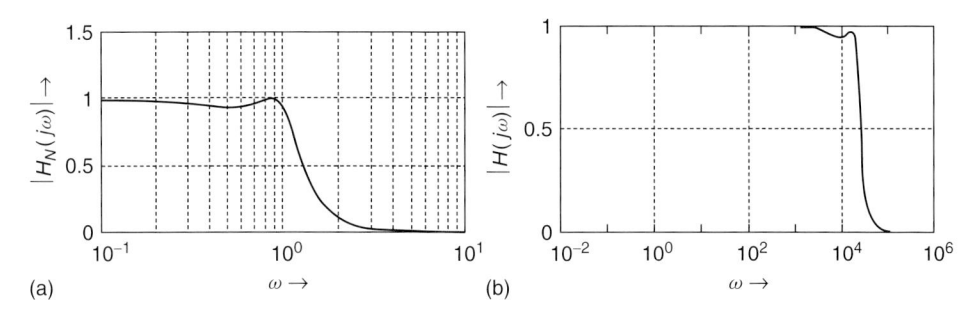

Figure 3.13 Magnitude responses of (a) the normalized and (b) the desired Chebyshev filter of Example 3.7.

frequencies. As far as the frequency response is concerned, this characteristic leads to a higher selectivity feature for a given order "*n*" of the rational function $R_n(\omega)$, compared with the selectivity obtained in case of MFM or CHEB approximating functions of the same order "*n*." For hardware design, this implies that less number of components will be needed to implement the filtering function. However, the mathematical complexity of the function is greatly increased.

When this rational function has the characteristic of a CHEB-type function, the TF will have ripple characteristics both in the PB and in the SB. Approximating the filter specifications with such a function that has poles and zeros at finite frequencies will introduce ripples in the PB as well as in the SB, and such an approximation is known as *elliptic magnitude approximation* for the filter TF, and the filter itself is called an *elliptic LP filter* (Huelsman, 1993). The magnitude-squared function for elliptic approximation is given by

$$|H_N(j\omega)|^2 = \frac{H^2}{1 + \varepsilon^2 R_n^2(\omega)} \qquad (3.57)$$

where the CHEB rational function $R_n(\omega)$ is represented by

$$R_n(\omega) = \left\{ M \prod_{k=1}^{n/2} \frac{\omega^2 - \omega_{p_k}^2}{\omega^2 - \omega_{z_k}^2} \quad \text{for } n \text{ even} \right. \qquad (3.58a)$$

$$= \left\{ M\omega \prod_{k=1}^{(n-1)/2} \frac{\omega^2 - \omega_{p_k}^2}{\omega^2 - \omega_{z_k}^2} \quad \text{for } n \text{ odd} \right. \qquad (3.58b)$$

where

$$\omega_{p_k} \omega_{z_k} = \omega_s^2 \qquad (3.58c)$$

The following points may be noted regarding $|H_N(j\omega)|$:

1) The PB is defined for $0 \leq \omega \leq 1$. The constants M are chosen so that in this region $0 \leq R_n^2(\omega) \leq 1$.
2) The values $\omega = \omega_{p_k}$ at which $R_n^2(\omega) = 0$ represent the PB peaks at which $|H_N(j\omega)| = H$.
3) In the PB, the values of ω at which $R_n^2(\omega) = 1$ correspond to the PB valleys at which $|H_N(j\omega)| = H/\sqrt{1+\varepsilon^2}$.
4) The SB is defined as $\omega \geq \omega_s$. In this region, the minimum value of $R_n^2(\omega)$ is R_{stop}^2, where $R_{stop}^2 \geq \eta = [(10^{0.1A_a} - 1)/(10^{0.1A_p} - 1)]$.
5) In the SB, at values of $\omega = \omega_{z_k}$, $R_n^2(\omega)$ approaches infinity so that $|H_N(j)|$ approaches zero. These frequencies are therefore called the *transmission zeros of the TF*.
6) In the SB, at frequencies where $R_n^2(\omega) = R_{stop}^2$, the attenuation corresponds to the SB peaks, and at these frequencies $|H_N(j\omega)|$ has a minimum attenuation of A_s decibels.

Figure 3.14 shows a plot representing a typical elliptic magnitude approximation function.

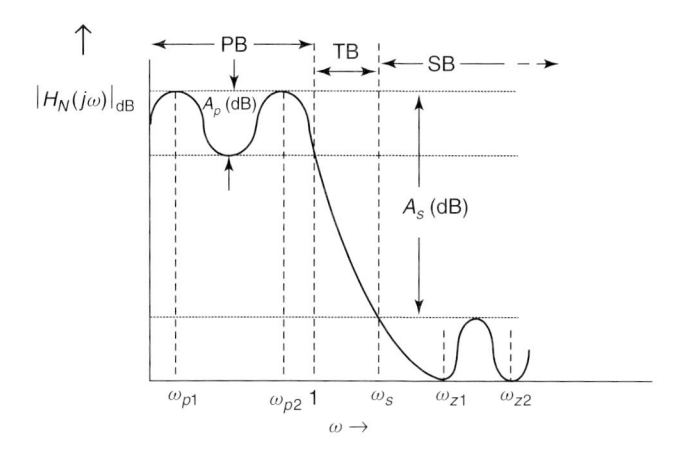

Figure 3.14 A typical elliptic magnitude approximation function.

As with the MFM and CHEB function approximations, we need to determine the order n of the function $R_n(\omega)$. For this purpose, we first calculate from the specifications,

$$\eta = \frac{10^{0.1A_s} - 1}{10^{0.1A_p} - 1} \tag{3.59a}$$

Then, we let

$$u(\eta) = \frac{1}{16\eta}\left(1 + \frac{1}{2\eta}\right) \text{ and } v(\omega_s) = \frac{\sqrt{\omega_s} - 1}{2(\sqrt{\omega_s} + 1)} \tag{3.59b}$$

Then, the order n is given by (Huelsman, 1993)

$$n = F(u)F(v) \tag{3.60a}$$

where

$$F(x) = \frac{1}{\pi}\ln(x + 2x^5 + 15x^9) \tag{3.60b}$$

Once the value of n is obtained, one can obtain the TF of the elliptic filter by using tables, which are available in many books on filter design (Huelsman, 1993; Schaumann, Ghausi, and Laker, 1990; Zverev, 1967; Christian and Eisermann, 1977). Explicit expressions are available to calculate the coefficients of the elliptic filter, given the specifications of the desired LP filter (Antoniou, 2006). For a few cases, the TFs of the elliptic filter are given in Appendix A.

Example 3.8. Consider the following specifications. PB edge frequency $f_c = 1000$ Hz, $A_p = 3$ dB, SB edge frequency $f_s = 1300$ Hz, $A_s = 22$ dB. Find the minimum order of the LP filters satisfying these specifications using (i) MFM approximation, (ii) CHEB approximation, and (iii) elliptic (ELLIP) approximation.

Using Eqs. (3.35), (3.50), and (3.60), we can derive the orders to be $n_{MFM} = 10$, $n_{CHEB} = 5$, and $n_{ELLIP} = 3$. Clearly, the elliptic approximation provides the same frequency response selectivity with a substantially lower order of implementation. A MATLAB code to perform the calculations is given in Program 3.6. Once the order of the filter is known, we can get the complete TF for the elliptic filter using tables.

Program 3.6 MATLAB code to calculate the orders of the filters in Example 3.8

```
%Filter order calcul
ap=3;as=22;fc=1000;fa=1300;
ws=fa/fc;
d=(10^(.1*as)-1)/(10^(.1*ap)-1);
%MFM approximation
nmfm=(log10(d))/(2*log10(ws))
%CHEB approximation
dr=sqrt(d);
ncheb=(acosh(dr))/acosh(ws)
%Elliptic approximation
cd=(1/(16*d))*(1+1/(2*d));
dws=(sqrt(ws)-1)/(sqrt(ws)+1);
dws=dws/2;
x1=(1/pi)*log(cd+2*cd^5+15*cd^9);
x2=(1/pi)*log(dws+2*dws^5+15*dws^9);
nelp=x1*x2
end
```

3.9.4
Inverse-Chebyshev (ICHEB) Magnitude Approximation

As the name suggests, in this approximation, the response in the PB is monotonic while the response in the SB has equal ripples, which is the inverse of what happens in a CHEB filter, and hence the approximation is called *Inverse-Chebyshev (ICHEB) approximation*. Such characteristics will usually produce somewhat inferior selectivity relative to the CHEB approximation because of the resemblance of the PB response to MFM approximation in the PB. Higher selectivity in magnitude characteristic, however, is attended to with large fluctuations in the phase and delay characteristics of the filter through the TB. Thus, in ICHEB, the delay characteristic will have less fluctuation as the signal frequencies traverse through the PB to the SB. Hence, filters with ICHEB characteristics are preferred in situations where small delay variations are important, such as in video or data transmission (Schaumann, Ghausi, and Laker, 1990).

In ICHEB approximation, the magnitude-squared function of the normalized LP filter has the form

$$|H_N(j\omega)|^2 = \frac{\varepsilon^2 C_n^2(1/\omega)}{1+\varepsilon^2 C_n^2(1/\omega)} \tag{3.61}$$

It can be shown (Huelsman, 1993) that the CHEB and ICHEB approximation functions have the same order for same set of loss parameters (A_p and A_s) and same SB to PB edge frequency ratio. The ICHEB approximation function can be directly obtained from the CHEB approximation function by using the relation

$$|H_N^{ICHEB}(j\omega)|^2 = 1 - |H_N^{CHEB}(j/\omega)|^2 \qquad (3.62)$$

It should be observed that the cutoff frequency of the ICHEB filter is no longer at $\omega = 1$, but is the beginning of the SB.

3.10 Frequency Transformations

Frequency transformations are used to synthesize HP, BP, BR, and other filters from a normalized LP filter. The synthesis of lossless LP filter networks, whose magnitude specifications are approximated by different types of functions such as Butterworth, CHEB, and elliptic, has been treated in detail by many of the authors dealing with passive filter design (Weinberg, 1962; Chen, 1964), and the element values are readily available in the literature (Huelsman, 1993; Weinberg, 1962; Christian and Eisermann, 1977; Zverev, 1967). Thus, the frequency transformations are very useful in obtaining realizations for the HP, BP, and BR filters from those of the LP filters. The motivation behind a frequency transformation is to find a function $F(s)$ such that the PB of the LP filter is transformed into the PB of the required filter, and, at the same time, to have the SB of the original LP filter transformed into the SB of the required filter.

3.10.1 LP to HP Transformation

Consider a normalized LP filter $H_{NLP}(S)$ for which the cutoff frequency is normalized to unity. Consider the transformation

$$S(=j\Omega) \to \frac{\omega_c}{s} \qquad (3.63a)$$

or equivalently

$$\Omega \to -\frac{\omega_c}{\omega} \qquad (3.63b)$$

Then, it is very clear that whatever be the magnitude at $-\Omega$ in the LP filter, the same would be the magnitude at ω_c/ω for the transformed filter. Since the magnitude characteristic is symmetric with respect to Ω, we see that the PB of the latter filter would be from ω_c to infinity and the SB from DC to ω_c. Thus, the transformed filter would correspond to a HP filter with cutoff frequency at ω_c, TB from ω_c/Ω_s to ω_c, and PB for $\omega_c > 1$, with PB ripple and SP attenuation being unchanged. In fact

$$H_{HP}(s) = H_{NLP}(\omega_c/s) \qquad (3.64)$$

It is also clear that if the LP filter has been realized by RLC elements, then the HP filter is realized from the LP filter just by replacing inductors by capacitors and vice versa. The actual element values are given in Table 3.2.

3 Network Theorems and Approximation of Filter Functions

Table 3.2 Frequency transformations and transformed elements.

Normalized LP filter	Corresponding HP filter	Corresponding BP filter	Corresponding BR filter
$H_{LP}(s)$	$H_{HP}(s) = H_{NLP}(\omega_c/s)$	$H_{BP}(s) = H_{NLP}\left(\frac{s^2+\omega_0^2}{Bs}\right)$	$H_{BR}(s) = H_{NLP}\left(\frac{Bs}{s^2+\omega_0^2}\right)$
$L_i \Rightarrow$	$1/L_i\omega_c$	$L_B = L_i/B$, $\quad L'_B = B/C_j\omega_0^2$	$L_e = BL_i/\omega_0^2$, $\quad L'_e = 1/BC_j$
$C_j \Rightarrow$	$1/C_j\omega_c$	$C_B = B/L_i\omega_0^2$, $\quad C'_B = C_j/B$	$C_e = 1/BL_i$, $\quad C'_e = BC_j/\omega_0^2$

From the above discussion, it is clear that given a set of specifications A_p, A_s, ω_c, and ω_s for an HP filter, we can convert the specifications to that of a normalized LP filter by using Eq. (3.63a). The corresponding specifications of the LP filter would then be A_p, A_s, $\Omega_c = 1$, and $\Omega_s = \omega_c/\omega_s$. We can then approximate this LP filter by any of the approximations considered in Section 3.9 and obtain the LP TF. The corresponding HP TF is then obtained using Eq. (3.64). To illustrate this, let us consider the following example.

Example 3.9. Derive the TF for a HP filter with a -3-dB frequency of 100 kHz and an attenuation of at least 100 dB for $f \leq 10$ kHz. Use MFM approximation method.

In this case, $\omega_c = 2\pi(10^5)$ rad s^{-1} and $\omega_s = 2\pi(10^4)$ rad s^{-1}. Hence, $\Omega_s = \omega_c/\omega_s = 10$ and $\Omega_c = 1$. Further $A_p = 3$ dB and $A_s = 100$ dB. These specifications are the same as that of the normalized LP considered in Example 3.5. Thus, the TF of the normalized LP filter is

$$H_{NLP}(S) = \frac{1}{(S+1)(S^2+0.618S+1)(S^2+1.618S+1)} \qquad (3.65)$$

Applying the LP to HP transformation (Eq. (3.64)), that is, changing $S \to (2\pi \times 10^5)/s$, we get the TF of the required HP filter to be

$$H_{HP}(s) = \frac{6.25 \times 10^{27} s^5}{(3.1416 \times 10^{15} + 5 \times 10^9 s)(4.9348 \times 10^{31} + 4.854 \times 10^{14} s + 1.250 \times 10^9 s^2)(3.9478 \times 10^{31} + 1.0166 \times 10^{15} s + 10^9 s^2)}$$

3.10.2
LP to BP Transformation

Consider the transformation

$$S \rightarrow \frac{s^2 + \omega_0^2}{Bs} \qquad (3.66)$$

on an LP filter given by $H_{LP}(s)$, where $S = j\Omega$ and $s = j\omega$ are the complex frequencies for the LP filter and the transformed filter. Then, we have on the imaginary axes,

$$\Omega = -\frac{-\omega^2 + \omega_0^2}{B\omega} \qquad (3.67)$$

It is seen that the frequency $\Omega = 0$ is mapped to the point $\omega = \omega_0$. Also, the cutoff frequencies $\Omega_c = \pm 1$ map to the points ω_{c2} and ω_{c1}, where ω_{c2} and ω_{c1} satisfy the equation

$$\omega^2 \pm B\omega - \omega_0^2 = 0 \qquad (3.68)$$

From these it can be shown that

$$B = \omega_{c2} - \omega_{c1}, \quad \omega_{c2}\omega_{c1} = \omega_0^2 \qquad (3.69)$$

Hence, the transformation (Eq. (3.66)) converts a normalized LP filter into a BP filter with lower and upper cutoff frequencies at ω_{c1} and ω_{c2}, with a bandwidth B and with the center frequency of the PB at ω_0, which is the geometric mean of its lower and upper cutoff frequencies. The TF of the BP filter is given by

$$H_{BP}(s) = H_{LP}\left(\frac{s^2 + \omega_0^2}{Bs}\right) \qquad (3.70)$$

It is to be observed that the BP filter that we obtain by this transformation is always symmetric, in the sense that the upper and lower SBs have the same attenuation requirement and the center frequency is always the geometric mean of the upper and lower cutoff frequencies. The quality factor of the BP filter is defined as

$$Q = \frac{\omega_0}{B} \qquad (3.71)$$

It may also be verified that the frequencies ω_{s1} and ω_{s2} corresponding to the SB edges of the BP filter satisfy the relation $\omega_{s1}\omega_{s2} = \omega_0^2$, and further that $\omega_{s2} - \omega_{s1} = B\Omega_s$, where Ω_s corresponds to the SB edge of the LP filter. Table 3.2 shows how the inductors and capacitors in the LP filter are transformed in the BP filter. It may be mentioned that the above method can be adapted to nonsymmetrical BP requirements, if need be (see Example 3.10).

From the above discussion, it is clear that we can obtain the TF of a symmetrical BP filter for a given set of specifications, A_p, A_s, $B = \omega_{c2} - \omega_{c1}$, ω_{s1}, ω_{s2}, (where $\omega_{c2} \omega_{c1} = \omega_{s1} \omega_{s2} = \omega_0^2$), by using the following steps.

1) Obtain the specifications of the corresponding normalized LP filter:

$$A_p, A_s, \Omega_c = 1, \Omega_s = (\omega_{s2} - \omega_{s1})/B = (\omega_{s2} - \omega_{s1})/(\omega_{c2} - \omega_{c1})$$

2) Approximate the normalized LP filter using one of the approximations considered in Section 3.9.
3) Then, the required BP TF is obtained by using the LP–BP transformation given by Eq. (3.70).

We now illustrate this by an example.

Example 3.10. Given that $\omega_{c1} = 430$ rad s^{-1}, $\omega_{c2} = 600$ rad s^{-1}, $\omega_{s1} = 350$ rad s^{-1}, $\omega_{s2} = 700$ rad s^{-1}, $A_p = 0.5$dB, $A_s = 40$ dB, find the BP filter TF. CHEB approximation is to be used.

It is clear that for this set of specifications, $\omega_{c1}\omega_{c2} = 258\,000$ and $\omega_{s1}\omega_{s2} = 245\,000$; thus, $\omega_{c2}\omega_{c1} = \omega_{s1}\omega_{s2}$ is not satisfied. However, by choosing $\omega'_{s1} = 368.571$, we can satisfy the condition $\omega_{c2}\omega_{c1} = \omega'_{s1}\omega_{s2} = (507.937)^2$, thus making $\omega_0 = 507.937$. The change we have made for ω_{s1} has not in any way changed the conditions of the PB and SB requirements, but only reduced the TB on the lower side. This is illustrated in Figure 3.15.

(a) The specifications of the corresponding normalized LP filter are

$$A_p = 0.5 \text{ dB}, A_s = 40 \text{ dB}, \Omega_c = 1, \Omega_s = (\omega'_{s1} - \omega_{s1})/(\omega_{c2} - \omega_{c1}) = 1.95$$

(b) To approximate the above LP filter by CHEB, we use Eq. (3.50b) to calculate

$$\eta = \frac{10^{0.1 A_s} - 1}{10^{0.1 A_p} - 1} = 81946.62$$

We now calculate the required value of n, the order of the filter, using Eq. (3.50a)

$$n \geq \frac{\cosh^{-1}\sqrt{\eta}}{\cosh^{-1}\Omega_s} = 4.703$$

Hence, we choose $n = 5$.

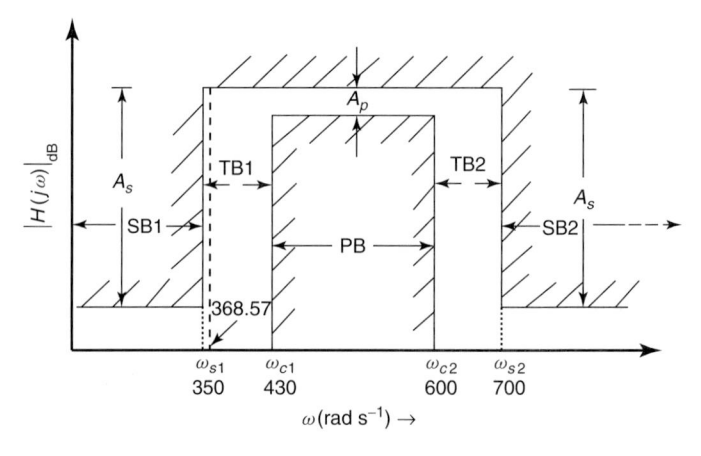

Figure 3.15 Characteristics of the BP filter of Example 3.10.

Now, using the tables for the CHEB function with $A_p = 0.5$ dB, we get $\varepsilon = 0.3493$, and the TF for the normalized LP filter to be

$$H_{LP}(S) = \frac{1}{2^{n-1}\varepsilon} \frac{1}{D(S)} = \frac{1}{2^4(0.3493)} \frac{1}{D(S)}$$

where

$$D(S) = S^5 + 1.1725 S^4 + 1.9374 S^3 + 1.3096 S^2 + 0.7525 S + 0.1789$$

(c) We now obtain the TF of the required BP filter by using the LP–BP transformation given by Eq. (3.70), that is, by changing $S \to (s^2 + \omega_0^2)/(Bs)$, where $\omega_0 = 507.937$ and $B = \omega_{c2} - \omega_{c1} = 170$. The details are left to the student.

3.10.3
LP to BR Transformation

In a similar way it can be shown that the transformation

$$S \to \frac{Bs}{s^2 + \omega_0^2} \tag{3.72}$$

transforms a normalized LP filter into a BR filter with PB from DC to ω_{c1} and ω_{c2} to infinity, the rejection bandwidth being $B = \omega_{c2} - \omega_{c1}$ and the center frequency of the rejection band being ω_0 which again is the geometric mean of ω_{c1} and ω_{c2}. Thus,

$$H_{BR}(s) = H_{NLP}\left(\frac{Bs}{s^2 + \omega_0^2}\right) \tag{3.73}$$

Also, Table 3.2 shows how the inductors and capacitors in the LP filter are transformed in the BR filter.

It should be mentioned that the order of the BP or the BR filter will always be double that of the normalized LP filter in view of the nature of LP to BP or LP to BR transformation.

3.11
Phase Approximation

Magnitude characteristic of a filter is important in cases where the signal perception does not depend critically on the phase of the processed signal. This is the case, for example, in voice communication. This is because human ear is insensitive to small errors in the phase of the received signal. However, for digital and video signals, phase relations among the various frequency components are to be carefully preserved for faithful reproduction. In such applications, the phase characteristics of the filter assume importance.

3.11.1
Phase Characteristics of a Transfer Function

The phase function associated with a signal is easily recognized by considering the phasor representation for a time-varying signal. Thus, in polar coordinates $V = V_M \exp(-j\varphi)$ represents a signal of magnitude V_M having a phase delay of φ radians. For a system, such as a filter, the frequency-dependent TF can be similarly expressed in polar form with a magnitude part and a phase part. The magnitude approximation problem discussed earlier is associated with this magnitude part. An understanding about the phase part can be obtained by writing the TF in the form

$$H(j\omega) = \frac{N(j\omega)}{D(j\omega)} = \frac{m_1(j\omega) + jn_1(j\omega)}{m_2(j\omega) + jn_2(j\omega)} \quad (3.74)$$

where m_1 and n_1 are the even and odd parts (i.e., terms involving even and odd powers in ω) of $N(\omega)$, and m_2 and n_2 are the even and odd parts of $D(\omega)$. The magnitude of $H(j\omega)$ is given by $\sqrt{(m_1^2 + n_1^2)/(m_2^2 + n_2^2)}$, while the phase angle is given by $\phi(j\omega) = \arctan(n_1/m_1) - \arctan(n_2/m_2)$. Using basic trigonometric identities, $\phi(j\omega)$ can be expressed as

$$\phi(j\omega) = \arctan\left[\frac{n_1 m_2 - n_2 m_1}{m_1 m_2 + n_1 n_2}\right] \quad (3.75)$$

In phase approximation problems, this phase function $\phi(j\omega)$ will be required to have a desirable characteristic.

3.11.2
The Case of Ideal Transmission

For digital/video signals, the goal of signal processing is very different from the concept of frequency-selective filtering (i.e., rejection of certain frequencies relative to other frequencies). The signal transmission has to be ideal, that is, the magnitude can be changed only by a constant factor K irrespective of the frequency, and there can only be a constant delay in time so that the relative timing (phase) among the various frequency contents of the signal remains unchanged. Analytically, if $f(t)$ is a given signal, the processed signal can be $f(t) = Kf(t - t_o)$, where t_o is a fixed delay. Taking Laplace transform and setting $s = j\omega$, the processed signal would be $F'(j\omega) = K F (j\omega)\exp(-j\omega t_o)$. Thus the transfer characteristic of the processing function is required to be $H(j\omega) = F'(j\omega)/F(j\omega) = K \exp(-j\omega t_o)$. Since $\exp(-j\omega t_o)$ represents only a delay, the TF $H(j\omega)$ has a magnitude equal to K, which implies an AP characteristic in the frequency domain. The phase $\varphi = -\omega t_o$ implies a constant delay $\tau = t_o$. The phase is a linear function of frequency ω leading to a constant delay in the time domain.

3.11.3
Constant Delay (Linear Phase) Approximation

From the above discussion, it is clear that the phase characteristic of the filter which preserves the time domain shape of the signal has to be such that a constant delay to the signal is produced. In practice, it will be required that this delay be constant for as large a range of ω as possible. Hence, the problem reduces to the synthesis of a filter TF with maximally flat delay characteristic, or $H(s) = K \exp(-\tau s)$. This problem was solved by Thomson (1949), who showed that the filter TF in such cases can be approximated by

$$H(s) = \frac{B_n(0)}{B_n(y)} \tag{3.76}$$

where $B_n(y)$ is the Bessel polynomial of order n in the normalized frequency variable $y = s\tau$, τ being the delay of signal propagation through the filter. This filter function is known as the *Bessel–Thomson (BT) filter*. We now derive the TF (Eq. (3.76)) approximating $H(s) = K \exp(-\tau s)$. Letting $y = s\tau$, we can write the normalized TF of the filter to be

$$H(y) = \frac{1}{\cosh(y) + \sinh(y)} = \frac{1/(\sinh y)}{1 + (\cosh y)/(\sinh y)} \tag{3.77}$$

Using power series expansions for the hyperbolic functions, we have

$$\cosh y = 1 + \frac{y^2}{2!} + \frac{y^4}{4!} + \cdots \tag{3.78}$$

$$\sinh y = y + \frac{y^3}{3!} + \frac{y^5}{5!} + \cdots \tag{3.79}$$

Expanding $(\cosh y / \sinh y)$ by continued fraction expansion,

$$\frac{\cosh y}{\sinh y} = \frac{1}{y} + \cfrac{1}{\cfrac{3}{y} + \cfrac{1}{\cfrac{5}{y} + \cdots}} \tag{3.80}$$

and truncating the expansion after n terms, we can rewrite the TF in the form

$$H(y) = \frac{c_0}{B_n(y)} \tag{3.81}$$

where c_0 is chosen such that $H(0) = 1$ (i.e., DC gain $= 1$) and

$$B_n(y) = c_n y^n + c_{n-1} y^{n-1} + \cdots + c_1 y + c_0 \tag{3.82}$$

Higher the value of n, closer will be the phase characteristic to the ideal situation of constant delay. The coefficients c_n in $B_n(y)$ are given by

$$c_k = \frac{(2n-k)!}{2^{n-k} k! (n-k)!}, \quad k = 0, 1, 2, \ldots, n \tag{3.83}$$

Using Eq. (3.83), it can be shown that $B_n(y)$ satisfies the recurrence relation

$$B_n(y) = (2n - 1)B_{n-1}(y) + y^2 B_{n-2}(y) \text{ with } B_0(y) = 1 \text{ and } B_1(y) = y + 1 \tag{3.84}$$

which can be used to find the higher-order polynomials. Note that the form of $H(y)$ shows that it has an all-pole LP characteristic, where y is the normalized complex frequency, $y = s\tau$.

3.11.4
Graphical Method to Determine the BT Filter Function

Determination of the BT TF for a given set of magnitude and delay approximations is easily carried out using two sets of curves, which show the magnitude attenuation and percentage delay error as functions of the normalized frequency $\omega = \Omega\tau_o$, where τ_o is the desired delay (in seconds) and Ω is the radian frequency over which the magnitude and delay specifications are to be satisfied. Figures 3.16a and 3.16b present several of these graphs. Filter orders up to 15 are included. MATLAB program listings associated with these graphs are presented in Appendix B.

Example 3.11. Consider the case of producing a multidimensional image from a TV signal by delaying it by 70 ns and superimposing on the master signal. The bandwidth of the TV image signal is 4.5 MHz. The magnitude error should not exceed 1.5 dB and the delay error should be within 1%.

Since the bandwidth for the signal is 4.5 MHz, consider the normalized frequency variable $\Omega\tau_o = p = 2\pi \times 4.5 \times 10^6 \times 70 \times 10^{-9} = 1.98$. From the magnitude attenuation set of graphs, we find that for attenuation of 1.5 dB and with $p = 1.98$, one requires an order $n = 6$ (approximately). From the delay error set of graphs, the corresponding value is $n = 4$. So, a conservative design strategy will be to choose $n = 6$. This will ensure a magnitude attenuation of no more than 1.5 dB and a delay error of less than 1%. Thus, the BT filter function will be

$$H(y) = \frac{B_6(0)}{B_6(y)} \tag{3.85}$$

Using Eq. (3.84) repeatedly, we get

$$y^6 + 21y^5 + 210y^4 + 1260y^3 + 4725y^2 + 10395y + 10395$$

Hence,

$$H(s) = \frac{10395}{y^6 + 21y^5 + 210y^4 + 1260y^3 + 4725y^2 + 10395y + 10395} \tag{3.86}$$

where

$$y = s\tau_o = 70 \times 10^{-9}s$$

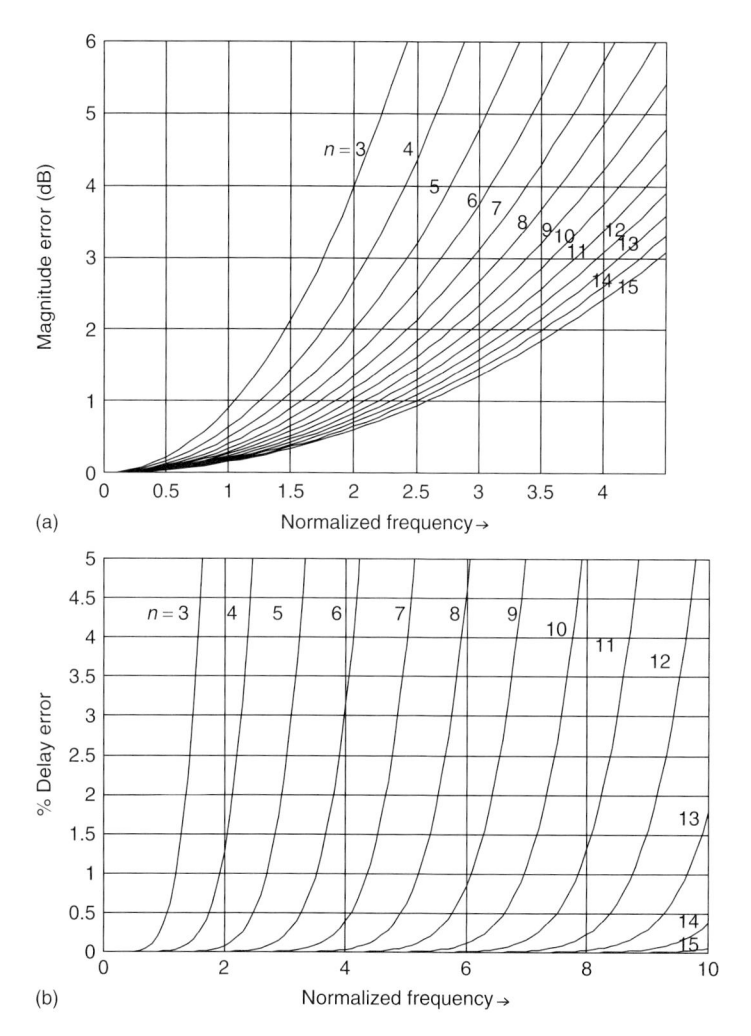

Figure 3.16 (a) Magnitude error in decibels and (b) percentage delay error, as functions of the normalized frequency.

3.12
Delay Equalizers

In practical signal processing, one, however, requires some frequency selectivity (especially to guard against noise) together with a constant delay feature in order to preserve the ideal transmission property. This can be achieved by cascading an AP network (such as a BT filter discussed above) with a frequency selective network such as an LP or a BP filter. The AP network, however, need not have a phase characteristic as $\phi(\omega) = -\omega\tau_o$. Instead, its phase characteristic should be such that the overall phase function $\phi(\omega)$ of the AP network and the frequency-selective network together is $\phi(\omega) = \phi_{APN} + \phi_{FSN} = -\omega\tau_0$. Thus, ϕ_{APN} should be adjusted to work with ϕ_{FSN} so that the overall $\phi(\omega)$ becomes a linear function of the radian frequency ω. Such special AP networks are known as *delay equalizers*. When the frequency-selective network is an external physical channel, such as a telephone or a coaxial line, the characteristics of the AP network are to be adjusted in an adaptive way with changing characteristics of the frequency-selective network to meet the criterion of linear phase characteristic. Such an equalizer is known as *adaptive delay equalizer*. Numerical computations are extensively used to derive the characteristics of delay equalizers. Hardware implementation of delay equalizers may be facilitated using logical computations via a microprocessor.

Practice Problems

3.1 Consider the Sallen–Key LP filter shown in Figure P3.1, where $R_1 = (1/3)$ Ω, $R_2 = 1$ Ω, $C_1 = C_2 = 1$ F, and $K = 2$. (a) Find the transfer function of the LP filter as well as its ω_p and Q_p. (b) Apply the LP to HP transformation $s \to (1/s)$ to obtain the corresponding HP function, and the corresponding filter structure. Find the values of the different components. (c) Transform all the impedances in the HP structure so obtained by the impedance

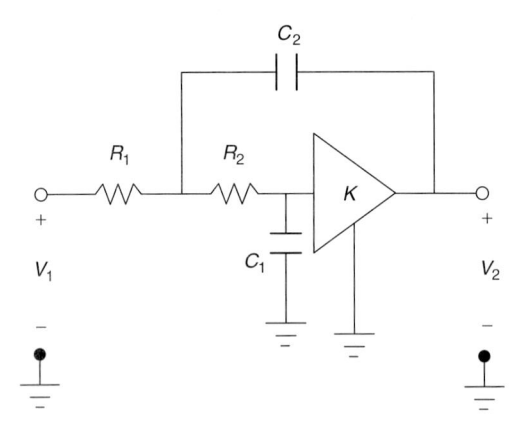

Figure P3.1

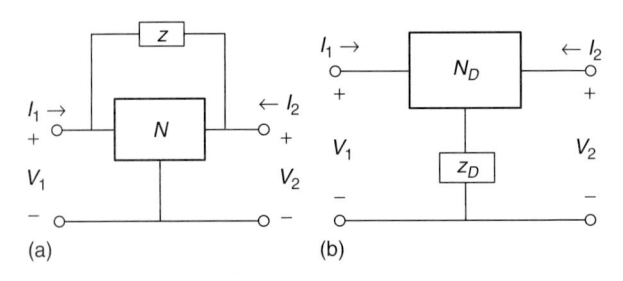

Figure P3.2

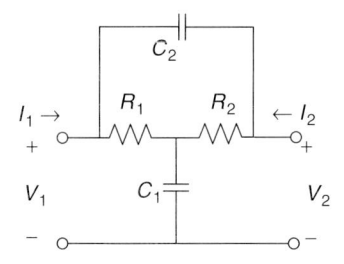

Figure P3.3

transformation $z^*(s) = (1/s) z(s)$. What is the resulting transfer function? Also, show the resulting filter structure with its component values. (d) What are the ω_p and Q_p values of the resulting filter? (Note: This example shows that an OA-RC LP filter can be converted to an OA-RC HP filter by a simple replacement of resistors by capacitors and vice versa without changing the OA. This is also called the *RC:CR transformation* (Mitra, 1969) and is useful in deriving HP structures from LP structures realized using VCVS (or CCCS) and RC elements.)

3.2 Consider the two-ports shown in Figures P3.2a and b, where the three-terminal two-ports N and N_D, as well as the one-ports z and z_D are duals of each other w.r.t. $f(s)$. Show that in such a case, the two-ports themselves are duals of each other w.r.t. $f(s)$.

3.3 Using the result of Problem 3.2, find the capacitive dual of the twin-T network of Figure P3.3.

3.4 Consider the cascade network N, as shown in Figure P3.4, which consists of a cascade of a unity-gain VCVS, a series resistor r_m, and a unity-gain CCCS. (a) Evaluate the overall chain matrix of N, and hence show that the overall network N corresponds to a VCCS of transadmittance $g_m = 1/r_m$. (b) Find N_D, the dual of the network N, w.r.t. $f(s) = r_m r_n$, by applying Theorems 3.1 and 3.2 and show that it corresponds to a CCVS of transresistance r_n. (c) Show that the transpose of N is nothing but a VCCS of transadmittance g_m, but with its input and output ports reversed. (d) Show that the transpose of N_D, that is, $(N_D)^T$, is nothing but a CCVS of transresistance r_n, but with its input and output ports reversed.

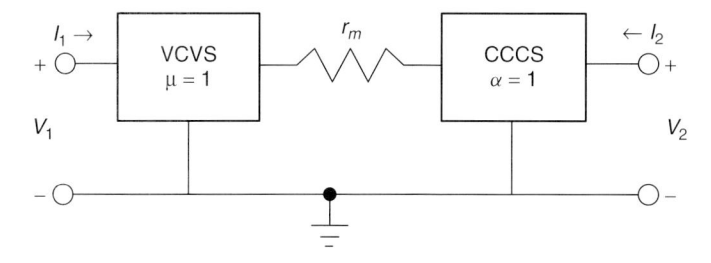

Figure P3.4

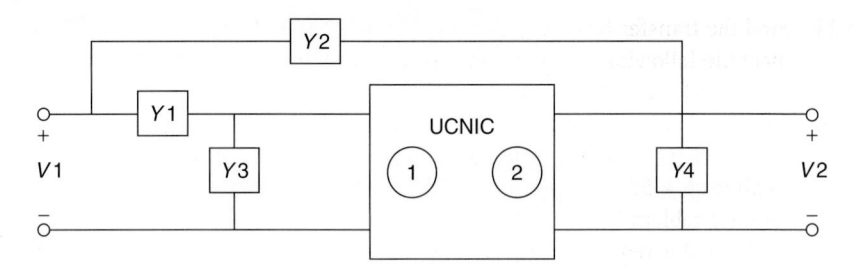

Figure P3.6

3.5 (a) Show that a cascade of a VCCS of transadmittance g_m followed by CCVS of transresistance r_n corresponds to a VCVS of gain (r_m/r_n). (b) Show that if the positions of the VCVS and the CCVS are reversed in (a), then the cascaded network would correspond to a CCCS of gain (r_m/r_n).

3.6 Consider the network N shown in Figure P3.6, where the chain matrix of the unity-gain current inversion type negative impedance converter (UCNIC) is given by $\begin{bmatrix} 1 & 0 \\ 0 & -1 \end{bmatrix}$. (a) Show that the open circuit VTF is given by $\frac{V_2}{V_1} = \frac{Y_1-Y_2}{(Y_1-Y_2)+(Y_3-Y_4)}$. (b) Obtain the network N_R^T corresponding to N, and find its short circuit CTF. (c) Obtain the network N_D corresponding to N and find its short circuit CTF. (d) Finally, obtain the network $\overline{N} = (N_R^T)_D = (N_D)_R^T$ and find its open circuit VTF. (Note: The structures N and $\overline{N}$ were proposed by Yanagisawa (1957) as alternate structures for realizing VTFs, while the structures N_R^T and N_D were proposed by Thomas (1959) to realize CTFs (Mitra, 1969). As one can see, using the concepts of duals, transposes, and dual transposes, they are all mutually related.)

3.7 Derive the transfer function of a maximally flat LP filter which has an attenuation of 25 dB at $f = 4f_c$ where f_c is the -3 dB frequency. Repeat the case when f_c is the -1 dB frequency.

3.8 Find the transfer function of a Butterworth LP filter with $f_c = 10^3$ Hz and where the attenuation increases at 25 dB per octave.

3.9 An LP filter is required to meet the following specifications: (i) maximum flat PB in $0 \leq f \leq 8$ kHz, (ii) maximum loss in PB, $A_p = 1$ dB, and (iii) minimum SB attenuation, $A_s = 15$ dB at 12 kHz. Find the transfer function for the filter.

3.10 Find the transfer function of an LP filter such that (i) the PB is equiripple with $A_p = 0.5$ dB in $0 \leq f \leq 6$ kHz and (ii) monotonic SB with $A_s \geq 30$ dB for $f > 15$ kHz.

3.11 Find the transfer function of an HP filter that has (i) an equiripple PB for $f > 15$ kHz with $A_p = 0.5$ dB and (ii) a monotonic SB for $f < 7.8$ kHz with $A_s = 35$ dB.

3.12 Find the Chebyshev LP filter function which has 0.5-dB loss ripple in the PB and 60 dB per decade attenuation increase in the attenuation band.

3.13 Find the transfer function of an all-pole BP filter with maximally flat PB to meet the following specifications: (i) $A_p < 1$ dB in 15 kHz $\leq f \leq 20$ kHz. (ii) $A_s > 40$ dB for $f \leq 8.6$ kHz and $f \geq 35$ kHz.

3.14 Find the BR filter transfer function which satisfies (i) equiripple PB with $A_p = 1$ dB in $f \leq 40$ kHz and $f \geq 100$ kHz and (ii) monotonic SB with $A_s \geq 15$ dB for $f \geq 50$ kHz, and $f \leq 80$ kHz.

3.15 Repeat Problem 3.7 where the PB is equiripple with $A_p < 1$ dB.

3.16 (a) Find the required order for an elliptic function, which has a cutoff frequency of $\omega_c = 1$ rad s^{-1}, with $\omega_s/\omega_c = 1.2$, $A_p = 0.5$ dB, and $A_s = 45$ dB. (b) What will be the corresponding orders for MFM and CHEB function characteristics?

3.17 A sixth-order MFM filter has $A_p = 1$ dB. What will be the exact attenuation for $\omega = 10\omega_c$, where ω_c is the LP band-edge frequency?

3.18 Find the elliptic filter transfer function which satisfies the following specifications: $A_p = 1$ dB in the PB $0 \leq \omega \leq 1$ rad s^{-1}, $A_s \geq 46$ dB in the SB $\omega > 1$.

3.19 Find an all-pole LP transfer function which has an attenuation <3 dB up to $\omega = 1000$ rad s^{-1} and the delay is maximally flat at $\tau_o = 2.5$ ms with a delay error of less than 3% up to $\omega = 700$ rad s^{-1}.

3.20 Find an AP function with constant delay of 300 µs. The delay error must be no greater than 1.5% up to $f = 4.0$ kHz.

4
Basics of Passive Filter Design

As mentioned in Chapter 3, lossless networks terminated by resistors at one or both ends are usually used in communication circuits to couple energy source to a load. A general coupling network may be excited by a voltage or current source, which may be ideal or nonideal. Further, the network may be working into a load which is finite, zero, or infinite. All these situations may be represented by the arrangements shown in Figures 3.7a and 3.7b. Such a network is called a *doubly terminated network*. It is called a *singly terminated network* if (i) $R_S = 0$ in Figure 3.7a or $R_S = \infty$ in Figure 3.7b, or (ii) $R_L = 0$ in Figure 3.7a or $R_L = \infty$ in Figure 3.7b. Before we consider the realization of terminated lossless networks, we deal with some general properties of singly terminated networks, similar to what was done with doubly terminated networks in Section 3.6.

4.1
Singly Terminated Networks

Various possible situations in singly terminated networks along with the specifications necessary to realize them are tabulated in Table 4.1. The arrangement shown in (a) in the table corresponds to a case where the voltage source has negligible internal resistance, while that in (b) to a situation where the current source has infinite internal resistance. An example of the latter is the output from a field effect transistor, while for the former case, it is the output from an OA. The two network arrangements shown in (c) are equivalent in view of Thevenin's and Norton's theorems, a similar statement holding true for the two networks in (d). The arrangement in (c) corresponds to the case when the source resistance is finite and the output is connected to a device or network whose impedance is so small that it is virtually a short circuit. Finally, the arrangement in (d) applies when the source resistance is finite and the output is connected to a device or network that essentially is an open circuit; an example of such a situation is when the output is fed to an OA. It can be shown by elementary analysis of the above arrangements that

$$T_V^{(a)}(s) = Y_{21}^{(a)}(s) = \frac{1}{A+B} \tag{4.1}$$

Modern Analog Filter Analysis and Design: A Practical Approach. Rabin Raut and M. N. S. Swamy
Copyright © 2010 WILEY-VCH Verlag GmbH & Co. KGaA, Weinheim
ISBN: 978-3-527-40766-8

4 Basics of Passive Filter Design

Table 4.1 Various arrangements of singly terminated networks and the corresponding specifications in terms of the chain parameters of the network.

Network	Suitable specification
(a) $V_s = V_1$, N, $R_L = 1$, output V_2	$\text{VTF} = T_V^{(a)}(s) = \dfrac{V_2(s)}{V_1(s)}$ $\text{TAF} = Y_{21}^{(a)}(s) = -\dfrac{I_2(s)}{V_1(s)}$ $\Bigg\} = \dfrac{1}{A+B}$
(b) $I_s = I_1$, N, $R_L = 1$, output V_2	$\text{CTF} = T_I^{(b)}(s) = -\dfrac{I_2(s)}{I_1(s)}$ $\text{TIF} = z_{21}^{(b)}(s) = \dfrac{V_2(s)}{I_1(s)}$ $\Bigg\} = \dfrac{1}{C+D}$
(c) V_s, $R_s = 1$, N, then I_s, $R_s = 1$, N	$\text{CTF} = T_I^{(c)}(s) = -\dfrac{I_2(s)}{I_s(s)}$ $\text{TAF} = Y_{21}^{(c)}(s) = -\dfrac{I_2(s)}{V_s(s)}$ $\Bigg\} = \dfrac{1}{B+D}$
(d) V_s, $R_s = 1$, N, output V_2, then I_s, $R_s = 1$, N, output V_2	$\text{VTF} = T_V^{(d)}(s) = \dfrac{V_2(s)}{V_s(s)}$ $\text{TIF} = Z_{21}^{(d)}(s) = \dfrac{V_2(s)}{I_s(s)}$ $\Bigg\} = \dfrac{1}{A+C}$

$$T_I^{(b)}(s) = Z_{21}^{(b)}(s) = \frac{1}{C+D} \tag{4.2}$$

$$T_I^{(c)}(s) = Y_{21}^{(c)}(s) = \frac{1}{B+D} \tag{4.3}$$

$$T_V^{(d)}(s) = Z_{21}^{(d)}(s) = \frac{1}{A+C} \tag{4.4}$$

where $T_V(s)$ and $T_I(s)$ correspond to the VTF and the CTF, respectively, $Y_{21}(s)$ to the TAF and $Z_{21}(s)$ to the TIF. These are tabulated in Table 4.1. It is seen from this table that there are only four different specifications that may be assumed for a singly terminated network, and these are $T_V^{(a)}(s)$, $T_I^{(b)}(s)$, $T_I^{(c)}(s)$, and $T_V^{(d)}(s)$. We will now show that if we know how to realize $T_V^{(a)}(s)$, then the other three functions may be synthesized directly from the realization of $T_V^{(a)}(s)$ using the concepts of N_D, N_R^T, and $\overline{N}$, introduced in Chapter 3.

Let us assume that we know how to synthesize the network N of (a) in Table 4.1 for a given VTF $T_V^{(a)}$, that is, $T_V^{(a)} = \frac{1}{A+B}$. Now suppose that the network N in (b) of the table is replaced by its dual $(N_b)_D$ w.r.t. $f(s) = 1$. Since it is known that if

$$[a] = \begin{bmatrix} A & B \\ C & D \end{bmatrix} \tag{4.5}$$

then

$$[a_D] = \begin{bmatrix} A_D & B_D \\ C_D & D_D \end{bmatrix} = \begin{bmatrix} D & Cf(s) \\ B/f(s) & A \end{bmatrix} = \begin{bmatrix} D & C \\ B & A \end{bmatrix} \tag{4.6}$$

resulting in

$$\left(T_I^{(b)}\right)_D = \frac{1}{C_D + D_D} = \frac{1}{A+B} = T_V^{(a)} \tag{4.7}$$

Thus, if the specification given is the CTF with a load resistor $R_L = 1\,\Omega$, we can first synthesize the network N of (a) in Table 4.1, realizing the desired specification as the VTF $T_V^{(a)}$. The reason for this is that most of the times, the specifications are given in terms of a VTF realization and there exist many methods dealing with the realization of a VTF. Once the network N is found, its dual N_D can be found and used in the structure of (b) in Table 4.1 to realize the given system function as a CTF. Similarly, given a $T_I^{(c)}$, it may be first realized as the $T_V^{(a)}$ of (a) in Table 4.1 and the corresponding network N found; then the N_R^T of this network is used as the two-port in the structure of (c) to obtain the desired $T_I^{(c)}$. Finally, it can be shown that a given $T_V^{(d)}$ may be realized by first realizing the specifications as the $T_V^{(a)}$ of (a) and the corresponding network N found; then, $\overline{N} = \left(N_R^T\right)_D = (N_D)_R^T$ of this network is used as the two-port in the structure of (d) to get the desired $T_V^{(d)}$.

We will first concentrate on the realization of an LP filter in a singly terminated form using only LC elements for the coupling network, wherein we assume the LP filter is an all-pole function (that is, all the zeros of the filter are at infinity). Once we have realized an LP filter, we know how to transform an LP filter to an HP, a BP, or a BR filter by using appropriate frequency transformations (see Chapter 3). Hence, we assume that the specifications of the given filter (LP, HP, BP, BR) have been transformed to that of the associated normalized LP filter, and that the normalized LP filter specifications have been approximated using Butterworth, Chebyshev, or Bessel–Thomson approximations, all of which lead to an all-pole transfer function.

Since we are dealing with ladder structures containing only LC elements, whether these structures are singly or doubly terminated, let us first consider some basic properties of such networks.

4.2 Some Properties of Reactance Functions

A network that consists of only LC elements is called a *reactance network*. Consider a reactance one-port network and let $Z_{LC}(s)$ be its DPI. Since the real part of

$Z_{LC}(j\omega) = 0$, $Z_{LC}(s)$ must be a ratio of an even polynomial $n(s)$ to an odd polynomial $m(s)$, or vice versa. A very fundamental property of a reactance network is that it is lossless, and hence does not dissipate any energy. Hence, all the poles of the DPI of such a network must be on the imaginary axis; otherwise, the impulse response would contain decaying factors. For a similar reason, the zeros of a reactance function (which are the poles of the DPA function) should also be on the imaginary axis. Thus, the DPI $Z_{LC}(s)$ (or its DPA $Y_{LC}(s) = 1/Z_{LC}(s)$) is of the form

$$Z_{LC}(s) = \frac{n(s)}{m(s)} \text{ or } Z_{LC}(s) = \frac{m(s)}{n(s)} \tag{4.8}$$

It can be shown that a reactance function, that is, the DPI or the DPA of a reactance network, has the following properties (Van Valkenburg, 1960; Weinberg, 1962):

1) A reactance function is the ratio of an even to odd polynomial or vice versa.
2) The degrees of the numerator and denominator polynomials differ exactly by unity.
3) The zeros and poles lie on the imaginary axis, alternate, and are simple with positive residues at its poles.
4) The slope of $dX(\omega)/d\omega$ is positive, where $X(\omega) = (1/j)Z(j\omega)$.

The impedance $Z_{LC}(s)$ or its inverse $Y_{LC}(s)$ can be realized by *Cauer* networks either by successively removing the pole at infinity or the pole at the origin using continued fraction (CF) expansion around $s = \infty$ or $s = 0$, respectively. They are respectively called *Cauer-I* and *Cauer-II* forms of realization of $Z_{LC}(s)$ (or $Y_{LC}(s)$ as the case may be). We illustrate these forms through the following example.

Example 4.1. Find the Cauer-I and Cauer-II forms for the reactance function given by

$$Z_{LC}(s) = \frac{(s^2 + 1)(s^2 + 4)}{s(s^2 + 2)} \tag{4.9}$$

We first obtain the Cauer-I form by a CF expansion of $Z_{LC}(s)$ around $s = \infty$ as follows:

$$s^3 + 2s \overline{\smash{\big)}\, s^4 + 5s^2 + 4} \,(s \longleftarrow \text{inductor}, L = 1\text{H}$$

$$\underline{s^4 + 2s^2}$$

$$3s^2 + 4 \overline{\smash{\big)}\, s^3 + 2s} \,(s/3 \longleftarrow \text{capacitor}, C = 1/3\text{F}$$

$$\underline{s^3 + \tfrac{4}{3}s}$$

$$\tfrac{2}{3}s \overline{\smash{\big)}\, 3s^2 + 4} \,(\tfrac{9}{2}s \longleftarrow \text{inductor}, L = \tfrac{9}{2}\text{H}$$

$$\underline{3s^2}$$

4.2 Some Properties of Reactance Functions

$$4)\ \frac{2}{3}s\left(\frac{1}{6}s \leftarrow \text{capacitor},\ C = \frac{1}{6}\text{F}\right.$$
$$\underline{\frac{2}{3}s}$$
$$--$$

Thus,

$$Z_{LC}(s) = s + \cfrac{1}{\cfrac{1}{3}s + \cfrac{1}{\cfrac{9}{3}s + \cfrac{1}{\cfrac{1}{6}s}}}} \quad (4.10)$$

The above $Z_{LC}(s)$ may be realized in the Cauer-I form as shown in Figure 4.1a. The Cauer-II form is obtained by CF expansion of $Z_{LC}(s)$ around $s = 0$ and is given below.

$$2s + s^3)\ 4 + 5s^2 + s^4\ (2/s \leftarrow \text{capacitor},\ C = 1/2\text{F}$$
$$\underline{4 + 2s^2}$$
$$3s^2 + s^4)\ 2s + s^3\ \left(\frac{2}{3}s \leftarrow \text{inductor},\ L = 3/2\text{H}\right.$$
$$\underline{2s + \frac{2}{3}s^3}$$
$$\frac{1}{3}s^3)\ 3s^2 + s^4\ \left(\frac{9}{s} \leftarrow \text{capacitor},\ C = 1/9\text{F}\right.$$
$$\underline{3s^2}$$
$$s^4)\ \frac{1}{3}s^3\ \left(\frac{1}{3}s \leftarrow \text{inductor},\ L = 1/3\text{H}\right.$$
$$\underline{\frac{1}{3}s^3}$$
$$--$$

Hence,

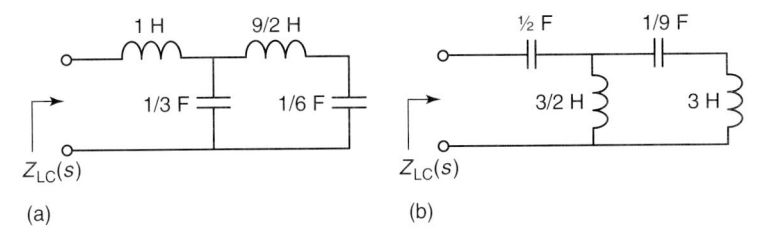

$$Z_{LC}(s) = \frac{2}{s} + \cfrac{1}{\cfrac{2}{3}\cfrac{1}{s} + \cfrac{1}{\cfrac{9}{s} + \cfrac{1}{\cfrac{1}{3}\cfrac{1}{s}}}}} \quad (4.11)$$

Figure 4.1 Realization of $Z_{LC}(s)$ of Example 4.1: (a) Cauer-I form and (b) Cauer-II form.

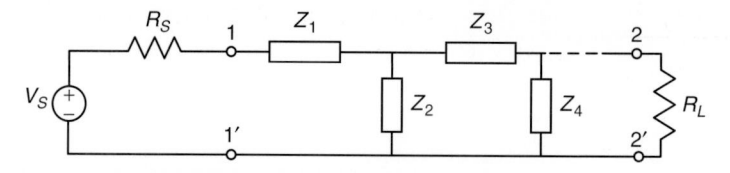

Figure 4.2 A doubly terminated lossless ladder network.

The corresponding Cauer-II realization is shown in Figure 4.1b.

Now consider a doubly terminated lossless ladder as shown in Figure 4.2, driven by a voltage source. It is noted that the conclusions we are going to be drawing with such a ladder are equally true when the ladder is driven by a current source or when the ladder is singly terminated. It is obvious that there is no transmission from port 1 to port 2, if and only if one of the series arms is an open circuit or one of the shunt arms is a short circuit. That is, the zeros of transmission, which are the zeros of the transfer function, are nothing but the poles of the series elements and the zeros of the shunt elements.

Since we are going to be dealing with ladder network realization of LP all-pole filters in singly or doubly terminated form, it is clear that all the series elements should be inductors and the shunt elements be capacitors, as all the zeros of the transfer function are at infinity. If LP to HP transformation is employed to convert the LP filter into an HP filter, it is known from Table 3.2 (see Chapter 3) that the inductors become capacitors and vice versa. Thus, in the HP filter, all the capacitors would be in the series arms and the inductors in the shunt arms, thereby producing all the transmission zeros at the origin. When the LP to BP transformation is applied to convert an LP filter into a BP filter, then each series arm L is converted to a series combination of an L and a C, while each shunt arm C is transformed to a parallel combination of a C and an L. Hence, the n transmission zeros at infinity for the LP filter are converted into n transmission zeros at $s = 0$ and n transmission zeros at $s = \infty$. This is due to the fact that at $s = \infty$, the n inductors in the series arm become open circuits along with the n capacitors in the shunt arms becoming short circuits, while at $s = 0$, the n capacitors in the series arm become open circuits along with the n inductors in the shunt arms becoming short circuits.

We now consider the design of a singly terminated lossless ladder realizing an LP filter.

4.3
Singly Terminated Ladder Filters

Consider a lossless ladder network terminated in a load of $1\,\Omega$, as shown in Figure 4.3, and let $Z_{LC}(s)$ be the DPI looking into the network N at port 1. This network can be driven by a voltage or a current source. In the former case, the appropriate specification would be the VTF $T_V(s) = V_2(s)/V_1(s)$, V_1 being the

4.3 Singly Terminated Ladder Filters

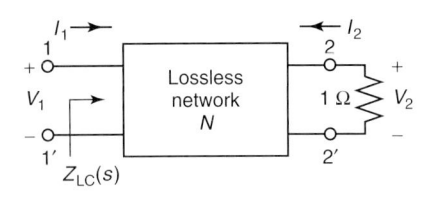

Figure 4.3 A singly terminated lossless network.

driving voltage, while in the latter case the appropriate specification would be the TIF, $Z_{21}(s) = V_2(s)/I_1(s)$, I_1 being the driving current. Let us first consider the case of the realization of a VTF. Using Eq. (3.22), and letting $R_s = 0$ and $R_L = 1$, we get

$$T_V(s) = \frac{V_2(s)}{V_1(s)} = \frac{1}{A+B} \tag{4.12}$$

where

$$[a] = \begin{bmatrix} A & B \\ C & D \end{bmatrix} \tag{4.13}$$

is the chain matrix of N. Using the interrelations between the chain matrix $[a]$ and the short circuit admittance matrix $[y]$, we can rewrite Eq. (4.12) as

$$T_V(s) = \frac{1}{-\frac{y_{22}}{y_{21}} - \frac{1}{y_{21}}} = \frac{-y_{21}}{1 + y_{22}} \tag{4.14}$$

The transfer function to be realized is of the form

$$H(s) = \frac{k}{s^n + a_1 s^{n-1} + \cdots + a_0} = \frac{k}{D(s)} \tag{4.15}$$

since we are assuming $H(s)$ to be an all-pole function. Also, $D(s)$ is known to be a strictly Hurwitz polynomial; that is, all its zeros are in the LH of the s-plane. We may rewrite $H(s)$ as

$$H(s) = \frac{k}{m(s) + n(s)} = \frac{\frac{k}{n(s)}}{1 + \frac{m(s)}{n(s)}} \tag{4.16}$$

where $m(s)$ and $n(s)$ are the even and odd parts of $D(s)$. It is known that if $D(s)$ is strictly Hurwitz, then $m(s)/n(s)$ is a reactance function (Van Valkenburg, 1960). From Eqs. (4.14) and (4.16) we have

$$T_V(s) = \frac{-y_{21}}{1 + y_{22}} = H(s) = \frac{\frac{k}{n(s)}}{1 + \frac{m(s)}{n(s)}} \tag{4.17}$$

Thus,

$$-y_{21} = \frac{k}{n(s)}, \quad y_{22} = \frac{m(s)}{n(s)} \tag{4.18}$$

Any realization of y_{22} will automatically realize the poles of y_{21}, since the poles are determined by the network determinant and is the same for all the $[y]$ parameters; hence, we can concentrate on realizing y_{22} in such a way that all the transmission zeros for $T_V(s)$ are realized at $s = \infty$. This means that the ladder should have all the inductors in the series arms and capacitors in the shunt arms. Hence, Cauer-I form is utilized to realize y_{22}, which is a reactance function. Since y_{22} is being realized, we have to start at port 2 of N (i.e., at the load end) and move left towards port 1. If the degree of $m(s) = 1 +$ degree of $n(s)$, then the CF expansion around $s = \infty$ will lead to a term like αs as the first term. Since y_{22} is an admittance, αs will correspond to a capacitor of value α in the shunt arm of the ladder at port 2. However, if the degree of $m(s) =$ degree of $n(s) - 1$, then we have to invert and start the division with $n(s)/m(s)$ to obtain the CF expansion around $s = \infty$. Since we are now dealing with $(1/y_{22})$, the quotient of the first division, βs, will correspond to an inductor of value β, in the series arm of the ladder at port 2.

Since the ladder realization involves a single inductor in the series arm and a single capacitor in the shunt arm, the CF expansion will lead to exactly n elements. The last element must always be an inductor (in the series arm) and cannot be a capacitor in the shunt arm, since it would be shorted while computing y_{22}. Finally, the 1-Ω load resistor is inserted at port 2 to complete the realization of $H(s)$ as the VTF of the structure of Figure 4.3. The constant k in Eq. (4.15) can be determined by evaluating $H(0)$, that is, the value of the VTF at $s = 0$. If $H(s)$ corresponds to a normalized LP filter with a cutoff of 1 rad s^{-1}, then frequency scaling is used to obtain the required cutoff frequency. Once we have realized $H(s)$ as the VTF of a singly terminated ladder with voltage as the source, then from the discussion of Section 4.1, we know how to realize $H(s)$ for the other arrangements of the singly terminated networks.

We now illustrate the procedure we have discussed with the following example.

Example 4.2. Realize the transfer function given by

$$H(s) = \frac{k}{2s^4 + 2s^3 + 6s^2 + 5s + 2} = \frac{k}{D(s)} \qquad (4.19)$$

as the VTF of a singly terminated lossless ladder with a load of 100 Ω. Determine the value of k.

From Eq. (4.19), we see that $m(s) = 2s^4 + 6s^2 + 2$ and $n(s) = 2s^3 + 5s$. Since the degree of $m(s) = 1+$ degree of $n(s)$, we perform the CF expansion of $m(s)/n(s)$ around $s = \infty$ to realize y_{22}. The CF expansion is as follows:

$$2s^3 + 5s \overline{)\,2s^4 + 6s^2 + 2\,} (s \longleftarrow \text{capacitor, } C = 1F$$
$$\underline{2s^4 + 5s^2}$$
$$s^2 + 2 \overline{)\,2s^3 + 5s\,} (2s \longleftarrow \text{inductor, } L = 2H$$
$$\underline{2s^3 + 4s}$$
$$s) \,s^2 + 2\, (s \longleftarrow \text{capacitor, } C = 1F$$
$$\underline{s^2}$$
$$2) \, s \, (\frac{1}{2}s \longleftarrow \text{inductor, } L = \frac{1}{2}H$$
$$\underline{s}$$
$$-$$

4.3 Singly Terminated Ladder Filters

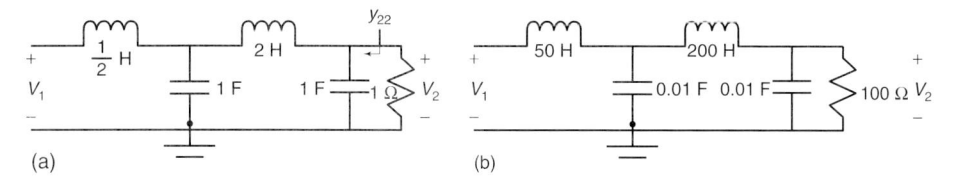

Figure 4.4 Realization of $H(s)$ of Example 4.2 when (a) $R_L = 1\ \Omega$ and (b) $R_L = 100\ \Omega$.

Hence, the realization of $H(s)$ is as shown in Figure 4.4a. At $s = 0$, all the inductors are short circuits and the capacitors are open circuits, and hence, $V_2/V_1 = 1 = H(0) = (k/2)$; thus $k = 2$. Since the given termination is $100\ \Omega$, we now scale all the impedances by $100\ \Omega$, which of course will not alter the transfer function. The scaled network is shown in Figure 4.4b.

Example 4.3. The function $H(s) = \dfrac{k}{s^3 + 2s^2 + 2s + 1}$ represents a third-order normalized LP Butterworth filter function. Realize $H(s)$ by a singly terminated network when the network is driven (a) by an ideal voltage source and (b) by an ideal current source, the terminating load resistance being $1\ \Omega$. (c) If the load resistance is $100\ \Omega$ and the cutoff frequency of the filter is $100\ \mathrm{rad\ s^{-1}}$, find the corresponding realizations.

(a) In this case, $m(s) = 2s^2 + 1$ and $n(s) = s^3 + 2s$. Since the degree of $m(s) <$ the degree of $n(s)$, we have to perform the CF expansion on $n(s)/m(s)$ around $s = \infty$ to realize $(1/y_{22})$. The CF expansion is as follows:

$$2s^2 + 1\ )\ s^3 + 2s\ \left(\dfrac{s}{2}\ \longleftarrow \text{inductor},\ L = \dfrac{1}{2}\mathrm{H}\right.$$

$$\underline{s^3 + \dfrac{1}{2}s}$$

$$\dfrac{3}{2}s\ )\ 2s^2 + 1\ \left(\dfrac{4}{3}s\ \longleftarrow \text{capacitor},\ C = \dfrac{4}{3}\mathrm{F}\right.$$

$$\underline{2s^2}$$

$$1\ )\ \dfrac{3}{2}s\ \left(\dfrac{3}{2}s\ \longleftarrow \text{inductor},\ L = \dfrac{3}{2}\mathrm{H}\right.$$

$$\underline{\dfrac{3}{2}s}$$

$$-$$

The realization of $H(s)$ as a VTF with $R_L = 1\ \Omega$ is shown in Figure 4.5a. The value of $k = 1$, since $H(0) = 1$.

(b) As discussed in Section 4.1, we may now realize the given $H(s)$ at the TIF, $Z_{21}(s)$, of a singly terminated network driven by an ideal current source by simply replacing the network N by N_D, and the corresponding realization is shown in Figure 4.5b.

(c) If the terminating resistance is changed to $100\ \Omega$, we shall use impedance scaling. Thus scale all $L \rightarrow 100L$ and all $C \rightarrow C/100$.

If the cutoff frequency is $100\ \mathrm{rad\ s^{-1}}$, that is, $\omega_c = 2\pi \times 100\ \mathrm{rad\ s^{-1}}$, then frequency scale all $L \rightarrow L/(2\pi \times 100)$, and all $C \rightarrow C/(2\pi \times 100)$.

92 | 4 Basics of Passive Filter Design

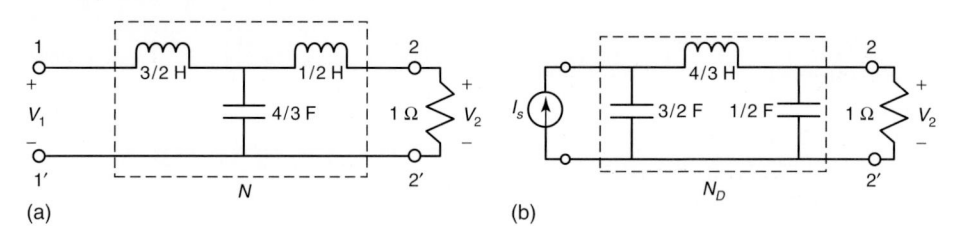

Figure 4.5 Realization of $H(s)$ of Example 4.3 when driven by (a) an ideal voltage source and (b) an ideal current source, with $R_L = 1\ \Omega$.

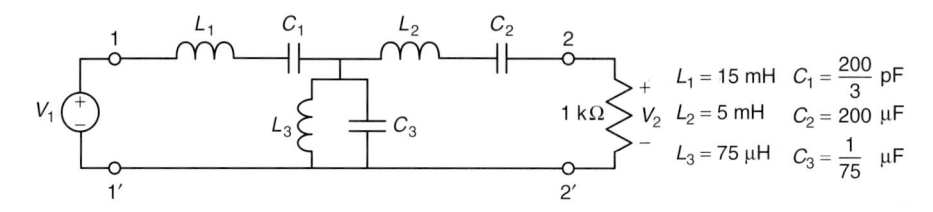

Figure 4.6 Realization of the BP filter of Example 4.4.

$L_1 = 15$ mH $C_1 = \dfrac{200}{3}$ pF
$L_2 = 5$ mH $C_2 = 200$ μF
$L_3 = 75$ μH $C_3 = \dfrac{1}{75}$ μF

We may also realize $H(s)$ as a singly terminated network with open circuit or short circuit at the output and driven by a nonideal voltage or current source using the results of Section 4.1; this is left to the reader to pursue.

Example 4.4. Design a third-order Butterworth BP filter with a center frequency $\omega_0 = 10^6$ rad s^{-1} and $Q = 10$ for a load of 1 kΩ, when the network is driven by a voltage source.

In Example 4.3, we have designed a normalized third-order Butterworth LP filter for $R_L = 1\ \Omega$. We convert this LP filter to the given BP filter by using the LP to BP transformation (see Section 3.10). In our case, $B = (\omega_0/Q) = 10^5$. Hence, each inductance L in the LP filter is to be replaced by an inductance (L/B) in series with a capacitance $B/(\omega_0^2 L)$, and each capacitance C in the LP network replaced by an inductance $B/(\omega_0^2 C)$ in parallel with a capacitance (C/B). Making these replacements and scaling the impedances by a factor of 1000, we get the required BP filter as shown in Figure 4.6.

Extensive tabulated results are available for singly terminated LC ladder filters corresponding to a cutoff frequency of 1 rad s^{-1} and a termination of 1 Ω. Some of these are included in Appendix C. We, therefore, conclude the discussion on singly terminated LC ladder filters and consider the subject of doubly terminated LC ladder filters.

4.4
Doubly Terminated LC Ladder Realization

When the voltage or current source is not ideal and the load is finite and nonzero, we have the situation of a doubly terminated lossless ladder network. Such an

4.4 Doubly Terminated LC Ladder Realization

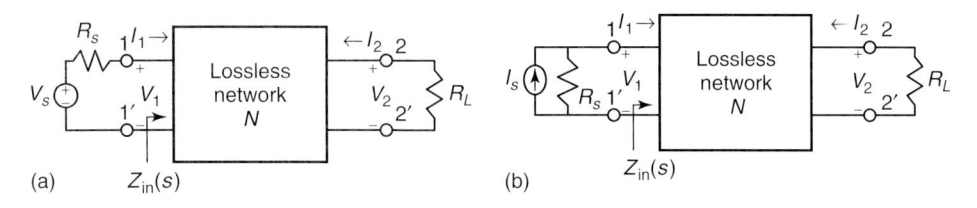

Figure 4.7 A doubly terminated lossless network driven by (a) a voltage source and (b) a current source.

arrangement is either the type shown in Figure 4.7a or b. It is observed that the VTF, $T_V = V_2/V_S$, and the TAF $Y_{21} = -I_2/V_S$ in Figure 4.7a, as well as the CTF $T_I = -I_2/I_S$ and the TIF $Z_{21} = V_2/I_S$, are all proportional to one another. Hence, without loss of generality, we assume the situation of Figure 4.7a and realize the given filter function as the VTF V_2/V_S. In a doubly terminated network, the performance is judged by the amount of power delivered to the load as compared to the maximum power that can be delivered by the source. We know that the maximum power is delivered if the load impedance is the complex conjugate of the source impedance. In the case of pure resistances, it means $R_S = R_L$; otherwise, maximum power is not delivered. The maximum power that the source can deliver is

$$P_{max} = \frac{|V_S(j\omega)|^2}{4R_S} \tag{4.20}$$

while the power delivered to the load is

$$P_L = \frac{|V_2(j\omega)|^2}{R_L} \tag{4.21}$$

We now define the transmission coefficient $t(s)$ as

$$|t(j\omega)|^2 = \frac{P_L}{P_{max}} = 4\frac{R_S}{R_L}\frac{|V_2(j\omega)|^2}{|V_S(j\omega)|^2} = 4\frac{R_S}{R_L}|H(j\omega)|^2 \tag{4.22}$$

Hence,

$$t(s)t(-s) = 4\frac{R_S}{R_L}H(s)H(-s)$$

or

$$t(s) = 4\frac{R_S}{R_L}H(s) \tag{4.23}$$

The reflection coefficient $\rho(s)$ is defined as

$$|\rho(j\omega)|^2 + |t(j\omega)|^2 = 1 \tag{4.24}$$

Let the input impedance looking into the port 1 with the load in place be $Z_{in}(s)$. If $Z_{in}(j\omega) = R_{in} + jX_{in}$, then the power delivered to the input terminals of N is $R_{in}|I_1(j\omega)|^2$. Since N is lossless, the power delivered to the load is also $R_{in}|I_1(j\omega)|^2$. Hence,

$$R_{in}|I_1(j\omega)|^2 = \frac{|V_2(j\omega)|^2}{R_L} \tag{4.25}$$

Therefore,

$$R_{in} \frac{|V_S(j\omega)|^2}{|R_S + Z_{in}(j\omega)|^2} = \frac{|V_2(j\omega)|^2}{R_L} \tag{4.26}$$

or

$$|H(j\omega)|^2 = \frac{|V_2(j\omega)|^2}{|V_S(j\omega)|^2} = \frac{R_L R_{in}}{|R_S + Z_{in}(j\omega)|^2} \tag{4.27}$$

Substituting Eq. (4.27) in Eq. (4.22) we get

$$|t(j\omega)|^2 = \frac{4 R_S R_{in}}{|R_S + Z_{in}(j\omega)|^2} \tag{4.28}$$

Hence, from Eqs. (4.24) and (4.28) we have

$$|\rho(j\omega)|^2 = 1 - \frac{4 R_S R_{in}}{|R_S + Z_{in}(j\omega)|^2} = \frac{(R_S + R_{in})^2 + X_{in}^2 - 4 R_S R_{in}}{|R_S + Z_{in}(j\omega)|^2}$$

$$= \frac{|R_S - Z_{in}(j\omega)|^2}{|R_S + Z_{in}(j\omega)|^2} \tag{4.29}$$

Equation (4.29) is satisfied if

$$\rho(s) = \pm \frac{R_S - Z_{in}(s)}{R_S + Z_{in}(s)} \tag{4.30}$$

Hence, the DPI $Z_{in}(s)$ is given by

$$\frac{Z_{in}(s)}{R_S} = \frac{1 \pm \rho(s)}{1 \mp \rho(s)} \tag{4.31}$$

Thus, there are two possible solutions for the normalized DPI, $Z_{in}(s)/R_S$. Darlington (1939) has shown that these normalized DPIs can always be realized by lossless networks terminated in a resistance, and that these two one-port networks are duals of each other.

Thus, the problem of synthesis of a doubly terminated lossless network consists of first finding $|t(j\omega)|^2$ from the given $H(s)$, and then finding $Z_{in}(s)/R_S$ and realizing it as a lossless two-port terminated in a resistor R_L. It is enough to get one realization for $Z_{in}(s)/R_S$, since the other realization is the dual of the first realization.

It is seen from Eq. (4.23) that the zeros of $H(s)$ are those of $t(s)$. Hence, if we are considering an all-pole LP function, then all the zeros are at infinity and the lossless two-port N will have inductors for all the series elements and capacitors for all the shunt elements. For an all-pole LP filter (such as the Butterworth or Chebyshev), we will then have

$$H(0) = \frac{R_L}{R_S + R_L} \tag{4.32}$$

Thus, if the given function $H(s)$ is of the form

$$H(s) = \frac{k}{D(s)} = \frac{k}{s^n + a_{n-1}s^{n-1} + \cdots + a_1 s + a_0} \tag{4.33}$$

then, from Eqs. (4.32) and (4.33),

$$H(0) = \frac{k}{a_0} = \frac{R_L}{R_S + R_L} \tag{4.34}$$

If the terminations are equal, that is, $R_S = R_L$, then from Eq. (4.34), k is constrained to be

$$\frac{k}{a_0} = \frac{R_L}{R_S + R_L} = \frac{1}{2} \tag{4.35}$$

The above condition assumes that $H(0) = H_{max}$ as in the case of Butterworth filter or a Chebyshev filter of odd order. If H_{max} does not occur at $\omega = 0$, like in the case of a Chebyshev filter of even order, then we require

$$H(j\omega) \leq H_{max} \text{ for all } \omega \tag{4.36}$$

and k has to be adjusted appropriately. This is why a Chebyshev filter of even order cannot be realized with equal terminations. Table 4.2 gives the systematic steps for the realization of an all-pole normalized LP filter with equal termination, normalized to 1 Ω.

Table 4.2 Various steps to be followed to design an all-pole doubly terminated LP filter.

Given the normalized transfer function $H(s)$ of an all-pole LP function (for example, MFM, CHEB, or BT) in the form of Eq. (4.33), and the source and load resistances:
↓
Find the value of k using Eq. (4.35)
↓
Derive $|H_n(j\omega)|^2$ using Eq. (4.33)
↓
Find $|t(j\omega)|^2$ using Eq. (4.22)
↓
Find $|\rho(j\omega)|^2$ using Eq. (4.24)
↓
Derive $\rho(s)$, from $\rho(s)\rho(-s) = |\rho(j\omega)|^2$, such that the poles of $\rho(s)$ are all in the LH of the s-plane; note that the zeros are not restricted to the LH of the s-plane
↓
Obtain the normalized DPI, $Z'_{in}(s) = Z_{in}(s)/R_S$, using Eq. (4.31)
↓
Choose one set of signs in Eq. (4.31) and express $Z'_{in}(s)$ as a ratio, $N(s)/D(s)$
↓
Expand $N(s)/D(s)$ in a CF around $s = \infty$ to obtain Cauer-I form realization for $Z'_{in}(s)$
↓
Then, the required normalized LP filter is obtained by driving this terminated network by a voltage source with a source resistance of 1 Ω, as shown in Figure 4.7a
↓
Impedance scale the filter so obtained by R_S.

If the target filter is an LP filter with a PB edge frequency ω_c, we need to apply frequency scaling to the components as $L \to L_n/\omega_c$ and $C \to C_n/\omega_c$, where L_n and C_n are the components in the normalized filter. If the target filter is not an LP filter, we need to derive the associated normalized LP filter transfer function by the methods illustrated in Chapter 3. One has to work according to the steps given in Table 4.2 to derive the LC network corresponding to the normalized LP filter transfer function. To realize the prescribed filter network, one then takes recourse to the component transformation relations as provided in Chapter 3 (Table 3.2, Section 3.10). Let us now consider a few examples.

Example 4.5. For the normalized Butterworth filter of order 3, obtain a doubly terminated LC ladder realization assuming $R_S = R_L = 1\,\Omega$.

We know that for $n = 3$, the transfer function $H(s)$ is given by

$$H(s) = \frac{k}{D(s)} = \frac{k}{s^3 + 2s^2 + 2s + 1}$$

Since $R_S = R_L$, we have from Eq. (4.35) that $k = (1/2)$. Hence, from Eq. (4.22) we have

$$|t(j\omega)|^2 = 4|H(j\omega)|^2 = 4\frac{1}{4}\frac{1}{|D(j\omega)|^2} = \frac{1}{1+\omega^6} \quad (4.37)$$

Hence,

$$|\rho(j\omega)|^2 = 1 - \frac{1}{1+\omega^6} = \frac{\omega^6}{1+\omega^6}$$

Therefore,

$$\rho(s)\rho(-s) = \frac{(s^3)(-s^3)}{(s^3 + 2s^2 + 2s + 1)(-s^3 + 2s^2 - 2s + 1)}$$

$$\rho(s) = \frac{s^3}{(s^3 + 2s^2 + 2s + 1)} = \frac{s^3}{D(s)} \quad (4.38)$$

Hence from Eq. (4.31), we have

$$Z_{in}(s) = \frac{2s^3 + 2s^2 + 2s + 1}{2s^2 + 2s + 1} \quad (4.39a)$$

or

$$Z_{in}(s) = \frac{2s^2 + 2s + 1}{2s^3 + 2s^2 + 2s + 1} \quad (4.39b)$$

It is seen that the two realizations we get using Eqs. (4.39a) and (4.39b) will be duals of each other w.r.t. $1\,\Omega^2$. Expanding $Z_{in}(s)$ given by Eq. (4.39a), by CF around $s = \infty$, we get

$$Z_{in}(s) = s + \cfrac{1}{2s + \cfrac{1}{s + \frac{1}{1}}} \quad (4.40)$$

Hence, $Z_{in}(s)$ can be realized by the ladder network N terminated in $R_L = 1\,\Omega$, as shown in Figure 4.8a. The required third-order Butterworth filter is realized with

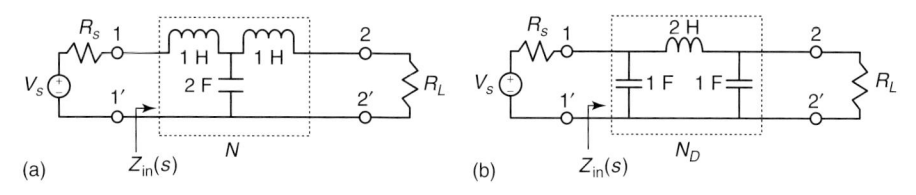

Figure 4.8 (a) Realization of the Butterworth filter of Example 4.5. (b) Alternate realization, which is the dual of the network of (a).

$k = (1/2)$ as the VTF $V_2(s)/V_S(s)$. Had we started with Eq. (4.39b) and realized it in Cauer-II form, we would have realized the VTF by the network of Figure 4.8b, which is nothing but the dual of that of Figure 4.8a w.r.t. $1\,\Omega^2$.

Example 4.6. Consider a LP Butterworth filter network with PB edge at 1590 Hz and an attenuation of 40 dB at 4000 Hz. Given that $R_s = R_L = 100\,\Omega$, find a realization of the filter using an equally terminated LC ladder structure.

From the specifications ($A_p = 3$ dB, $A_a = 40$ dB, $f_c = 1590$ Hz, $f_s = 4000$ Hz), one derives the order of the filter as $n = 5$. Then, from the table of standard Butterworth functions (Appendix A), the normalized transfer function is

$$H(s) = \frac{1}{s^5 + 3.2361s^4 + 5.2361s^3 + 5.2361s^2 + 3.2361s + 1} = \frac{k}{D(s)} \quad (4.41)$$

From Eq. (4.35), we see that $k = (1/2)$. For a fifth-order BUT function, we know that

$$|H(j\omega)|^2 = \frac{1}{1+\omega^{10}}$$

Hence,

$$|t(j\omega)|^2 = 4\,|H(j\omega)|^2 = 4\frac{1}{4}\frac{1}{|D(j\omega)|^2} = \frac{1}{1+\omega^{10}}$$

or

$$|\rho(j\omega)|^2 = 1 - |t(j\omega)|^2 = \frac{\omega^{10}}{1+\omega^{10}} \quad (4.42)$$

Therefore,

$$\rho(s) = \frac{s^5}{D(s)} \quad (4.43)$$

Then, the normalized input impedance function $Z_{in}(s)$ (i.e., with $R_S = 1\,\Omega$) is

$$Z_{in}(s) = \frac{1 + \rho(s)}{1 - \rho(s)}$$

$$= \frac{3.2361s^4 + 5.2361s^3 + 5.2361s^2 + 3.2361s + 1}{2s^5 + 3.2361s^4 + 5.2361s^3 + 5.2361s^2 + 3.2361s + 1} \quad (4.44)$$

Since the degree of the denominator is greater than the degree of the numerator, we expand $1/Z_{in}(s)$ by CF around $s = \infty$. The details are shown below. Only the coefficients of s^j ($j = 5, 4, 3, 2, 1$) are shown. The realized element values are shown by the side of the associated quotients.

$$3.2361, 5.2361, 5.2361, 3.2361, 1)\ 2, 3.2361, 5.2361, 5.2361, 3.2361, 1(0.618s \to \text{capacitance } C = 0.618F$$
$$2, 3.2361, 3.2361, 2, 0.618$$

- -

$$2, 3.2361, 2.618, 1)3.2361, 5.2361, 4.2361, 1.618(1.618s \to \text{inductance } L = 1.618H$$
$$3.2361, 5.2361, 4.2361, 1.618$$

- -

$$1, 1.618, 1)2, 3.2361, 2.618, 1(2s \to \text{capacitance } C = 2\ F$$
$$2, 3.2361, 2$$

- - - - - - - - - - - - - - -

$$0.618, 1)1, 1.618, 1(1.618s \to \text{inductance} L = 1.618H$$
$$1, 1.618$$

- - - - - - - - - - - -

$$1)0.618(0.618s \to \text{capacitance } C = 0.618F$$
$$0.618$$

- - - - - - - - - - -

$$0$$

The normalized LP filter ladder network obtained is shown in Figure 4.9a. An alternate realization would be one where the reactance two-port is replaced by its dual two-port, and is shown in Figure 4.9b.

For terminations of 100 Ω at each side and a PB frequency of $\omega_c = 2\pi(1590) \cong 10^4$ rad s^{-1}, the denormalized LC networks can be obtained by the substitutions $C \to C_n/(10^4.100) = C_n/(10^6)$ and $L \to L_n(100)/(10^4) = L_n/(10^2)$, where C_n and L_n are the components in the normalized filter networks. The denormalized filters are shown in Figures 4.10a and 4.10b.

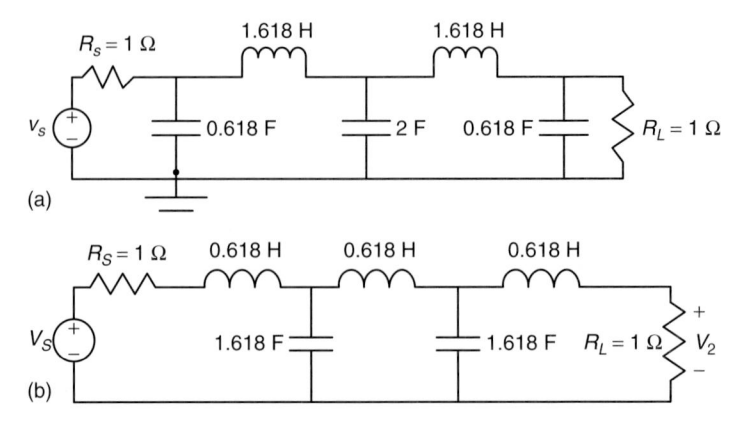

Figure 4.9 (a) Realization of the normalized Butterworth filter for Example 4.6. (b) Alternate realization.

4.4 Doubly Terminated LC Ladder Realization

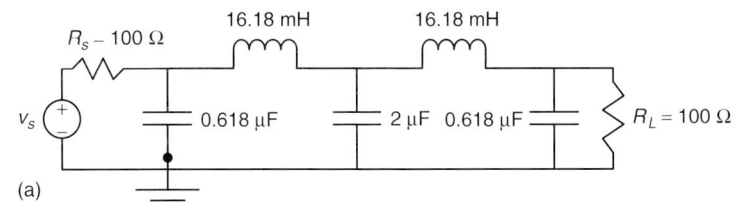

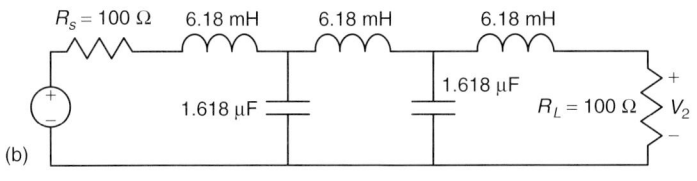

Figure 4.10 (a) Realization of the Butterworth filter for the specifications of Example 4.6. (b) Alternate realization.

We now consider the case of unequal terminations using the example of a third-order Butterworth filter.

Example 4.7. (a) Realize the third-order Butterworth filter of Example 4.5, with terminations of $R_S = 1\,\Omega$ and $R_L = 4\,\Omega$. Find the value of the gain k.

(b) Realize the same transfer function when the source and load resistances are reversed, and find the corresponding value of k.

(c) Obtain alternate realizations for the above two networks, and find the corresponding value of k.

Solution: (a) In this case, we have from Eq. (4.35) that

$$\frac{k}{a_0} = \frac{R_L}{R_S + R_L} = \frac{4}{5} \Rightarrow k = \frac{4}{5} \tag{4.45}$$

Hence, from Eqs. (4.22) and (4.24), we have

$$|\rho(j\omega)|^2 = 1 - |t(j\omega)|^2 = 1 - 4\frac{R_S}{R_L}|H(j\omega)|^2$$

$$= 1 - 4\frac{1}{4}\frac{(4/5)^2}{1+\omega^6} = \frac{\omega^6 + (9/25)}{\omega^6 + 1} \tag{4.46}$$

Therefore,

$$\rho(s)\rho(-s) = \frac{P(s)P(-s)}{D(s)D(-s)} = \frac{(s^3 + 0.6)(-s^3 + 0.6)}{(s^3 + 2s^2 + 2s + 1)(-s^3 + 2s^2 - 2s + 1)} \tag{4.47}$$

There is more than one choice for $P(s)$, since there is no restriction that the zeros of $P(s)$ have to be in the LHS of the s-plane, as those of $D(s)$. Choosing $P(s) = (s^3 + 0.6)$, we get

$$\rho(s) = \frac{P(s)}{D(s)} = \frac{(s^3 + 0.6)}{(s^3 + 2s^2 + 2s + 1)} \tag{4.48}$$

Hence from Eq. (4.32),

$$Z'_{in}(s) = \frac{Z_{in}(s)}{R_S} = \frac{1 + \rho(s)}{1 - \rho(s)} = \frac{2s^3 + 2s^2 + 2s + 1.6}{2s^2 + 2s + 0.4} \tag{4.49}$$

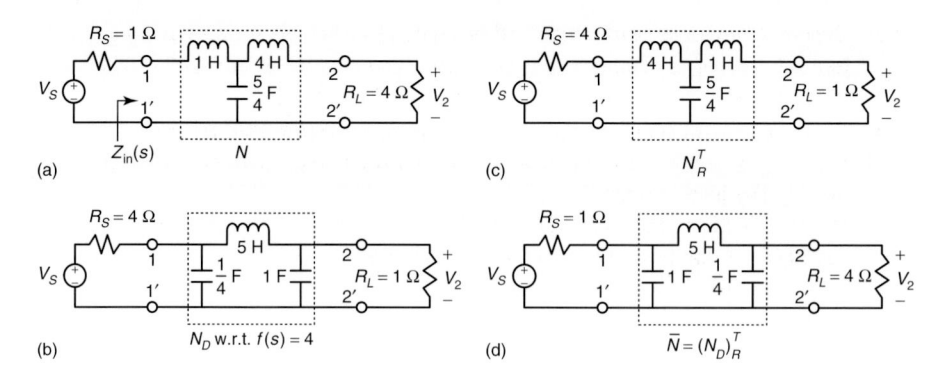

Figure 4.11 Various doubly terminated structures realizing a third-order Butterworth filter. (a) $R_S = 1\ \Omega$, $R_L = 4\ \Omega$, (b) $R_S = 4\ \Omega$, $R_L = 1\ \Omega$, (c) alternate realization for (a), and (d) alternate realization for (c).

which can be realized by a Cauer-I form ladder network N terminated in a 4-Ω resistor. The complete realization of the third-order Butterworth filter is shown in Figure 4.11a.

(b) We know from Chapter 3 that if the ladder network N is replaced by its dual network N_D w.r.t. $f(s) = R_S R_L = 4\ \Omega^2$, we realize the same VTF $H(s)$ with $k = (1/5)$, but with $R_S = 4\ \Omega$ and $R_L = 1\ \Omega$. This realization is shown in Figure 4.11b.

(c) Again from Chapter 3, we also know that we get two more realizations through N_R^T and $(N_D)_R^T$. However, since N is a reciprocal network, N^T is nothing but N itself. Hence, the realizations corresponding to N_R^T and $(N_D)_R^T$ may be obtained very easily from the realizations of Figures 4.11a and 4.11b and these are shown in Figures 4.11c and 4.11d, respectively. It is seen, as explained in Chapter 3, that realizations shown in Figures 4.11a and 4.11d are alternate structures with $R_S = 1\ \Omega$ and $R_L = 4\ \Omega$ (with $k = (4/5)$) while those in Figures 4.11b and 4.11c are alternate structures realizing the same VTF (with $k = (1/5)$), but with $R_S = 4\ \Omega$ and $R_L = 1\ \Omega$; that is, with the source and load terminations reversed. It should be noted that the former networks have a higher value of k compared to that of the latter two networks. A higher k value implies a higher gain at DC.

Just as in the case of the singly terminated LC ladder filters, extensive tables are available in the literature, giving the component values for doubly terminated LC ladder filters realizing Butterworth, Chebyshev, and other types (Christian and Eisermann, 1977; Zverev, 1967; Weinberg, 1962). Some of these are included in Appendix C. It should also be noted that closed-form solutions exist giving the component values for Butterworth and Chebyshev LP filters (Chen, 1986; Schaumann, Ghausi, and Laker, 1990).

Practice Problems

4.1 Synthesize using LC elements, a third-order Butterworth filter to work into a load of 1 Ω and excited by (a) an ideal voltage source and (b) an ideal current source.

4.2 Design a third-order Butterworth filter excited by a voltage source having an internal resistance of 50 Ω, if the filter is to be connected to an ideal OA. The input impedance of the OA can be assumed to be infinite.

4.3 Design an LC all-pole LP singly terminated filter with Butterworth response having $f_c = 1$ kHz and which produces an attenuation of 25 dB per octave in the SB. The load resistance is 100 Ω.

4.4 Design an LC all-pole LP singly terminated filter with an equiripple PB having $A_p = 0.5$ dB up to $f_c = 6$ kHz and a monotonic SB with $A_a \geq 30$ dB for $f \geq 15$ kHz. The termination resistance is 1 Ω.

4.5 Design a singly terminated LC HP filter having an equiripple PB with $A_p = 0.5$ dB for $f_c \geq 15$ kHz and a monotonic SB for $f \leq 7.8$ kHz with $A_a \geq 35$ dB.

4.6 Synthesize using LC elements, a CHEB LP filter of order 3 with 0.5 dB ripple in the PB. The load terminations are 100 Ω each.

4.7 Consider a CHEB LP filter of order 4 with a PB ripple of 1 dB. Synthesize the LC ladder filter for double termination with $R_s = 1 \Omega$.

4.8 Consider a CHEB LP filter of order 5 with passband ripple of 0.5 dB. The passband extends from DC to 1.2 kHz. The loss at 1.92 kHz is 23 dB. Synthesize the LC ladder filter.

4.9 Consider the LC ladder realization of a BP filter with Butterworth magnitude response. The passband edges are $f_{p1} = 1578$ Hz and $f_{p2} = 3168$ Hz. The loss is 40 dB at $f \geq 5$ kHz and at $f \leq 1$ kHz. The termination resistances are $R_L = R_s = 100 \Omega$. Synthesize the filter.

4.10 Obtain an LC ladder to be inserted between a source resistance of 1 Ω and a load resistance of 2 Ω given that

$$|t(j\omega)|^2 = \frac{k}{\omega^4 + 1}$$

Using the concepts of dual, transpose, and dual transpose, find an alternate structure realizing the same $|t(j\omega)|^2$. Also, find structures realizing the same $|t(j\omega)|^2$, but with a source resistance of 2 Ω and a load resistance of 1 Ω. Find the corresponding value of k.

4.11 Design an LP elliptic filter that has a maximum of 1 dB attenuation in the PB of DC to 1000 rad s^{-1} and a minimum of 33 dB attenuation in the SB, with SB edge being at 2000 rad s^{-1}. The filter should have equal terminations of 1000 Ω at the two ends.

5
Second-Order Active-RC Filters

Classically, the field of active filters arose out of a desire to realize filters without having to use inductances for low-frequency applications where the inductance is required to have a large value. Such large values implied large physical space, cost, and poor reliability. In the early 1960s, some innovative researchers came up with the desired alternative, that is, filters having no inductances.

In this chapter, we introduce second-order voltage-mode filters involving resistances, capacitances, and active devices. One of the advantages of realizing second-order active filters (also known as *biquadratic filters*) is that a general filter of a higher order can be designed by cascading a number of second-order ones, with an addition of a bilinear filter, if need be. In a voltage-mode system, the transfer function of interest is the ratio of the output voltage to the input signal voltage applied to the system. The active devices are usually voltage amplifiers (such as the OAs), OTAs, and CCs. We first introduce standard second-order transfer functions. The network theoretic background for realization of a frequency-selective network, such as a filter, using an active device embedded in an array of passive elements containing only resistances and capacitances is presented next. A general network containing a single-voltage amplifier and passive elements that can realize a biquadratic transfer function is then presented. This is used to illustrate the realization of several second-order filters using a single finite- or an infinite-gain voltage amplifier. The discussion is continued to realizations using several infinite-gain voltage amplifiers. Sensitivity considerations are introduced and several low-sensitivity second-order filters are presented. Second-order filters using OTA are considered next. The case of the frequency-dependent gain of an OA is considered and its effect on the realization of a second-order filter is discussed. Technological considerations regarding hardware implementation of the filters are presented at the end of the chapter. Before we present the realization of the second-order filters, we first briefly introduce some of the simple building blocks realizable using an OA and then discuss some general properties of second-order filters.

Modern Analog Filter Analysis and Design: A Practical Approach. Rabin Raut and M. N. S. Swamy
Copyright © 2010 WILEY-VCH Verlag GmbH & Co. KGaA, Weinheim
ISBN: 978-3-527-40766-8

5.1
Some Basic Building Blocks using an OA

In Chapter 2, we discussed how an OA could be used as a summer or integrator. There are a number of other simple building blocks, including bilinear filters, which are useful in the study of higher-order filters using OAs. These are tabulated in Table 5.1; these may be easily derived using the model of an ideal OA introduced in Chapter 2 and are left to the reader to derive them.

5.2
Standard Biquadratic Filters or Biquads

A second-order transfer function of the form

$$H(s) = \frac{N(s)}{D(s)} = \frac{b_2 s^2 + b_1 s + b_1}{s^2 + a_1 s + a_0} \tag{5.1}$$

is called a *biquadratic function*. Even though, in general, the poles and zeros may lie on the negative real axis, we will assume them to be complex conjugates, since poles and zeros on the negative real axis can be realized using passive-RC circuits (Van Valkenburg, 1960). In such a case, we may express

$$H(s) = \frac{N(s)}{D(s)} = H_0 \frac{(s+z)(s+z^*)}{(s+p)(s+p^*)} = H_0 \frac{s^2 + (\omega_z/Q_z)s + \omega_z^2}{s^2 + (\omega_p/Q_p)s + \omega_p^2} \tag{5.2}$$

where

$$\omega_z^2 = (\text{Re } z)^2 + (\text{Im } z)^2, \, Q_z = \omega_z/2\text{Re}(z) \tag{5.3a}$$

$$\omega_p^2 = (\text{Re } p)^2 + (\text{Im } p)^2, \, Q_p = \omega_p/2\text{Re}(p) \tag{5.3b}$$

and H_0 could be positive or negative. ω_p and Q_p are called the *pole frequency* and *pole Q*; sometimes ω_p is called the *undamped natural frequency* since there will be resonance at $s = j\omega_p$, if the s-term is not present in the denominator of Eq. (5.2). The poles are given by

$$p_{1,2} = -\frac{\omega_p}{2Q_p} \pm j\frac{\omega_p}{Q_p}\sqrt{4Q_p^2 - 1} \tag{5.4}$$

It is clear that, in order to have complex poles, $Q_p > 0.5$. For a highly selective filter, Q_p should be large, that is, the real part of the poles should tend to zero, and hence the poles will be close to the imaginary axis. The following properties can be observed from Eqs. (5.2) and (5.3):

1) The dc gain is $[H_0(\omega_z^2/\omega_p^2)]$.
2) The gain at $\omega = \infty$ is $|H_0|$.
3) The maximum value of $|H(j\omega)|$ occurs approximately at ω_p if $Q_p \gg 1$. This is particularly true if $\omega_z \gg \omega_p$ or $\omega_z \ll \omega_p$; otherwise, it is slightly moved away from ω_p.
4) The minimum value occurs at approximately $\omega = \omega_z$ for $Q_z \gg 1$.

Table 5.1 Some basic building blocks using an OA.

	Operational amplifier circuit	Transfer function	Remarks
A		$\frac{V_2}{V_1} = -\frac{R_2}{R_1}$	Inverting voltage amplifier
B		$\frac{V_2}{V_1} = 1 + \frac{R_2}{R_1}$	Noninverting voltage amplifier
C		$\frac{V_2}{V_1} = 1$	Voltage follower (unity gain voltage amplifier)
D		$V_o = -R \sum_{i=1}^{n} \frac{V_i}{R_i}$	Inverted summer

(*continued overleaf*)

Table 5.1 (continued).

	Operational amplifier circuit	Transfer function	Remarks
E	(op-amp with V_1 through R_1 to inverting input, V_2 through R_3 to non-inverting input, R_2 feedback, R_4 to ground)	$V_o = \frac{R_4}{R_1} \times \frac{R_1+R_2}{R_3+R_4} V_2 - \frac{R_2}{R_1} V_1$	Differential summer
F	(inverting op-amp with R input, C feedback)	$\frac{V_2}{V_1} = -\frac{1}{RCs}$	Inverting lossless integrator
G	(inverting op-amp with R_1 input, $R_2 \parallel C$ feedback)	$\frac{V_2}{V_1} = -\frac{(1/R_1)}{(1/R_2)+Cs}$	Lossy integrator
H	(inverting op-amp with R_1, C_1 in series at input, R_2, C_2 in series in feedback)	$\frac{V_2}{V_1} = -\frac{R_2+(1/C_2 s)}{R_1+(1/C_1 s)}$	Bilinear transfer function with a negative gain

Table 5.1 (continued).

	Operational amplifier circuit	Transfer function	Remarks
I		$\frac{V_2}{V_1} = -\frac{C_1}{C_2} \times \frac{s+(1/R_1 C_1)}{s+(1/R_2 C_2)}$	Bilinear transfer function with a negative gain
J		$\frac{V_2}{V_1} = \left[1 + \frac{R_2}{R_1}\right] \times \left[\frac{s+1/(R_1+R_2)C_1}{s+1/(R_1 C_1)}\right]$	Bilinear transfer function with a positive gain
K		$\frac{V_2}{V_1} = \left[\frac{s+\{1+(R_2/R_1)\}/(C_2 R_2)}{s+1/(R_2 C_2)}\right]$	Bilinear transfer function with a positive gain

5) Q_p is a measure of the sharpness of the maximum.
6) Q_z is a measure of the sharpness of minimum value.

These will become clearer when we sketch the magnitude response of the different types of filters. By choosing suitable values for the coefficients b_0, b_1, and b_2, that is, the positions of the zeros of $H(s)$, we can obtain different types of

5 Second-Order Active-RC Filters

Table 5.2 Standard biquadratic transfer functions.

Type of filter	N(s)
Low-pass (LP)	$H_o \omega_p^2$
Band-pass (BP)	$H_o(\omega_p/Q_p)s$
High-pass (HP)	$H_o s^2$
All-pass (AP)	$s^2 - (\omega_p/Q_p)s + \omega_p^2$
Notch	$H_o(s^2 + \omega_n^2)$

Note: For all filters, $\frac{V_2(s)}{V_1(s)} = H(s) = \frac{N(s)}{D(s)}$, with $D(s) = s^2 + (\omega_p/Q_p)s + \omega_p^2$.

biquadratic filters or biquads. These are listed in Table 5.2. It is seen that, except for the AP filter, the zeros are all on the imaginary axis and hence, $Q_z = \infty$.

The meaning of ω_p and Q_p will become more clear when we consider a BP filter of the form

$$H_{BP}(s) = H_o \frac{(\omega_p/Q_p)s}{s^2 + (\omega_p/Q_p)s + \omega_p^2} \tag{5.5}$$

At $s = j\omega_p$,

$$|H_{BP}(j\omega)| = |H_o| \tag{5.6}$$

Let us find the frequencies ω_1 and ω_2, where $|H_{BP}(j\omega)|$ is $(1/\sqrt{2})$ times that at the peak value, namely, $|H_o|$; then,

$$|H_{BP}(j\omega)|^2 = \frac{|H_o(\omega_p/Q_p)\omega|^2}{(\omega_p^2 - \omega^2)^2 + (\omega_p/Q_p)^2\omega^2} = \frac{1}{2}|H_o|^2 \tag{5.7}$$

From the above, we get

$$\omega^2 - \omega_p^2 = \pm(\omega_p/Q_p)\omega \tag{5.8}$$

Solving Eq. (5.8) and taking the positive roots ω_1 and ω_2 for ω, we get

$$\omega_1\omega_2 = \omega_p^2 \text{ and } \omega_2 - \omega_1 = (\omega_p/Q_p) \tag{5.9}$$

If $\omega_2 - \omega_1$ is defined as the bandwidth (BW) of the BP filter, we then have

$$Q_p = \frac{\omega_p}{BW} = \frac{\omega_p}{\omega_2 - \omega_1} \tag{5.10}$$

Thus, Q_p and BW are inversely related and hence, the higher the Q_p, the narrower the BW of the filter. The nature of the magnitude response of the BP filter is shown in Figure 5.1. Even though we cannot relate the BW in the same way as in a BP filter, the peak values for the LP and HP filters also increase with increasing value of Q_p. As for the notch filter, depending on whether $\omega_z > \omega_p$, $\omega_z < \omega_p$, or $\omega_z = \omega_p$, it is called an *LP notch*, an *HP notch*, or a *symmetric notch* filter. Figure 5.1 shows the magnitude response for typical biquad LP, HP, BP, and notch filters as well as the magnitude and phase response for the AP filter.

5.3 Realization of Single-Amplifier Biquadratic Filters

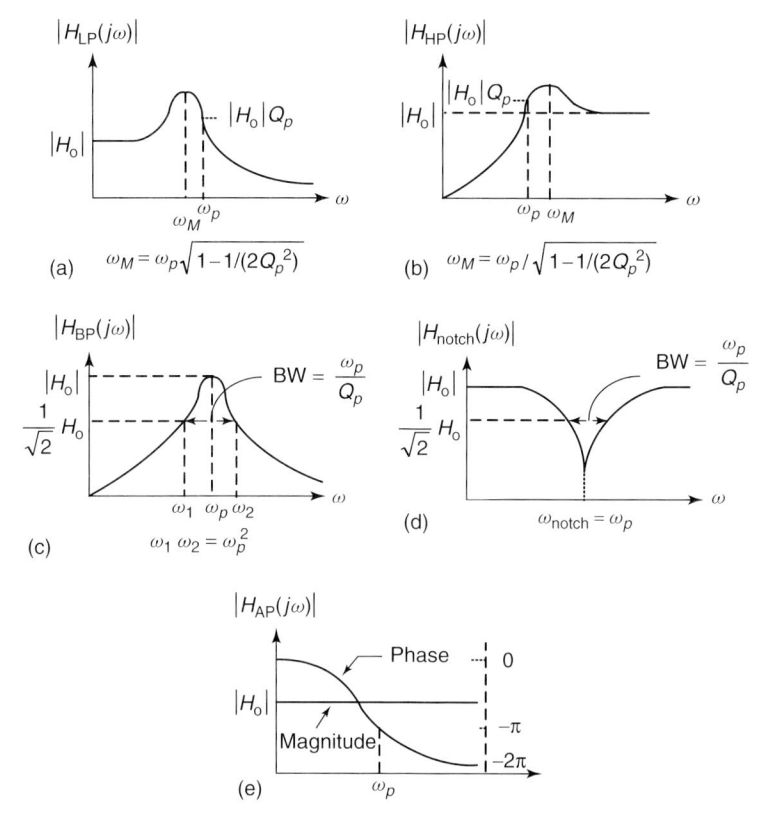

Figure 5.1 Magnitude response of (a) an LP, (b) an HP, (c) a BP, and (d) a notch filter. (e) Magnitude and phase responses of an AP filter.

For (a): $\omega_M = \omega_p\sqrt{1 - 1/(2Q_p^2)}$

For (b): $\omega_M = \omega_p/\sqrt{1 - 1/(2Q_p^2)}$

For (c): $BW = \omega_p/Q_p$, $\omega_1 \omega_2 = \omega_p^2$

For (d): $BW = \omega_p/Q_p$, $\omega_{notch} = \omega_p$

5.3
Realization of Single-Amplifier Biquadratic Filters

As mentioned earlier, a passive network consisting of only resistors and capacitors has all its poles on the negative real axis, and hence cannot have complex poles. Therefore, a passive-RC network cannot give rise to a frequency-selective transfer function. Imbedding an active device such as a voltage amplifier with a gain K in a passive-RC network, however, opens up the possibility of realizing a transfer function with complex poles. Consider the network of Figure 5.2, where a voltage amplifier of gain K is connected in a feedback structure with the three-port passive-RC network. Using the admittance parameters of the three-port network and the principle of constrained network analysis (see Chapter 2), the VTF can be expressed as

$$\frac{V_2(s)}{V_1(s)} = \frac{-Ky_{31}(s)}{y_{33}(s) + Ky_{32}(s)} \quad (5.11)$$

5 Second-Order Active-RC Filters

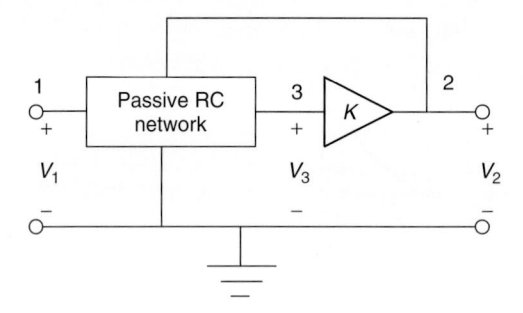

Figure 5.2 A general single-amplifier second-order filter configuration.

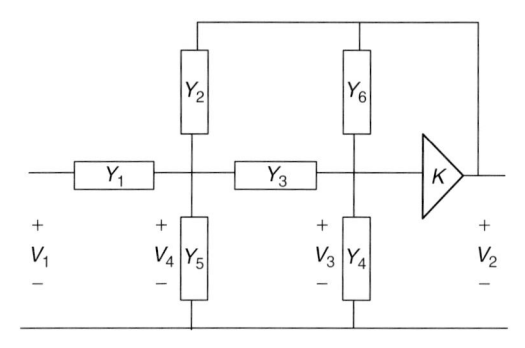

Figure 5.3 A specific single-amplifier biquad.

In Eq. (5.11), y_{31} and y_{32} are the short circuit TAFs and y_{33} is the short circuit DPA at port 3 of the passive network. While these functions cannot have poles in the complex plane, the gain K can be suitably adjusted to make $D(s) = y_{33}(s) + Ky_{32}(s)$ to possess zeros in the complex plane. Thus, the VTF in Eq. (5.11) can have complex poles. When the real parts of these poles lie in the LH of the s-plane, the VTF can produce a stable frequency-selective filter function.

A general configuration for producing a biquadratic transfer function using a single amplifier of finite gain K, popularly referred to as *single-amplifier biquad* (SAB) is shown in Figure 5.3. Using the method of analysis of constrained networks in conjunction with the nodal suppression technique (see Chapter 2), we can show that the transfer function $V_2(s)/V_1(s)$ is given by

$$\frac{V_2(s)}{V_1(s)} = \frac{KY_1Y_3}{(Y_1+Y_2+Y_5)(Y_3+Y_4+Y_6)+Y_3(Y_4+Y_6)-K\{Y_6(Y_1+Y_2+Y_3+Y_5)+Y_2Y_3\}} \quad (5.12)$$

Choosing the admittances appropriately, it is possible to realize second-order filters with LP, BP, and HP characteristics. Filters using a single *positive gain* (K is positive) ideal voltage amplifiers are also known as *Sallen and Key* (SK) filters (Sallen and Key, 1955).

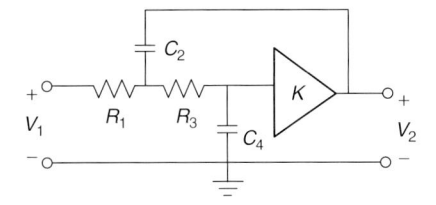

Figure 5.4 Sallen and Key LP filter.

5.4
Positive Gain SAB Filters (Sallen and Key Structures)

5.4.1
Low-Pass SAB Filter

In the general structure of Figure 5.3, if we choose $Y_1 = G_1 = 1/R_1$, $Y_2 = sC_2$, $Y_3 = G_3 = 1/R_3$, $Y_4 = sC_4$, and $Y_5 = Y_6 = 0$, we get the following transfer function, which is that of an LP filter. Specifically, the transfer function is given by

$$H_{LP}(s) = \frac{V_2(s)}{V_1(s)} = \frac{K(G_1 G_3/C_2 C_4)}{s^2 + s\left\{\frac{G_1}{C_2} + \frac{G_3}{C_2} + (1-K)\frac{G_3}{C_4}\right\} + \frac{G_1 G_3}{C_2 C_4}} \quad (5.13)$$

The schematic of the SAB LP filter is shown in Figure 5.4.

The network elements are related to the filter parameters ω_p, Q_p, and H_0 through the following design equations:

$$\omega_p = \frac{1}{\sqrt{R_1 R_3 C_2 C_4}} \quad (5.14a)$$

$$\frac{1}{Q_p} = \sqrt{\frac{R_3 C_4}{R_1 C_2}} + \sqrt{\frac{R_1 C_4}{R_3 C_2}} + (1-K)\sqrt{\frac{R_1 C_2}{R_3 C_4}} \quad (5.14b)$$

$$H_0 = K \quad (5.14c)$$

The above equations may be considered as the general design equations. In a practical case, simplifying assumptions are used to ease the task of design. The rationale behind such simplification lies in the fact that the general design equation contains more parameters (more component values) than are required to satisfy the three conditions given by Eqs. (5.14a)–(5.14c). Hence, some of these components can be assigned suitable values. Depending upon the simplifications, several alternative design equations may be arrived at. Consider the following design strategies.

Case 1: Equal-capacitor, gain of 2 design

It is given that $K = 2$. Let $C_2 = C_4 = C$. Then the design Eqs. (5.14a)–(5.14c) lead to

$$R_1 = R_3 Q_p^2, \quad R_3 = 1/(Q_p \omega_p C) \text{ and } H_0 = K = 2 \quad (5.15)$$

The advantages of this design are that the capacitors are of equal value and the gain $K = 2$ can be obtained using equal resistances (see Table 5.1). In

IC technology, it is easier to maintain ratios of capacitors or resistors more accurately than to maintain their absolute values precisely. However, the disadvantage is that the spread in the resistor value is rather large for large Q_p, since the ratio of the resistors is Q_p^2. Also, observe that gain at $\omega = 0$ is always 2. Let us consider an example.

Example 5.1. Design an LP filter with a pole Q of 4 and a pole frequency of 10^4 rad s^{-1}, using the above design procedure.

The transfer function of the LP filter is given by

$$H_{LP}(s) = H_0 \frac{\omega_p^2}{s^2 + \left(\frac{\omega_p}{Q}\right)s + \omega_p^2} \tag{5.16}$$

where the values of ω_p and Q_p are given by $\omega_p = 10^4$ rad s^{-1} and $Q_p = 4$, respectively. From Eq. (5.15), we see that $R_1 = 16R_3$, $C = \frac{1}{\{4(10^4)R_3\}}$, $K = 2$, and hence $H_0 = 2$. Assuming $R_3 = 1\,\Omega$, we get $R_1 = 16\,\Omega$ and $C = 25\,\mu F$. Impedance scaling by 1000, we have the component values as

$$R_3 = 1\,k\Omega,\ R_1 = 16\,k\Omega,\ C_1 = C_2 = 0.025\,\mu F,\ \text{and}\ K = 2.$$

Case 2: Equal-resistor and equal-capacitor design

Let $R_1 = R_3 = R$ and $C_2 = C_4 = C$. This leads to the design equations

$$C = 1/(R\omega_p^2)\ \text{and}\ K = 3 - (1/Q_p) \tag{5.17}$$

The advantage of this design is the resistor as well as the capacitors are of equal value, but it should be noted that the value of the gain K is dependent on Q_p.

Example 5.2. Design an LP filter for the same specifications as those in Example 5.1 using the above design equations.

Using Eq. (5.17), we get

$$C_2 = C_4 = C = 1/(10^8 R),\ K = 3 - \left(\tfrac{1}{4}\right) = 2.75$$

Choosing $R = 100\,\Omega$, we get the component values to be

$$R_1 = R_3 = 100\,\Omega,\ C_2 = C_4 = 100\,pF\ \text{and}\ K = 2.75$$

Case 3: Unity-gain design

Hence, $K = 1$. Let

$$m = C_4/C_2\ \text{and}\ n = R_3/R_1. \tag{5.18a}$$

Hence, from Eqs. (5.14a) and (5.14b), we get

$$R_1 C_2 = \frac{1}{\sqrt{mn}\,\omega_p} \qquad (5.18b)$$

$$\frac{1}{Q_p} = \sqrt{\frac{m}{n}}(n+1)$$

Solving the above equation, we get

$$n = \left(\frac{1}{2mQ_p^2} - 1\right) \pm \frac{1}{2mQ_p^2}\sqrt{1 - 4mQ_p^2} \qquad (5.18c)$$

For n to be real,

$$m \leq \frac{1}{4Q_p^2} \qquad (5.18d)$$

There are two solutions for n, and it is easy to see that they are reciprocals of each other, giving the same ratio for the resistors.

In this case, since $K = 1$, the gain can be achieved with a very high degree of accuracy by a voltage follower circuit (see Table 5.1). In addition, it saves two resistors (the ones used to obtain K). However, this design has the disadvantage of a large spread in capacitor values, since their ratio is Q_p^2.

Example 5.3. Design the LP filter for the specifications given in Example 5.1 using the above design procedure.

In this case, the design equations are given by Eqs. (5.18a)–(5.18d). From Eq. (5.18d), $m \leq (1/64) = 0.0156$. Choosing $m = 0.001$, we get from Eq. (5.18c), $n = 0.0329$ or 30.397. Hence from Eq. (5.18b), $R_1 C_2 = (5.7357) \times 10^{-4}$. If we choose $C_2 = 0.1\ \mu F$, then $C_4 = 100$ pF and $R_1 = 5.746$ kΩ. Hence, $R_3 = 174.7$ kΩ and $K = 1$.

5.4.2
RC:CR Transformation

Consider a biquadratic LP filter of pole frequency ω_p realized using an active-RC network. If we now apply the LP to HP filter transformation (see Chapter 3), namely, $s \to \omega_p^2/s$, then the LP filter of pole frequency ω_p would be transformed to a biquadratic HP filter having the same pole frequency ω_p; also, a capacitor of C Farads would be transformed into an inductor of value $(1/\omega_p^2 C)$ Henries, while a resistor of value $R\ \Omega$ would remain the same. Since we are avoiding inductors, and trying to realize filters by active-RC networks, this would not be useful. However, if we now perform an impedance transformation (see Chapter 2) on the new HP network, where every impedance $z(s)$ is transformed into another impedance of value $(1/s)\,z(s)$, then the transfer function of the HP filter will not be affected provided the active element is a VCVS or a CCCS; however, a resistor of value $R\ \Omega$ would become a capacitor of $(1/R)$ Farads and an inductor of value $(1/\omega_p^2 C)$ Henries

5 Second-Order Active-RC Filters

Table 5.3 RC:CR transformation and the transformed elements.

The original transfer function	Transformed transfer function
$H(s)$	$H(\omega_p^2/s)$
R Ω —/\/\/—	$1/R$ F —∣∣—
C F —∣∣—	$1/(\omega_p^2 C)$ Ω —/\/\/—
VCVS	VCVS (unaltered)
CCCS	CCCS (unaltered)

would become a resistor of value $(1/\omega_p^2 C)$ Ω. Thus, the resulting HP filter would then again be an active-RC filter using the same active elements as those in the LP filter. The combination of the frequency transformation $s \to \omega_p^2/s$ followed by the impedance transformation $z(s) \to (1/s)\, z(s)$ on the resulting network is also called the *RC:CR transformation* (Mitra, 1967, 1969), and is useful in converting an active-RC LP filter to an active-RC HP filter; it is noted that this is true only if the active element is a VCVS such as an OA, or a CCCS such as a current conveyor or a current OA. It does not hold good for an RC filter realized using OTAs. The RC:CR transformation and its effect on an active-RC filter using a VCVS or a CCCS is shown in Table 5.3.

As an application, consider a biquad active-RC filter designed using a VCVS. Let its transfer function be

$$H_{LP}(s) = H_0 \frac{\omega_p^2}{s^2 + \left(\frac{\omega_p}{Q_p}\right) s + \omega_p^2} \tag{5.19}$$

By applying the RC:CR transformation, we get the transfer function

$$H_0 \frac{s^2}{s^2 + \left(\frac{\omega_p}{Q}\right) s + \omega_p^2} \tag{5.20}$$

which can be readily seen as a biquad HP filter. Let us illustrate the procedure by the following example.

Example 5.4. Design an HP SAB filter with a pole Q of 4 and a pole frequency of $\omega_p = 10^4$ rad s^{-1}.

In Example 5.2, we have already designed an LP filter for the same specifications using the circuit of Figure 5.4, the transfer function being given by Eq. (5.16). One set of design values that we obtained were

$$R_1 = R_3 = 100\,\Omega,\ C_2 = C_4 = 100\,\text{pF and } K = 2.75$$

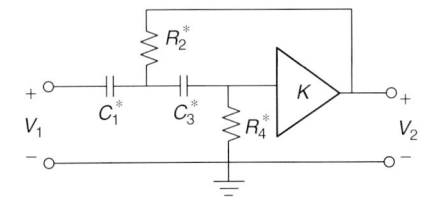

Figure 5.5 The HP filter obtained from Sallen and Key LP filter by RC:CR transformation.

Applying the RC:CR transformation, we get the transfer function

$$H_{LP}(s) = H_0 \frac{s^2}{s^2 + (10^4/4)s + \omega_p^2} \tag{5.21}$$

whose realization is shown in Figure 5.5. The component values are given by

$$C_1^* = C_3^* = \left(\frac{1}{R_1}\right) = 10^{-2}\,\text{F}, \quad R_2^* = R_4^* = \left(\frac{1}{\omega_p^2 C_2}\right) = \frac{1}{10^2 10^{-10}} = 100\,\Omega,$$

$$K = 2.75$$

Impedance scaling by 10^4, we get

$$C_1^* = C_3^* = 1\,\mu\text{F}, \quad R_2^* = R_4^* = 1\,\text{M}\Omega, \text{ and } K = 2.75$$

5.4.3
High-Pass Filter

We could, of course, have obtained the HP filter in the form given by Eq. (5.20) directly from the structure of Figure 5.3 by letting $Y_1 = sC_1$, $Y_3 = sC_3$, $Y_2 = (1/R_2)$, $Y_4 = (1/R_4)$, $Y_5 = Y_6 = 0$. The HP filter circuit along with the values for ω_p and Q_p in terms of the components R_2, R_4, C_1, C_3, and K, as well as a set of design equations, are given in Table 5.4.

5.4.4
Band-Pass Filter

By letting $Y_1 = (1/R_1)$, $Y_2 = (1/R_2)$, $Y_3 = sC_3$, $Y_4 = (1/R_4)$, $Y_5 = sC_5$, $Y_6 = 0$ in Figure 5.3, we get the SK BP filter. The BP filter along with the values of ω_p and Q_p in terms of the component values, as well as a set of design equations, is given in Table 5.4.

5.5
Infinite-Gain Multiple Feedback SAB Filters

In the previous sections, we have considered biquads designed using a finite gain amplifier. In this section, we consider a class of biquads that use a single OA as an infinite-gain voltage amplifier. We will see later that these biquads have lower

116 | 5 Second-Order Active-RC Filters

Table 5.4 Sallen and Key HP and BP filters and a set of design equations.

	HP filter	BP filter
Circuit diagram	(circuit with V_1, R_2, C_1, C_3, R_4, amplifier K, output V_2)	(circuit with V_1, R_1, R_2, C_3, C_5, R_4, amplifier K, output V_2)
ω_p^2	$\dfrac{1}{R_2 R_4 C_1 C_3}$	$\dfrac{1+(R_1/R_2)}{R_1 R_4 C_3 C_5}$
$\dfrac{1}{Q_p}$	$\sqrt{\dfrac{R_2 C_1}{R_4 C_3}} + \sqrt{\dfrac{R_2 C_3}{R_4 C_1}} + (1-K)\sqrt{\dfrac{R_4 C_3}{R_2 C_1}}$	$\dfrac{1+\left(\dfrac{R_1}{R_2}\right)(1-K)}{1+\left(\dfrac{R_1}{R_2}\right)}\sqrt{\dfrac{R_4 C_3}{R_1 C_5}} + \sqrt{\dfrac{R_1 C_3}{R_4 C_5}} + \sqrt{\dfrac{R_1 C_5}{R_4 C_3}}$
H_0	K	$\dfrac{\dfrac{K}{R_1 C_5}}{\dfrac{1}{R_1 C_5} + \dfrac{1}{R_4 C_5} + (1-K)\dfrac{1}{R_2 C_5}}$
A set of design equations	Equal-capacitor, gain of 2 design $C_1 = C_2 = C$, $K = 2$ $R_4 = m^2 R_2$ $R_2 = \dfrac{1}{m \omega_p C}$ $m = -\dfrac{1}{2Q_p} + \sqrt{\dfrac{1}{4Q_p^2} + 2}$ $H_0 = 2$.	Equal-capacitor, gain of 2 design $C_3 = C_5 = C$, $K = 2$ $R_1 = R_2 = R = \dfrac{1}{\omega_p Q_p C}$ $R_2 = \dfrac{2Q_p}{\omega_p C}$, $H_0 = 2Q_p^2$. Note: Hence, H_0 is dependent on Q_p. If we would like to design the filter for a specified H_0, then $R_1 \neq R_2$, and we have to assume $R_2 = mR_1$, $R_4 = nR_1$, and proceed.

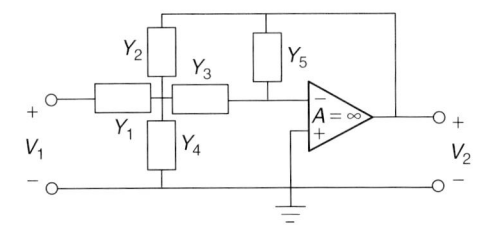

Figure 5.6 A general structure of an infinite-gain multiple feedback SAB filter.

sensitivities (i.e., variations of ω_p and Q_p with respect to changes in the values of the passive components) than the SABs considered in the previous sections. However, we will also see that the element spread will be large, thus resulting in its use only for low-Q filters. Figure 5.6 shows the general form of an infinite-gain multiple feedback (IGMFB) SAB filter.

Using the method of analysis of constrained networks (see Chapter 2), we can show that the VTF, V_2/V_1, is given by

$$H(s) = \frac{V_2}{V_1} = -\frac{Y_1 Y_4}{Y_5(Y_1 + Y_2 + Y_3 + Y_4) + Y_2 Y_3} \quad (5.22)$$

By choosing suitable values for the admittances in Eq. (5.22), we can generate LP, HP, and BP filters.

LP filter : Choose $Y_1 = G_1, Y_2 = sC_2, Y_3 = G_3, Y_4 = G_4, Y_5 = sC_5$ (5.23)

HP filter : Choose $Y_1 = sC_1, Y_2 = G_2, Y_3 = sC_3, Y_4 = sC_4, Y_5 = G_5$ (5.24)

BP filter : Choose $Y_1 = G_1, Y_2 = G_2, Y_3 = sC_3, Y_4 = sC_4, Y_5 = G_5$ (5.25)

where $G_i = (1/R_i)$ is the conductance. Comparing $H(s)$ for the above three filters with the standard biquad transfer functions listed in Table 5.2, we can determine the values of H_0, ω_p, and Q_p for each of these filters in terms of the various R's and C's. These are listed in Table 5.5 along with a set of design equations, which one can use to design an LP, HP, or BP filter. The values of the components determined using these design equations will have to be impedance scaled to bring the values to practically acceptable range.

5.6
Infinite-Gain Multiple Voltage Amplifier Biquad Filters

In the early era of active-RC filters, the OAs were expensive and hence attention was paid toward SAB designs. With the advancement in semiconductor technology, the OAs became more affordable. It was found that realizations of second-order filters using several OAs lead to a simpler implementation procedure and lend to additional desirable features, such as reducing total capacitance, enabling easy tuning procedure, and reduced sensitivity to component tolerances. In this section, we present several cases of biquad filter realization using multiple infinite-gain voltage amplifiers. The voltage amplifiers are in practice realized from high-gain OAs.

Table 5.5 Infinite-gain multiple feedback SAB filters and a set of design equations.

	LP filter	HP filter	BP filter
Circuit diagram			
ω_p^2	$\dfrac{1}{R_2 R_3 C_4 C_5}$	$\dfrac{1}{R_4 R_5 C_2 C_3}$	$\dfrac{1+(R_4/R_1)}{R_4 R_5 C_2 C_3}$
$\dfrac{1}{Q_p}$	$\sqrt{\dfrac{C_5}{C_4}}\left(\sqrt{\dfrac{R_2 R_3}{R_1}} + \sqrt{\dfrac{R_3}{R_2}} + \sqrt{\dfrac{R_2}{R_3}}\right)$	$\sqrt{\dfrac{R_4}{R_5}}\left(\sqrt{\dfrac{C_1}{C_2 C_3}} + \sqrt{\dfrac{C_3}{C_2}} + \sqrt{\dfrac{C_2}{C_3}}\right)$	$\dfrac{1}{\sqrt{1+\left(\dfrac{R_4}{R_1}\right)}}\dfrac{R_5/R_1}{1+(C_2/C_3)}\left(\sqrt{\dfrac{R_4 C_2}{R_5 C_3}} + \sqrt{\dfrac{R_4 C_3}{R_5 C_2}}\right)$
$\|H_0\|$	$\dfrac{R_2}{R_1}$	$\dfrac{C_1}{C_2}$	
A set of design equations	Let $C_4 = C$ and $C_5 = m^2 C$ Then, $R_2 = \dfrac{1 \pm \sqrt{1-4m^2 Q_p^2(1+\|H_0\|)}}{2\omega_p Q_p m^2 C}$ $R_1 = \dfrac{R_2}{\|H_0\|}$ $R_3 = \dfrac{1}{(m\omega_p C)^2}\dfrac{1}{R_2}$ Note: $m^2 < \dfrac{1}{4Q_p^2(1+\|H_0\|)}$	Let $C_3 = m^2 C_1$ Then, $C_2 = \dfrac{\dfrac{C_1}{\|H_0\|}}{\|H_0\|}$ $R_4 = \dfrac{\omega_p Q_p C(1+(m^2+1)\|H_0\|)}{\|H_0\|}$ $R_5 = \dfrac{Q_p[1+(m^2+1)\|H_0\|]}{m^2\omega_p C_1}$	Let $C_3 = C$ and $C_2 = m^2 C$ Then, $R_1 = \dfrac{Q_p}{m^2 \|H_0\|\omega_p C}$ $R_4 = \dfrac{Q_p}{(m^2+1)Q_p^2 - m^2\|H_0\|\omega_p C}$ $R_5 = \dfrac{m^2+1}{m^2}\dfrac{Q_p}{\omega_p C}$ Note: $\left(\dfrac{m^2+1}{m^2}\right)Q_p^2 > \|H_0\|$

5.6.1
KHN State-Variable Filter

This filter is named after the originators Kerwin, Huelsman, and Newcomb (KHN), and has very low sensitivities, good performance, and flexibility. It is called a *state-variable filter*, since state-variable methods of solving differential equations is used in the realization. This is best understood by taking the example of an LP filter and getting a state-variable realization for the same. Consider the standard LP filter function given by

$$\frac{V_2(s)}{V_1(s)} = \frac{H_0 a_0}{s^2 + a_1 s + a_0} \tag{5.26}$$

This can be written as

$$V_2(s) = \frac{(H_0 a_0)/s^2}{1 + \left(\frac{a_1}{s}\right) + \left(\frac{a_0}{s^2}\right)} V_1(s)$$

If we let

$$V_a(s) = \frac{1}{1 + \left(\frac{a_1}{s}\right) + \left(\frac{a_0}{s^2}\right)} V_1(s) \tag{5.27}$$

then

$$V_2(s) = \frac{H_0 a_0}{s^2} V_a(s) \tag{5.28}$$

If we now take inverse Laplace transforms of Eqs. (5.27) and (5.28), we get

$$v_a(t) = v_1(t) - a_1 \int v_a(t) dt - a_0 \int \left\{ \int v_a(t) dt \right\} dt \tag{5.29}$$

and

$$v_2(t) = H_0 a_0 \int \left\{ \int v_a(t) dt \right\} dt \tag{5.30}$$

In system theory, $v_a(t) = \ddot{x}(t)$, $\int v_a(t) dt = \dot{x}(t)$, and $\int\{\int v_a(t)dt\}dt = x(t)$ are called *state variables*, and hence this kind of a filter realization is termed a *state-variable realization*. It is easy to interpret Eqs. (5.29) and (5.30) by state-variable realization, as shown in Figure 5.7.

The block diagram shown in Figure 5.7 can be easily converted to an OA–RC circuit by using the OA as an integrator, and is shown in Figure 5.8. This circuit is known as the *Kerwin–Huelsamn–Newcomb* state-variable filter (Kerwin, Huelsman and Newcomb, 1967).

For the two integrators in Figure 5.8, we have

$$V_{o3} = -\frac{1}{sR_2 C_2} V_{o2} \text{ and } V_{o2} = -\frac{1}{sR_1 C_1} V_{o1} \tag{5.31}$$

For the differential summer

$$V_{o1} = -\frac{R_6}{R_5} V_{o3} + \frac{R_4}{R_3 + R_4} \frac{R_5 + R_6}{R_5} V_1 + \frac{R_3}{R_3 + R_4} \frac{R_5 + R_6}{R_5} V_{o2} \tag{5.32}$$

5 Second-Order Active-RC Filters

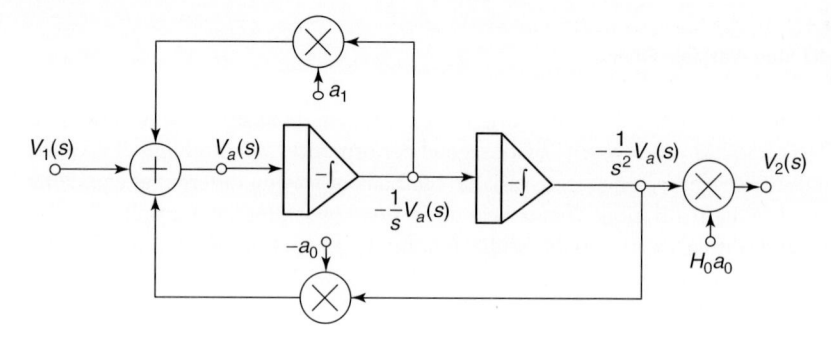

Figure 5.7 State-variable realization of the LP filter given by Eq. (5.26).

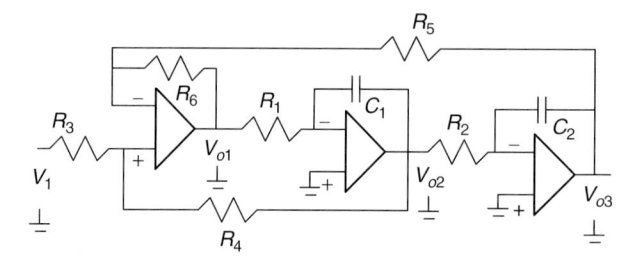

Figure 5.8 KHN state-variable filter.

From Eqs. (5.31) and (5.32), we can show that

$$\frac{V_{o3}}{V_1} = \frac{\left[\frac{1+R_6/R_5}{1+R_3/R_4}\right]\frac{1}{R_1 R_2 C_1 C_2}}{D(s)} \tag{5.33}$$

$$\frac{V_{o2}}{V_1} = -\frac{\left[\frac{1+R_6/R_5}{1+R_3/R_4}\right]\frac{s}{R_1 C_1}}{D(s)} \tag{5.34}$$

and

$$\frac{V_{o1}}{V_1} = \frac{\left[\frac{1+R_6/R_5}{1+R_3/R_4}\right]s^2}{D(s)} \tag{5.35}$$

where

$$D(s) = s^2 + \frac{s}{R_1 C_1}\frac{1+R_6/R_5}{1+R_4/R_3} + \frac{R_6/R_5}{R_1 R_2 C_1 C_2} \tag{5.36}$$

It is observed that the KHN state-variable filter can be used as an LP, BP, or HP filter depending on where the output is taken. It is also seen that the LP and HP outputs are noninverting, while that of the BP is inverting. The parameters ω_p

and Q_p of the filters are related to the network components as follows:

$$\omega_p = \sqrt{\frac{R_6/R_5}{R_1 R_2 C_1 C_2}} \tag{5.37}$$

$$\frac{1}{Q_p} = \frac{1 + R_6/R_5}{1 + R_4/R_3} \sqrt{\frac{R_5 R_2 C_2}{R_6 R_1 C_1}} \tag{5.38}$$

However, the value of H_0 for the three filters is different and is given by

$$H_0(\text{LP}) = \frac{1 + R_6/R_5}{1 + R_3/R_4} \frac{R_5}{R_6} \quad (\text{at } \omega = 0) \tag{5.39a}$$

$$H_0(\text{BP}) = -\frac{R_4}{R_3} \quad (\text{at } \omega = \omega_p) \tag{5.39b}$$

$$H_0(\text{HP}) = \frac{1 + R_6/R_5}{1 + R_3/R_4} \quad (\text{at } \omega = \infty) \tag{5.39c}$$

Example 5.5. Design a BP filter with $Q_p = 10$ and $\omega_p = 10^3$ rad s^{-1}, using the KHN biquad structure.

Equations (5.34), (5.37), (5.38), and (5.39b) show that there are eight variables (R_i, $i = 1, \ldots, 6$ and C_1, C_2), while there are only two specifications. Hence, we have sufficient flexibility in assigning arbitrary values to six of the components. An examination of Eqs. (5.37) and (5.34) reveals that the ratio R_4/R_3 occurs only in the expression for Q_p. Therefore, we determine this to satisfy the value of specified Q_p. We now assume that $R_1 = R_2 = R$, $C_1 = C_2 = C$, and $R_5 = R_6 = R_x$. Then we have two simple expressions for ω_p and Q_p:

$$\omega_p = \frac{1}{RC} \Longrightarrow R = \frac{1}{\omega_p C} \tag{5.40}$$

$$\frac{1}{Q_p} = \frac{2}{1+(R_4/R_3)} \Longrightarrow R_4 = (2Q_p - 1)R_3 \tag{5.41}$$

Assuming $C = 1$ F, we get $R = 10^{-3}$ Ω. Impedance scaling by 10^{-8}, we have $C = 0.01$ µF, $R_1 = R_2 = R = 100$ kΩ. Further, since for R_3, R_4, R_5, and R_6, only the ratio $(R_6/R_5) = 1$ and $(R_4/R_3) = 19$ are important, we can choose them to be $R_5 = R_6 = R_3 = 1$ kΩ, and $R_4 = 19$ kΩ. The value of H_0, given by Eq. (5.39b), is $H_0 = -(R_4/R_3) = -19$. It is noted that the same circuit gives LP and HP outputs at V_{o1} and V_{o3} for which $H_0 = (2Q_p - 1)/Q_p$, that is, $H_0 = 1.9$.

5.6.2
Tow–Thomas Biquad

If one examines the KHN biquad, one finds that the signal is fed at only one node of the system, while filters of different characteristics (viz., LP, HP, and BP) are obtained at distinct output nodes. Such a system is commonly referred to as a *single-in, multi-out* (SIMO) system. In contrast, there could be a system where the input signal is fed to several nodes in the system while only one node delivers the desired output. Filters of different characteristics are obtained by special choice of

5 Second-Order Active-RC Filters

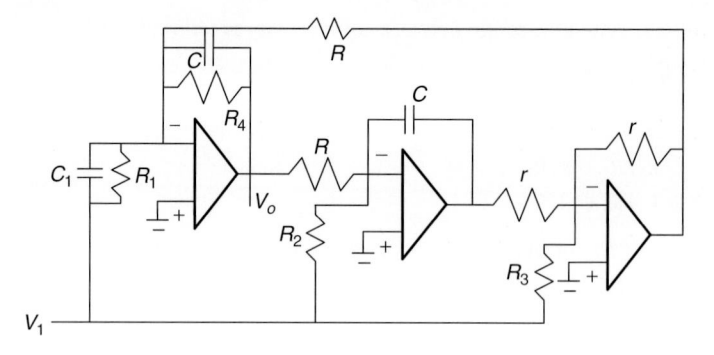

Figure 5.9 The Tow–Thomas universal biquad structure.

Table 5.6 Tow–Thomas universal biquad generating various filters.

Filter type	Component values	Remarks
LP	$C_1 = 0$, $R_1 = R_3 = \infty$, $R_2 = \frac{R}{H_0}$	H_0 is the gain at $\omega = 0$
BP (+ve gain)	$C_1 = 0$, $R_2 = R_3 = \infty$, $R_1 = \frac{Q_p}{H_0} R$	H_0 is the gain at $\omega = \omega_p$
BP (−ve gain)	$C_1 = 0$, $R_1 = R_2 = \infty$, $R_3 = \frac{Q_p}{|H_0|} r$	H_0 is negative and is the gain at $\omega = \omega_p$
HP	$C_1 = H_0 C$, $R_1 = R_2 = R_3 = \infty$	H_0 is the gain at $\omega = \infty$
Notch	$C_1 = H_0 C$, $R_1 = R_3 = \infty$, $R_2 = \frac{\omega_p^2}{\omega_n^2} \frac{R}{H_0}$	H_0 is the gain at $\omega = \infty$ and ω_n is the notch frequency
AP	$C_1 = H_0 C$, $R_1 = \infty$, $R_2 = \frac{R}{H_0}$, $R_3 = \frac{Q_p}{H_0} r$	H_0 is the flat gain for all ω

Note: For all cases, $R_4 = Q_p R$, $R = \frac{1}{\omega_p C}$; C and r are arbitrary.

the several input nodes. This system is known as *multi-in, single-out* (MISO) system. In Figure 5.9 we present a MISO biquad, originally due to Tow and Thomas (Tow, 1968, 1969; Thomas, 1971a, 1971b; Sedra and Smith, 1970, 2004).

The VTF, V_o/V_i, of Figure 5.9 is given by

$$\frac{V_o}{V_i} = \frac{\frac{C_1}{C}s^2 + \frac{1}{RC}\left[\frac{R}{R_1} - \frac{r}{R_3}\right]s + \frac{1}{R_2 RC^2}}{s^2 + \frac{1}{R_4 C}s + \frac{1}{R^2 C^2}} \tag{5.42}$$

Design guidelines for various standard filters are presented in Table 5.6.

Since all the five standard biquad filter functions are available, the Tow–Thomas structure may be considered to be a universal biquad filter structure. A second

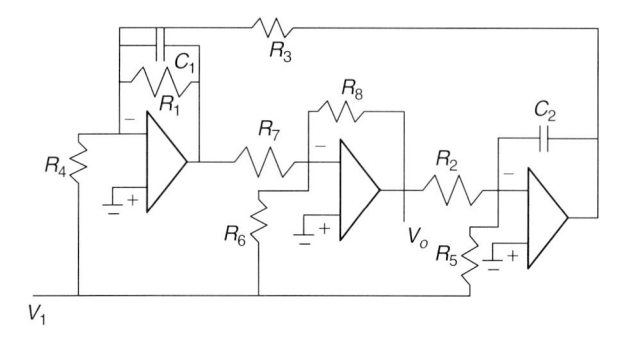

Figure 5.10 The Fleischer–Tow universal biquad structure.

structure that has similar potentials, but which does not use a capacitor (i.e., C_1), in the feed-forward path is presented next.

5.6.3
Fleischer–Tow Universal Biquad Structure

The schematic in Figure 5.10 presents the Fleischer–Tow biquad structure, which follows the MISO topology (Fleischer and Tow, 1973).

It can be shown that the transfer function of the Fleischer–Tow biquad structure is

$$\frac{V_o}{V_1} = -\frac{(R_8/R_6)s^2 + (1/R_1C_1)[R_8/R_6 - (R_1R_8/R_4R_7)]s + R_8/(R_3R_5R_7C_1C_2)}{s^2 + (1/R_1C_1)s + R_8/(R_2R_3C_1C_2R_7)} \tag{5.43}$$

By suitable choice of input nodes, we can generate the various types of filters and these are tabulated in Table 5.7. In view of this, the Fleischer–Tow biquad is also a

Table 5.7 Fleischer–Tow universal biquad generating various filters.

Filter type	Component values	Remarks
LP	$R_4 = R_6 = \infty, R_5 = \frac{R_a}{|H_0|}$	H_0 is negative and is the gain at $\omega = 0$
BP	$R_5 = R_6 = \infty, R_4 = \frac{Q_p}{H_0}R_a$	H_0 is the gain at $\omega = \omega_p$
HP	$R_5 = \infty, R_4 = \frac{Q_p}{|H_0|}R_a, R_6 = \frac{R}{|H_0|}$	H_0 is negative and is the gain at $\omega = \infty$
Notch	$R_4 = \frac{Q_p}{|H_0|}R_a, R_5 = \frac{\omega_p^2}{\omega_n^2}\frac{1}{|H_0|}R_a, R_6 = \frac{R}{|H_0|}$	H_0 is negative and is the gain at $\omega = \infty$ and ω_n is the notch frequency
AP	$R_4 = \frac{Q_p}{2|H_0|}R_a, R_5 = \frac{R_a}{|H_0|}, R_6 = \frac{R}{|H_0|}$	H_0 is negative and is the flat gain for all ω

Note: For all cases, it is assumed that $R_2 = R_3 = R_a$, $R_7 = R_8 = R$, $C_1 = C_2 = C$, $R_a = \frac{1}{\omega_p C}$, and $R_1 = Q_p R_a$; R and C being arbitrary.

universal biquad structure, and the pole frequency and pole Q are given by

$$\omega_p = \sqrt{R_8/R_2 R_3 C_1 C_2 R_7} \text{ and } \omega_p/Q_p = 1/R_1 C_1 \tag{5.44}$$

5.7
Sensitivity

Sensitivity is a measure of the variability of the performance of a filter as a result of changes in the values of the components and in the characteristics of the active device(s) used to implement the filter. Such changes may occur due to aging, manufacturing tolerances, environmental (i.e., temperature and power supply) variations, and so on. In the previous sections, we have discussed several different architectures for the realization of a second-order active-RC filter. There are many more structures that have been proposed in this area and an interested reader may consult the additional references (Ghausi and Laker, 1981; Schaumann, Ghausi, and Laker, 1990; Schaumann and Van Valkenburg, 2001; Chen, 1986) provided at the end of this book. Sensitivity is one of the aspects that can be used to compare the various filter structures with regard to their robustness toward changes in the component values and the characteristics of the active devices used to implement the filters. Various researchers have aimed their efforts toward minimizing such sensitivity and thereby have come up with novel filter structures. In the following, we shall introduce the basic definition of sensitivity and illustrate the use of this definition by considering some known filter structures.

5.7.1
Basic Definition and Related Expressions

The sensitivity of a function Y with respect to a parameter (or a variable) x is denoted by S_x^Y and is defined as

$$S_x^Y = \frac{\% \text{ change in Y}}{\% \text{ change in x}} = \frac{\Delta Y}{\Delta x} = \frac{dY/Y}{dx/x} = \frac{d(\ln Y)}{d(\ln x)} \tag{5.45}$$

If Y is a function of several variables, $Y = f(x_1, x_2, \ldots, x_n)$, then the sensitivity of Y w.r.t. x_i is given by

$$S_{x_i}^Y = \frac{\partial Y/Y}{\partial x_i/x_i} = \frac{\partial (\ln Y)}{\partial (\ln x_i)} \tag{5.46}$$

In the above, the symbol ∂ implies the partial derivative. Specifically, a sensitivity of 1/2 means that a 5% change in x would bring a 2.5% change in Y. A sensitivity of $-1/2$ means that a change of +5% in x would cause a change of -2.5% in Y. Several identities that are useful in calculating the sensitivity for complex functional relations are listed in Table 5.8.

In general, the filter parameters, such as ω_p, will depend upon several components that are used to implement the filter. In such cases, the concept of total differential is applicable. Thus, if the characteristic F of the filter depends upon several

Table 5.8 Some basic sensitivity formulas.

Sensitivity	Equivalent expression
S_x^{kY}, k is a constant	S_x^Y
$S_x^{Y_1 Y_2}$	$S_x^{Y_1} + S_x^{Y_2}$
$S_x^{Y_1/Y_2}$	$S_x^{Y_1} - S_x^{Y_2}$
$S_{1/x}^Y$	$-S_x^Y$
$S_x^{1/Y}$	$-S_x^Y$
$S_x^{aY^n}$	nS_x^Y
S_x^Y	$S_z^Y S_x^z$

parameters (i.e., constituent components) $x_1, x_2, \ldots, x_n$ of the filter, we can write $F = \psi(x_1, x_2, \ldots, x_n)$; then, $dF = \sum_{i=1}^{n} \frac{\partial F}{\partial x_i} dx_i$. Accordingly, the relative change in F for all the constituent component variations will be

$$\frac{dF}{F} = \sum_{i=1}^{n} \frac{\partial F}{\partial x_i} \frac{dx_i}{F} = \sum S_{x_i}^F \frac{dx_i}{x_i} \qquad (5.47)$$

Thus, the sensitivities with respect to the individual components x_i need to be calculated.

Example 5.6. Consider the SK LP filter of Section 5.4.1 with the following expressions concerning the pole frequency and pole Q:

$$\omega_p = 1/\sqrt{R_1 R_3 C_2 C_4}, \quad \omega_p/Q_p = 1/R_3 C_4 + 1/R_1 C_2 + 1/R_3 C_2 - K/R_3 C_4$$

To calculate $S_{R_1}^{\omega_p}$, we can proceed with

$$\ln(\omega_p) = -(1/2)\ln R_1 - (1/2)\ln R_3 - (1/2)\ln C_2 - (1/2)\ln C_4.$$

Taking the partial derivative with respect to R_1, we get $\partial \omega_p/\omega_p = -(1/2)(\partial R_1/R_1)$. Thus,

$$S_{R_1}^{\omega_p} = \frac{\partial \omega_p/\omega_p}{\partial R_1/R_1} = -(1/2) \qquad (5.48a)$$

It can be similarly shown that

$$S_{R_3}^{\omega_p} = S_{C_2}^{\omega_p} = S_{C_4}^{\omega_p} = -(1/2) \qquad (5.48b)$$

To evaluate the sensitivity of Q_p, one can start with $Q_p = \omega_p/P$, where $P = 1/R_3 C_4 + 1/R_1 C_2 + 1/R_3 C_2 - K/R_3 C_4$. Then, the sensitivity of Q_p, say, with respect to R_1, can be expressed as $S_{R_1}^{Q_p} = S_{R_1}^{\omega_p} - S_{R_1}^P$. From the previous results, $S_{R_1}^{\omega_p} = -(1/2)$. Now, $\partial P/\partial R_1 = -(1/R_1 C_2)(1/R_1)$. Hence,

$$S_{R_1}^P = \frac{\partial P/P}{\partial R_1/R_1} = -(1/P)(1/R_1 C_2)$$

$$= -(Q_p/\omega_p)(1/R_1 C_2) = -(Q_p/R_1 C_2)\sqrt{R_1 R_3 C_2 C_4} = -Q_p\sqrt{R_3 C_4/R_1 C_2}.$$

Thus,

$$S_{R_1}^{Q_p} = -(1/2) + Q_p\sqrt{R_3 C_4 / R_1 C_2}. \tag{5.49}$$

The sensitivities of Q_p with respect to other parameters can be similarly calculated. It is noted that the sensitivity of Q_p is proportional to Q_p. So, for a high-Q filter, the SK structure will have a large sensitivity for the pole Q. This information is important in deciding if one should adopt this architecture or not when the specifications demand for a design with a high Q_p value.

One can consider the clue that the above expressions provide toward a design with low sensitivity to component variations. Thus, in an ideal situation, we would like to make $S_{R_1}^{Q_p}$ equal to zero. Considering the expression of $S_{R_1}^{Q_p}$ above, we see that one has to adjust the components R_1, C_2, R_3, C_4 to achieve $0 = -(1/2) + Q_p\sqrt{R_3 C_4 / R_1 C_2}$, that is, $1/2 = Q_p\sqrt{R_3 C_4 / R_1 C_2}$. Since Q_p and ω_p are dependent on the R,C components, achieving the above condition may be quite challenging, if not impossible. However, the above discussion simply provides a clue for an approach toward a low-sensitivity design for a specific case.

5.7.2
Comparative Results for ω_p and Q_p Sensitivities

In Table 5.9, we present three different biquad filter structures with expressions for the ω_p and Q_p sensitivities. The structure in (a) is the SK LP SAB filter of Section 5.4.1, (b) is the IGFMB LP filter of Section 5.5, and (c) corresponds to the KHN state-variable filter. A discussion regarding the relative merits of the structures for low-sensitivity design follows.

The entries in Table 5.9 can be used to draw a comparison among the filter structures in (a)–(c) in that table. We have already noted that the network in the SK filter of (a) has a high sensitivity for Q_p. For the IGMFB filter shown in (b), the expressions for Q_p sensitivities with respect to the resistances R_1, R_2, R_3, and Q_p are multiplied by factors that are less than $1/Q_p$; hence, the products are less than 1. Thus, the Q_p sensitivities of the IGMFB filter is less than those in the SK filter. So, the IGMFB filter of (b) in Table 5.9 should be preferred to that of (a) for a low Q_p-sensitivity design. For the KHN state-variable filter of (c), all the sensitivities are less than unity. This reveals the superiority of the state-variable filter structure, when a low-sensitivity design is required. In particular, the Q_p sensitivities with respect to R_5 and R_6 could be made zero if R_5 and R_6 are made equal. In general, state-variable filters have low sensitivity. However, they require more number of active devices and network elements (especially resistances), which may add to the cost.

5.7.3
A Low-Sensitivity Multi-OA Biquad with Small Spread in Element Values

In Figure 5.11, we present a multi-OA general biquad filter network proposed by Mikhael and Bhattacharyya (1975). The sensitivities of ω_p, ω_z, Q_p, and Q_z to the

Table 5.9 Three different LP structures and their ω_p and Q_p sensitivities.

Structure	Expressions for ω_p and Q_p	ω_p and Q_p sensitivities
(a) Sallen and Key LP structure	$\omega_p = \dfrac{1}{\sqrt{R_1 R_3 C_2 C_4}}$ $\dfrac{1}{Q_p} = \sqrt{\dfrac{R_3 C_4}{R_1 C_2}} + \sqrt{\dfrac{R_1 C_4}{R_3 C_2}} + (1-K)\sqrt{\dfrac{R_1 C_2}{R_3 C_4}}$	$-S_{R_3}^{Q_p} = S_{R_1}^{Q_p} = -(1/2) + Q_p\sqrt{\dfrac{R_3 C_4}{R_1 C_2}}$ $S_{C_2}^{Q_p} = -S_{C_4}^{Q_p} = -(1/2) + Q_p\left(\sqrt{\dfrac{R_1 C_4}{R_3 C_2}} + \sqrt{\dfrac{R_3 C_4}{R_1 C_2}}\right)$ $S_{R_1,R_3,C_2,C_4}^{\omega_p} = -(1/2)$
(b) IGMFB LP structure	$\omega_p = \dfrac{1}{\sqrt{R_2 R_3 C_4 C_5}}$ $\dfrac{1}{Q_p} = \sqrt{\dfrac{C_5}{C_4}}\left(\sqrt{\dfrac{R_2 R_3}{R_1}} + \sqrt{\dfrac{R_3}{R_2}} + \sqrt{\dfrac{R_2}{R_3}}\right)$	$S_{R_1}^{Q_p} = Q_p\left(\dfrac{1}{R_1}\sqrt{\dfrac{R_2 R_3 C_5}{C_4}}\right)$ $S_{R_2}^{Q_p} = -\dfrac{Q_p}{2} - \dfrac{1}{R_1}\sqrt{\dfrac{R_2 R_3 C_5}{C_4}} - \sqrt{\dfrac{R_3 C_5}{R_2 C_4}} + \sqrt{\dfrac{R_2 C_5}{R_3 C_4}}$ $S_{R_3}^{Q_p} = -\dfrac{Q_p}{2} - \dfrac{1}{R_1}\sqrt{\dfrac{R_2 R_3 C_5}{C_4}} + \sqrt{\dfrac{R_3 C_5}{R_2 C_4}} - \sqrt{\dfrac{R_2 C_5}{R_3 C_4}}$ $S_{C_4}^{Q_p} = -S_{C_5}^{Q_p} = (1/2)S_{R_1}^{\omega_p} = 0$, $S_{R_2,R_3,C_4,C_5}^{\omega_p} = -(1/2)$
(c) KHN state-variable filter	$\omega_p = \sqrt{\dfrac{R_6/R_5}{R_1 R_2 C_1 C_2}}$ $\dfrac{1}{Q_p} = \dfrac{1+R_6/R_5}{1+R_4/R_3}\sqrt{\dfrac{R_5 R_2 C_2}{R_6 R_1 C_1}}$	$S_{R_1,C_1}^{Q_p} = -S_{R_2,C_2}^{Q_p} = (1/2)$ $S_{R_3}^{Q_p} = -S_{R_4}^{Q_p} = -1/\{1+(R_3/R_4)\}$ $S_{R_5}^{Q_p} = -S_{R_6}^{Q_p} = -\dfrac{Q_p}{2}\dfrac{R_5-R_6}{1+R_4/R_3}\sqrt{\dfrac{R_2 C_2}{R_5 R_6 R_1 C_1}}$ $S_{R_1,R_2,R_5,C_1,C_2}^{\omega_p} = -(1/2), S_{R_3,R_4}^{\omega_p} = 0$

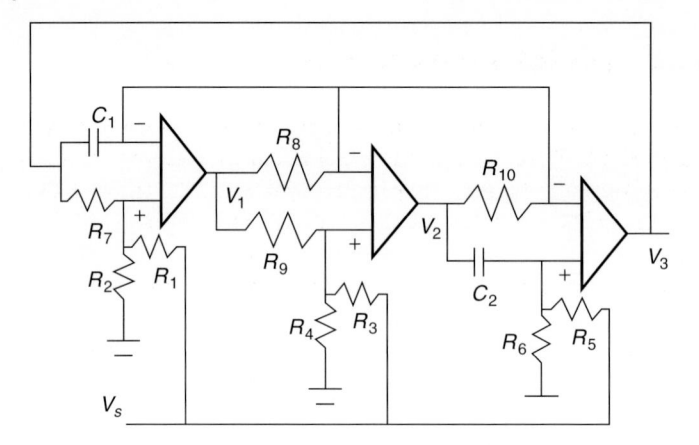

Figure 5.11 Mikhael–Bhattacharyya multiple amplifier biquad structure.

passive components are all less than unity. The network uses both the inverting and noninverting inputs of the OA. The transmission zeroes are realized with feed-forward resistive network. For a given pole Q, the spread in the resistance values (using equal capacitance values) is proportional to only $Q^{1/2}$, while most other biquad structures require resistance spreads of the order of Q or Q^2.

The transfer functions for the two most useful output terminals, assuming ideal OA, are

$$\frac{V_1}{V_s} = \left\{s^2\left[G_1\left(1+\frac{G_4}{G_9}\right) - \frac{G_2 G_3}{G_9}\right] + s\frac{G_3 G_7 G_8}{G_9 C_1}\right.$$
$$\left. + \frac{G_7 G_{10}}{C_1 C_2}\left[\left(1+\frac{G_4}{G_9}\right)G_5 - \frac{G_3 G_6}{G_9}\right]\right\}\bigg/ D(s) \quad (5.50)$$

$$\frac{V_3}{V_s} = \left\{s^2 G_1 + s\frac{G_7 G_8}{G_9 C_1}\left[\left(1+\frac{G_2}{G_7}\right)G_3 - \frac{G_1 G_4}{G_7}\right]\right.$$
$$\left. + \frac{G_7 G_{10}}{C_1 C_2}\left[\left(1+\frac{G_2}{G_7}\right)G_5 - \frac{G_4 G_6}{G_7}\right]\right\}\bigg/ D(s) \quad (5.51)$$

where $G_i = 1/R_i$, and

$$D(s) = s^2(G_1 + G_2) + s\frac{G_7 G_8}{G_9 C_1}(G_3 + G_4) + \frac{G_7 G_{10}}{C_1 C_2}(G_5 + G_6) \quad (5.52)$$

Thus,

$$\omega_p = \sqrt{\frac{G_7 G_{10}}{C_1 C_2}\frac{G_5 + G_6}{G_1 + G_2}} \quad \text{and} \quad Q_p = \frac{R_3 R_4 R_8}{R_9(R_3 + R_4)}\sqrt{\frac{R_7 C_1(R_1 + R_2)(R_5 + R_6)}{R_6 R_{10} C_2 R_1 R_2 R_5}} \quad (5.53)$$

For details regarding this circuit, we refer the reader to Mikhael and Bhattacharyya (1975).

5.7.4
Sensitivity Analysis Using Network Simulation Tools

In Section 5.7.1, we have dealt with finding analytical expressions for the sensitivity functions for a given filter network. When numerical values are required, one can plug in appropriate element values in the pertinent expression to arrive at the numerical value. If, however, the analytical expressions are not readily available, network simulations tools such as SPICE, PSpice, or HSPICE can be used to arrive at the numerical value of the sensitivity. In such a case, the component with respect to which the sensitivity is required can be assigned a variable parameter value and the filter performance function (such as ω_p and Q_p) can be evaluated as a function of this variable parameter. The sensitivity value can then be calculated by using the results (i.e., postprocessing) of pertinent simulations. Figure 5.12 shows the schematic of a fourth-order Chebychev BP filter realized using a cascade of two biquadratic filters. Each biquad is realized using the multiple feedback infinite-gain amplifier configuration, as discussed in Section 5.5 of this chapter. We are interested in evaluating the sensitivity of the center frequency ω_o of the filter with respect to the resistance R_{16} whose nominal value is 6.34 kΩ. The variation in the value of R_{16} is assumed to be 10%. Thus, its value could lie between 5.71 and 6.97 kΩ. In a typical network simulation using PSpice, one can assign R_{16} as a variable parameter with values such as 5.71, 6.34, and 6.97 kΩ. The center frequency of the BP response can be evaluated as a function of this parameter and thereafter the ratio $\frac{\Delta \omega_o / \omega_o}{\Delta R_{16}/R_{16}(\text{nominal})}$ can be calculated. This ratio is the sensitivity of the center frequency ω_o with respect to R_{16}. A plot of the variation of ω_o as

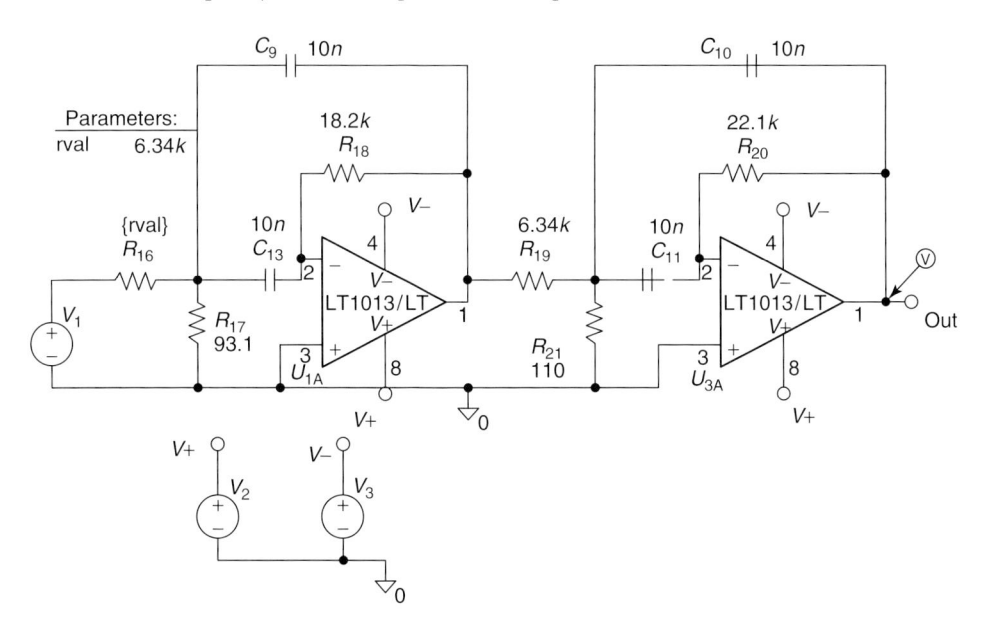

Figure 5.12 Schematic of a fourth-order filter using a cascade of two biquad filters.

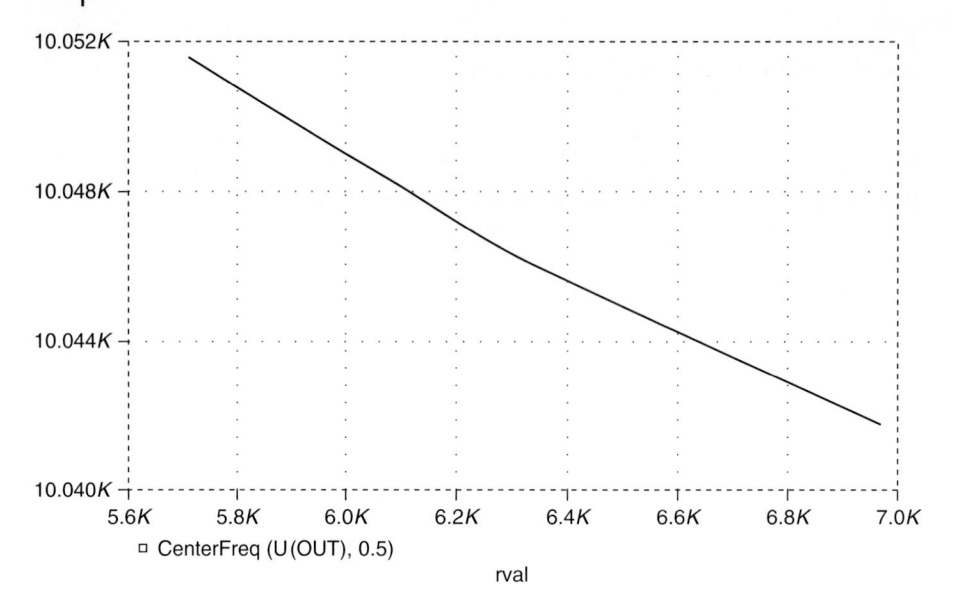

Figure 5.13 Variation of the center frequency ω_o of the BP filter of Figure 5.12 as a function of the resistor R_{16}.

a function of the variable parameter R_{16} obtained from the Pspice simulation is presented in Figure 5.13.

5.8
Effect of Frequency-Dependent Gain of the OA on the Filter Performance

Until now, we have assumed the OA used to realize the filters as ideal. It is well known that a practical OA is far from ideal. The open-loop gain is finite and changes with frequency. Since the filter is a frequency-selective device, the frequency-dependent gain of the OA can have a considerable impact on the performance of an active-RC filter. In this section, we present an analytical technique to take care of the frequency-dependent gain of an OA used as a building block for a filter. The technique will be illustrated by considering the case of a biquadratic filter.

5.8.1
Cases of Inverting, Noninverting, and Integrating Amplifiers Using an OA with Frequency-Dependent Gain

5.8.1.1 Inverting Amplifier
Consider the amplifier shown in Figure 5.14. Application of KCL leads to

$$(V_i - V_x)G_1 = (V_x - V_o)G_2$$

5.8 Effect of Frequency-Dependent Gain of the OA on the Filter Performance

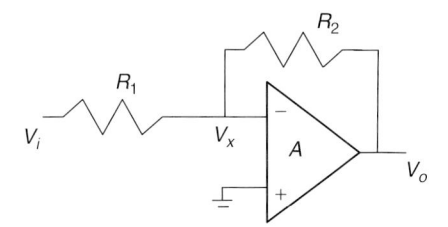

Figure 5.14 An inverting voltage amplifier whose gain is a function of the OA finite gain A.

But, $V_x = -V_o/A$. Substituting this in the above equation, we get $V_i G_1 = -V_o \left[G_2 + \frac{G_1 + G_2}{A} \right]$.

The amplifier voltage gain then becomes

$$K_- = \frac{V_o}{V_i} = -\frac{G_1}{G_2 + \frac{1}{A}(G_1 + G_2)} = K_o \frac{1}{1 + \frac{1}{A}\left(1 + \frac{R_2}{R_1}\right)}$$

or

$$K_- = K_o \frac{1}{1 + \frac{K_1}{A}} \tag{5.54a}$$

where $K_o = -R_2/R_1$ is the inverting amplifier gain when the OA has the ideal open-loop gain of infinity; K_1 is the gain of the noninverting amplifier realized with an ideal OA.

5.8.1.2 Noninverting Amplifier

Figure 5.15 presents the noninverting amplifier using an OA of gain A. The KCL equations $V_x G_1 + (V_x - V_o)G_2 = 0$ and $V_x(G_1 + G_2) = V_o G_{2x}(G_1 + G_2) = V_o G_2$ together with $V_o = A(V_i - V_x)$ will lead to

$$K_+ = \frac{V_o}{V_i} = \frac{G_1 + G_2}{G_2} \frac{1}{1 + \frac{1}{A}\frac{G_1+G_2}{G_2}}$$

Finally, since $G_i = 1/R_i$, we have

$$K_+ = K_1 \frac{1}{1 + \frac{K_1}{A}} \tag{5.54b}$$

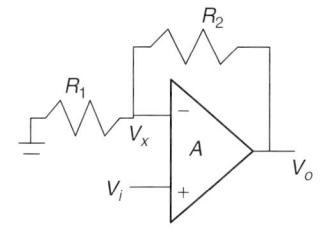

Figure 5.15 A noninverting voltage amplifier whose gain is a function of the OA finite gain A.

where K_1 is the gain of the noninverting amplifier with ideal OA.

5.8.1.3 Inverting Integrating Amplifier

Following methods similar to the above, one can derive the gain of an inverting integrating amplifier realized with an OA of gain A, as

$$T(s) = \frac{V_o(s)}{V_i(s)} = T_I(s) \frac{1}{1 + \frac{1}{A}(1 - T_I(s))} \tag{5.55}$$

where $T_I(s) = -\frac{1}{sCR}$, the integrator gain function with an ideal OA.

From the above derivations, it is clear that when the OA has a frequency-dependent gain $A = A(s)$, all the gain functions K_-, K_+, and $T(s)$ become frequency dependent to a certain degree. One can therefore use, in general,

$$\begin{aligned} K_- &= K_0 M_1(\omega) \exp(-j\phi_1) \\ K_+ &= K_1 M_2(\omega) \exp(-j\phi_2) \\ T(j\omega) &= T_I(j\omega) M_3(\omega) \exp(-j\phi_3) \end{aligned} \tag{5.56}$$

In the above, the explicit natures of the M and ϕ functions will depend upon the specific model used to represent the frequency-dependent gain $A(s)$ of the OA. Thus, with an integrator model for $A(s)$, that is, $A(s) = \omega_t/s$, $s = j\omega$, one can derive

$$M_1(\omega) = \frac{1}{\sqrt{1 + (K_1\omega/\omega_t)^2}} = M_2(\omega), \; \phi_1(\omega) = \tan^{-1}\left(\frac{K_1\omega}{\omega_t}\right) = \phi_2(\omega), \text{ and}$$

$$M_3(\omega) = \frac{1}{\sqrt{1 + (\omega/\omega_t)^2}}, \; \phi_3(\omega) = \tan^{-1}\left(\frac{\omega}{\omega_t}\right) \tag{5.57}$$

In the above, ω_t is the gain-bandwidth (GB) value of the OA. In deriving M_3 and ϕ_3, the assumption $1/(\omega_t CR) \ll 1$ has been used.

The technique to analyze an active filter realized with an OA that has a frequency-dependent gain function involves (i) identification of the basic mode of operation of the OA in the filter (i.e., inverting or noninverting amplifier or integrating amplifier) and then (ii) employing the pertinent gain function as discussed in Sections 5.8.1.1–5.8.1.3 and summarized in the form presented in Eq. (5.56). The technique is illustrated by considering a specific case.

5.8.2
Case of Tow–Thomas Biquad Realized with OA Having Frequency-Dependent Gain

Consider Figure 5.16, which shows the Tow–Thomas filter in its basic form. By inspection, we can understand that OA 1 and 2 are functioning as inverting integrators and OA 3 is working as an inverting amplifier.

Considering the OA to be ideal, one can write the following equation for the signal V_2:

$$V_2 = -\frac{V_s}{sC_1 R_4} - \frac{V_2}{sC_1 R_1} - \frac{V_1}{sC_1 R_3} \tag{5.58a}$$

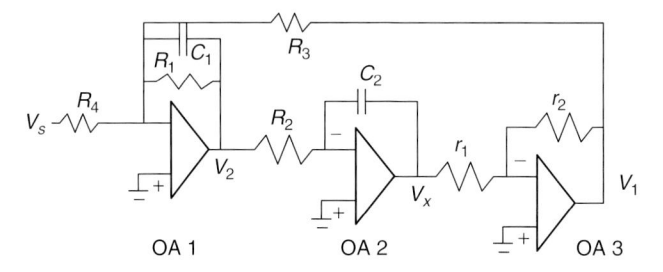

5.8 Effect of Frequency-Dependent Gain of the OA on the Filter Performance

Figure 5.16 The Tow–Thomas filter in its basic form.

But

$$V_1 = -\frac{r_2}{r_1} V_x \text{ and } V_x = -\frac{1}{sC_2 R_2} V_2 \quad (5.58b)$$

Now, considering the frequency-dependent gain function of the OA, we rewrite the above equation as

$$V_2 = -\frac{V_s}{j\omega C_1 R_4} M_3 \exp(-j\phi_3) - \frac{V_2}{j\omega C_1 R_1} M_3 \exp(-j\phi_3)$$

$$-\frac{V_1}{j\omega C_1 R_3} M_3 \exp(-j\phi_3) \quad (5.59a)$$

But

$$V_1 = -\frac{r_2}{r_1} M_1 \exp(-j\phi_1) V_x \text{ and } V_x = -\frac{1}{j\omega C_2 R_2} V_2 M_3 \exp(-j\phi_3) \quad (5.59b)$$

In the above, we have assumed that all the OAs are identical. M_1, M_3, ϕ_1, and ϕ_3 have the same significance as explained in Eq. (5.56). Writing $K_o = (r_2/r_1)$ and collecting all terms involving V_2 on one side, we have

$$V_2 \left[1 + \frac{M_3 \exp(-j\phi_3)}{j\omega C_1 R_1} + \frac{K_o}{-\omega^2 C_1 R_3 C_2 R_2} M_3^2 M_1 \exp(-j(2\phi_3 + \phi_1)) \right]$$

$$= -\frac{V_s}{j\omega C_1 R_4} M_3 \exp(-j\phi_3) \quad (5.60)$$

On further simplification, we get the transfer function, $V_2(j\omega)/V_s(j\omega) = N(j\omega)/D(j\omega)$, where

$$N(j\omega) = -\frac{j\omega}{C_1 R_4} M_3 \exp(-j\phi_3) \quad (5.61a)$$

$$D(j\omega) = -\omega^2 + \frac{j\omega}{C_1 R_1} M_3 \exp(-j\phi_3) + \frac{K_o}{C_1 C_2 R_2 R_3} M_3^2 M_1 \exp(-j(2\phi_3 + \phi_1)) \quad (5.61b)$$

In order to find the pole frequency and pole Q under this new condition, one has to expand $D(j\omega)$ by writing $\exp(jx) = \cos(x) + j\sin(x)$. Then,

$$D(j\omega) = -\omega^2 + \frac{\omega M_3}{C_1 R_1} \sin(\phi_3) + \frac{K_o M_3^2 M_1}{C_1 C_2 R_2 R_3} \cos(2\phi_3 + \phi_1) + j \left[\frac{\omega M_3}{C_1 R_1} \cos(\phi_3) \right.$$

$$\left. -\frac{K_o}{C_1 C_2 R_2 R_3} M_3^2 M_1 \sin(2\phi_3 + \phi_1) \right] \quad (5.62)$$

When ω equals the pole frequency $\hat{\omega}_p$, one must have $\mathrm{Re}(D(j\omega)) = 0$. The new pole frequency is obtained by solving the quadratic equation in ω

$$0 = -\omega^2 + \frac{\omega M_3}{C_1 R_1}\sin(\phi_3) + \frac{K_o M_3^2 M_1}{C_1 C_2 R_2 R_3}\cos(2\phi_3 + \phi_1) \tag{5.63}$$

If the GB of the OA is very large (20 times or more) compared to the ideal pole frequency $\omega_p = \sqrt{\frac{K_o}{C_1 C_2 R_2 R_3}}$, one can expect the phase angles ϕ_1 and ϕ_3 to be small. Then, approximately

$$\hat{\omega}_p \cong \omega_p M_3 \sqrt{M_1 \cos(2\phi_3 + \phi_1)} \tag{5.64}$$

The new pole frequency will thus be smaller than the nominal (i.e., with ideal OA) pole frequency ω_p. To find the new $Q_p \to \hat{Q}_p$, we recall that the coefficient of $s = j\omega$ term in $D(s)$ of the standard biquadratic transfer function is ω_p/Q_p. Then, in Eq. (5.62), we have to pay attention to the imaginary part of $D(s)$, since this could be put in the form

$$j\omega\left[\frac{M_3}{C_1 R_1}\cos(\phi_3) - \frac{K_o}{\omega C_1 C_2 R_2 R_3}M_3^2 M_1 \sin(2\phi_3 + \phi_1)\right]$$

and be related to $j\omega\left(\frac{\hat{\omega}_p}{\hat{Q}_p}\right)$. Again, in case of an ideal OA, $\omega_p/Q_p = 1/C_1 R_1$. Thus,

$$\frac{\hat{\omega}_p}{\hat{Q}_p} = \frac{\omega_p}{Q_p}M_3\cos(\phi_3) - \frac{\omega_p^2 M_3^2 M_1 \sin(2\phi_3 + \phi_1)}{\omega} \tag{5.65}$$

The above leads to (assuming $\omega \cong \omega_p$)

$$\hat{Q}_p \cong \frac{\hat{\omega}_p}{(\omega_p/Q_p)M_3\cos(\phi_3) - \omega_p M_3^2 M_1 \sin(2\phi_3 + \phi_1)} \tag{5.66}$$

Finally, using the ratio of $\hat{\omega}_p/\omega_p$ from Eq. (5.69), we get

$$\hat{Q}_p = Q_p \frac{M_3\sqrt{M_1 \cos(2\phi_3 + \phi_1)}}{M_3\cos(\phi_3) - Q_p M_3^2 M_1 \sin(2\phi_3 + \phi_1)} \tag{5.67}$$

Equation (5.67) indicates that the new pole Q (i.e., $\hat{Q}_p$) could become higher than the ideal (i.e., with ideal OA) Q_p if the denominator becomes less than the numerator. This is known as *Q-enhancement* caused due to the frequency-dependent gain function of the OA. Further, $\hat{Q}_p$ can even become negative if the denominator in Eq. (5.67) becomes negative. A negative pole Q implies a positive real part for the poles of the second-order filter. A positive real part will produce an unstable system. Thus, when the OA has a frequency-dependent gain, the pole Q will in general show an increase from the nominal value and the system may even become unstable when the filter requires a high Q_p design and if the GB of the OA is not substantially higher than the nominal pole frequency ω_p.

A rule of thumb for avoiding the above degradations due to a finite, frequency-dependent gain of the OA is to choose an OA whose GB is at least 20 times higher than the design ω_p. Further, before embarking on the design task, one must estimate the enhancement in the value of Q_p by employing Eq. (5.67). A known model for the OA gain function $A(s)$ has to be used for this purpose.

Example 5.7. Consider the design of a second-order filter using the architecture shown in Figure 5.16. Assume the model $A(s) = \omega_t/s$ for the OA gain function, where $\omega_t = 2\pi \times 10^6$ rad s^{-1} is the GB value of the OA. The design goals are as follows: a nominal ω_p of $2\pi \times 50 \times 10^3$ rad s^{-1} and a Q_p of 10. Calculate the ω_p and Q_p values that can be realized under this case.

We can use Eq. (5.57) to calculate the values of M_1, M_3, ϕ_1, and ϕ_3. We shall assume that $K_o = 1$. Then $K_1 = 2$. Now we can find $M_3 = 0.99875$, $M_1 = 0.99503$, $\cos(2\phi_3 + \phi_1) = 0.9801$, $\sin(2\phi_3 + \phi_1) = 0.1983$, and $\cos(\phi_3) = 0.9987$. These values, when substituted in Eqs. (5.64) and (5.67), give $\hat{f}_p = \hat{\omega}_p/2\pi = 49.315$ kHz and $\hat{Q}_p = -1.08793$. This reveals that the realized pole frequency (49.315 kHz) will be slightly less than the nominal pole frequency f_p of 50 kHz. It also shows that the realized Q_p will be negative, implying that the system might oscillate after being designed.

If the design is revised for a lower Q_p value, say, $Q_p = 2$, the above calculations will produce $\hat{Q}_p = 3.297$. This shows an enhancement in the Q_p value from 2 to a value of 3.297.

5.9
Second-Order Filter Realization Using Operational Transconductance Amplifier (OTA)

OTAs accept voltage signals at the input, as in OAs. Unlike OA, the output resistance of an OTA is very high (100 kΩ or more) and hence at the output it behaves more like a current source. Thus, the OTA produces a current signal as response to a voltage signal at its input. Hence, the name *transconductance*. If the output of an OTA is terminated in a high resistance, a large signal voltage can be produced at the output and hence the OTA can also be made to function as a voltage amplifier. An OTA can be configured to produce several special network response functions, similar to the OA. Hence, an OTA can be conveniently used to produce frequency-selective transfer functions such as a second-order filter. A few examples of these have been considered in Chapter 2. The symbol for an OTA is shown in Figure 5.17a and the AC equivalent circuit is shown in Figure 5.17b. In Figure 5.17a, the terminal marked V_c (or I_c) represents a control terminal that can be used to change the transconductance value g_m of the OTA. Table 5.10 presents several building blocks and networks realizable with OTAs, including special

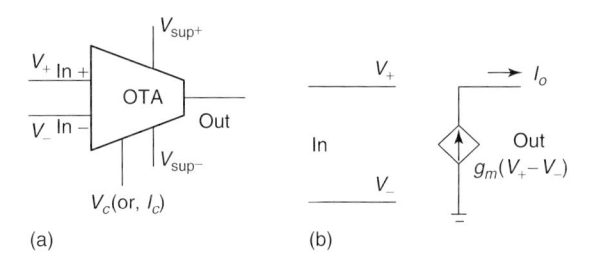

Figure 5.17 (a) The symbol for an OTA and (b) equivalent circuit of an OTA.

5 Second-Order Active-RC Filters

Table 5.10 Building blocks and network elements realizable using OTAs (Geiger and Sanchez-Sinencio, 1985; Su, 1996).

	Network	Function	Comment
A		$Z_{in} = \dfrac{1}{g_m}$	Grounded resistance of $\dfrac{1}{g_m}$ Ω. If the input and ground terminals are reversed, we get a negative resistance of $\dfrac{1}{g_m}$ Ω
B		$g_{m1} = g_{m2} = g_m$ $Z_{in} = \dfrac{1}{g_m}$	Floating resistance of $\dfrac{1}{g_m}$ Ω. If the input and ground terminals are reversed, we get a negative floating resistance of $\dfrac{1}{g_m}$ Ω
C		$\dfrac{V_o}{V_i} = \dfrac{g_m}{sC_L}$	Voltage integrator
D		$\dfrac{I_o}{I_i} = \dfrac{g_m}{sC_i}$	Current integrator
E		$\dfrac{V_o}{V_i} = \dfrac{g_{m1}}{sC+g_{m2}}$	Lossy voltage integrator

5.9 Second-Order Filter Realization Using Operational Transconductance Amplifier (OTA)

Table 5.10 (continued).

	Network	Function	Comment
F		$\dfrac{V_o}{V_i} = \dfrac{g_{m1}}{g_{m2}}$	Voltage amplifier
G		$V_o = \dfrac{1}{g_m}\sum\limits_{i=1}^{n} g_{mi} V_i$	Weighted summer
H		$\dfrac{V_o}{V_i} = \dfrac{sC_1 + g_{m1}}{s(C_1+C_2)+g_{m2}}$	First-order section
I		$Z_{in} = \dfrac{1}{g_{m1}g_{m2}}\dfrac{1}{Z_L}$	Grounded positive impedance inverter. If Z_L is a capacitor of C Farads, then Z_{in} reflects a grounded inductor of value $C/(g_{m1}g_{m2})$ H.
J		$g_{m2} = g_{m3}Z_{in} = \dfrac{1}{g_{m1}g_{m2}}\dfrac{1}{Z_L}$	Floating positive impedance inverter. If Z_L is a capacitor of C Farads, then Z_{in} reflects a grounded inductor of value $C/(g_{m1}g_{m2})$ H.

network elements such as a *gyrator*, FDNR, and so on. These building blocks and functional elements can be conveniently utilized to produce fully monolithic continuous-time filters using an available IC technology. In the ac equivalent circuit, the OTA has been assumed as ideal, that is, with an infinite input resistance and an output resistance. Also, it is assumed that no parasitic capacitances are present.

5.9.1
Realization of a Filter Using OTAs

On examining the networks in Table 5.10, one may appreciate that an OTA can be converted to a resistance by simply connecting the output to the input (entries A, B in Table 5.10). This is the same principle as that used to implement an active resistance from a BJT or MOS transistor. Thus, the RC-active filters realized with OAs can now be realized using capacitors, OTAs connected as resistors, and OTA connected as a voltage amplifier. This gives rise to what is popularly known as OTA-C filters or $g_m - C$ filters. It may also be noted that since an OTA and a capacitor are easily available in an IC technology, OTA-C filters easily render themselves to monolithic realizations. It should be noted that OTA-C filters can be designed to work at frequencies of the order of tens of hundreds of megahertz; they are physically small and consume low power.

5.9.2
An OTA-C Band-Pass Filter

Consider Figure 5.18a, whose equivalent circuit is shown in Figure 5.18b.
  KCL at node b gives

$$(V_i - V_b)sC_1 - g_{m1} V_o = 0 \tag{5.68}$$

and KCL at node a gives

$$g_{m2} V_b - g_{m3} V_o = V_o sC_2 \tag{5.69}$$

Eliminating V_b from the above two equations, we have the VTF $V_o(s)/V_i(s)$ to be

$$\frac{V_o(s)}{V_i(s)} = \frac{sg_{m2} C_1}{s^2 C_1 C_2 + sg_{m3} C_1 + g_{m1} g_{m2}} \tag{5.70}$$

This represents a BP filter. On comparing with the standard biquadratic transfer function (Table 5.2), we can identify

$$\omega_p = \sqrt{\frac{g_{m1} g_{m2}}{C_1 C_2}}, \quad \frac{\omega_p}{Q_p} = \frac{g_{m3}}{C_2}, \text{ therefore } Q_p = \sqrt{\frac{C_2}{C_1}} \frac{\sqrt{g_{m1} g_{m2}}}{g_{m3}} \tag{5.71}$$

It should be noted that in an OTA-C filter the design parameters are the capacitance C (an electrical element) and g_m (a network function). The g_m has to be realized by using appropriate electrical control voltage or current signals.

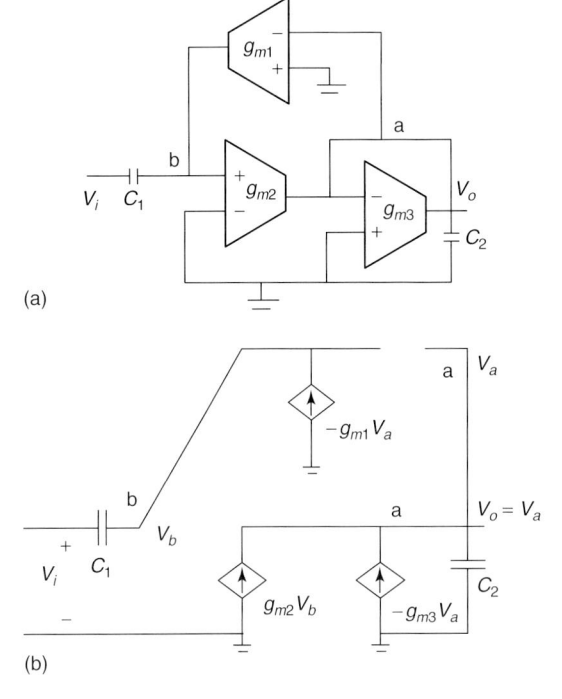

5.9 Second-Order Filter Realization Using Operational Transconductance Amplifier (OTA)

Figure 5.18 An OTA-C BP filter; (a) schematic, (b) AC equivalent circuit.

5.9.3
A General Biquadratic Filter Structure

Figure 5.19a presents a structure, which can be used to realize any of the five standard biquadratic transfer functions. It consists of five OTAs. The analysis can be carried out by considering the AC equivalent circuit shown in Figure 5.19b. By inspection, we have

$$i_{C_1} = g_{m5} V_A - g_{m1} V_o \qquad (5.72a)$$

$$V_x = \frac{i_{C_1}}{sC_1} = \frac{g_{m5} V_A - g_{m1} V_o}{sC_1} \qquad (5.72b)$$

KCL at node V_y leads to

$$g_{m2} V_x + g_{m4} V_B - g_{m3} V_o = sC_2(V_o - V_C) \qquad (5.73)$$

Substituting for V_x from Eq. (5.77b), and simplifying, we get

$$V_o = \frac{s^2 V_C + (g_{m4}/C_2)s V_B + (g_{m2}g_{m5}/C_1 C_2) V_A}{s^2 + (g_{m3}/C_2)s + (g_{m1}g_{m2}/C_1 C_2)} \qquad (5.74)$$

Comparing the denominator of Eq. (5.79) with the standard biquadratic form, we get

$$\omega_p = \sqrt{\frac{g_{m1}g_{m2}}{C_1 C_2}}, \frac{\omega_p}{Q_p} = \frac{g_{m3}}{C_2} \qquad (5.75a)$$

5 Second-Order Active-RC Filters

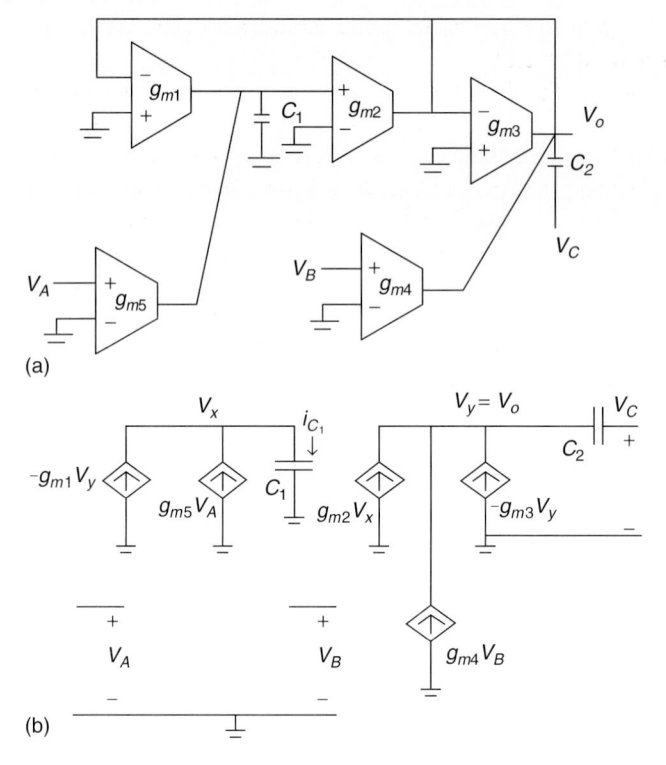

Figure 5.19 (a) A universal biquad filter using five OTAs and (b) equivalent circuit of (a).

and

$$Q_p = \frac{1}{g_{m3}} \sqrt{\frac{g_{m1} g_{m2} C_2}{C_1}} \tag{5.75b}$$

On examining Eq. (5.74), it is obvious that, by appropriately choosing the signal voltages V_A, V_B, or V_C, we can realize different transfer functions. Thus

For an LP filter: choose $V_B = V_C = 0$ (5.76)

For an HP filter: choose $V_A = V_B = 0$ (5.77)

For a BP filter: choose $V_A = V_C = 0$ (5.78)

For a notch filter: choose $V_B = 0$, $V_A = V_C$ (5.79)

For an AP filter: choose $V_A = V_B = V_C = V_i$, $g_{m1} = g_{m5}$, and $g_{m2} = g_{m4}$. (5.80)

5.10
Technological Implementation Considerations

In the early stages of active-RC filters, the filters were made using OAs and discrete RC elements mounted together on a printed circuit board (PCB). As the impetus toward miniaturization gained momentum, *hybrid technology* was adopted for the implementation of active-RC filters.

In hybrid technology, the OA dies and the discrete components are bonded in a single package. Special materials and techniques are used to implement the resistances and capacitances. In thick film technique, resistances are prepared from a paste of conducting particles and glass, deposited on an insulating surface. The physical sizes can be a lot smaller than that of the discrete resistances. The sheet resistance of a thick film paste ranges between 1 Ω/square and 10 MΩ/square (Chen, 1995). Thus, a large range (0.5 Ω to 1 GΩ) of resistances can be realized. The tolerance (20–30%) and the temperature coefficient (50–100 ppm/$^\circ$K) of the resistances are, however, rather large. The tolerance could be greatly reduced in thin film technology, where the resistances are made from thin films of metal alloys deposited on an insulating surface (i.e., glass). Thin film circuits are much smaller than thick film circuits. The available resistance range is, however, smaller. The sheet resistance value is of the order of 10–200 Ω/square, leading to a nominal resistance value ranging from 10 Ω to 10 MΩ. The temperature coefficient of resistance (TCR) is of the order of 15–150 ppm/$^\circ$K and the absolute tolerance is typically between 5 and 10%. To obtain high nominal resistance, the resistive layer is specially shaped by laser trimming. This process, however, increases the cost of production of the resistance. Capacitances are also available in hybrid technology, but mostly resistances using either thick- or thin film techniques have been used in hybrid technological production of an active-RC filter.

Advancements in semiconductor process IC technology leading to on-chip OTA devices opened up the possibilities of implementing an active filter entirely on a monolithic substrate. In this respect, it is useful to know about the R and C elements as are available in a given IC technology. Brief discussions on these follow.

5.10.1
Resistances in IC Technology

In its simplest form, a resistance is formed by a rectangular layer of semiconductor material, with a specified sheet resistance (resistance per square). The total resistance is equal to the number of squares multiplied by the sheet resistance. For a resistance of large value, an economy in substrate area can be achieved by adopting a serpentine (meandered) structure. An accurate estimation of the resistance requires use of conformal mapping technique.

5.10.1.1 Diffused Resistor

Usually, the diffusion layers in a transistor process can be used to implement resistors. Thus, in a BJT process, the base, emitter, or collector diffusion can be employed to build resistors. The most preferred one is the base-diffused resistor. A typical geometry and the electrical model are shown in Figure 5.20 (Grebene, 1984). For good matching tolerance, a large width should be used for the diffusion layer. Base-diffused resistors with 50-µm resistor width can achieve a matching

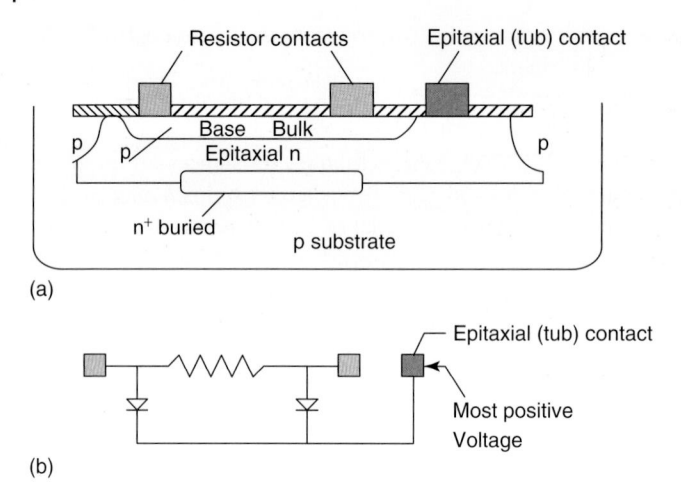

Figure 5.20 (a) Cross section and (b) electrical circuit model of a typical diffused resistor.

tolerance of $\pm 0.2\%$. The sheet resistance ranges between 100 and 200 Ω/square. The TCR ranges between 1500 and 2000 ppm/°C.

5.10.1.2 Pinched Resistor

To obtain a relatively high value of the sheet resistance (and, hence, to build a resistor of high value with less area), the base diffusion area is pinched, leading to pinched diffused resistors. The sheet resistance that can be realized falls in the range of 2–10 kΩ/square. The pinching is obtained, for example, by diffusing an n+ diffusion layer over a p-type base diffusion. The emitter diffusion greatly reduces the effective cross section (i.e., pinches) of the p-type resistor, thereby raising its sheet resistivity. Figure 5.21 shows a geometrical sketch of a p-type base pinched resistor. The process tolerance is very poor, ranging to a maximum of $\pm 50\%$.

5.10.1.3 Epitaxial and Ion-Implanted Resistors

Large values of resistance can also be obtained by using low doping concentration that forms the bulk of the resistor. Thus, an epitaxial resistor uses the bulk resistance of the n-type epitaxy in the BJT process. The p-type base diffusion can be

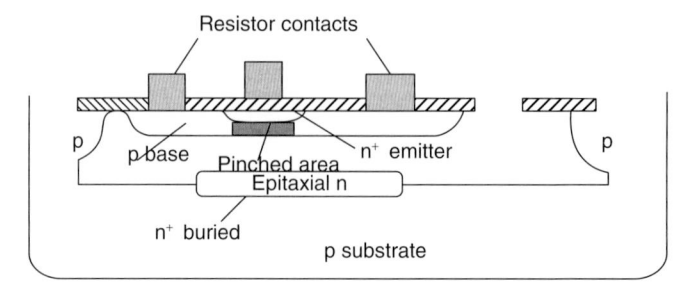

Figure 5.21 Cross section of a typical pinched resistor.

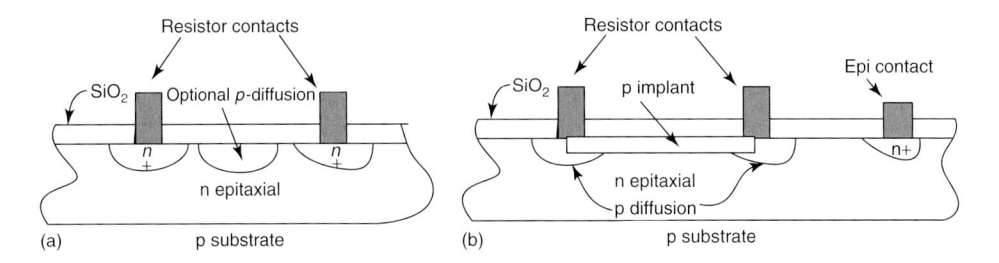

Figure 5.22 Cross section of (a) epitaxial and (b) ion-implanted resistors.

used to pinch the cross section of the epitaxial resistor, thereby leading to a pinched epitaxial resistor. For an epi thickness of 10 μm and a doping concentration of 10^{15} donor atoms per cubic centimeter, an effective sheet resistance of 5 kΩ/square can be obtained. The TCR value for epitaxial resistor is rather high, being of the order of 3000 ppm/°C.

An ion-implanted resistor can be built using a very thin layer of implant (0.1–0.8 μm), which leads to a very high value of sheet resistivity compared to that of the ordinary diffused resistors. Commonly used impurities are the p-type boron atoms. A sheet resistance of 100–1000 Ω/square can be obtained. The matching tolerance is very good, being typically within ±2%. The TCR is controllable to ±100 ppm/°C. Figures 5.22a and 5.22b, respectively, show the geometrical structures of epi and ion-implanted type of resistors.

5.10.1.4 Active Resistors

Active resistance implies resistance obtained from active devices, that is, transistors. Considering the I–V characteristics of well-known active devices, that is, BJT and MOS transistors, one can easily see that the devices exhibit the behavior of a resistance in certain areas of the I–V characteristics. For the BJT, it is the saturation zone, while for the MOS it is the linear zone. This is illustrated in Figure 5.23.

The advantages of an active resistance compared with a passive semiconductor resistance are twofold: (i) the available resistance can be very large without requiring large area of the substrate and (ii) the value of the resistance can be changed

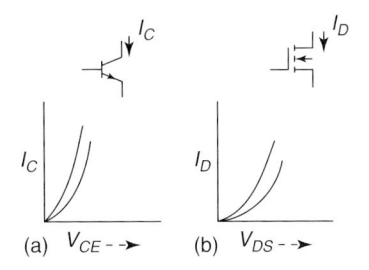

Figure 5.23 Current–voltage characteristics of (a) BJT and (b) MOS transistors pertaining to implementation of resistors.

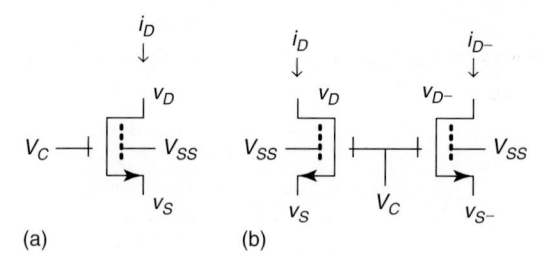

Figure 5.24 Active resistor (a) with a single MOS transistor and (b) with MOS transistors in differential configuration.

by a control voltage or current. In the modern MOS and CMOS technology, MOS transistors working in the linear region are frequently used to produce a resistance. Considering the N-type metal-oxide semiconductor (NMOS) transistor in Figure 5.24a, the instantaneous (i.e., AC plus DC) drain–source current, assuming operation in the linear (ohmic) region, can be modeled by the level = 1 equation:

$$i_D = K_n[2(V_C - v_S - V_{Thn})(v_D - v_S) - (v_D - v_S)^2] \tag{5.81}$$

In the above, K_n is the transconductance parameter, V_{Thn} is the threshold voltage, and V_C is the DC control voltage applied to the gate of the NMOS transistor. The operation of the transistor is in the linear (ohmic) region so that $v_D - v_S \leq V_C - v_S - V_{Thn}$ is true. The drain–source conductance is given by $\frac{\partial i_D}{\partial v_{DS}}$ and the associated resistance (R_{MOS}) is the inverse of this value. Since in an MOS transistor, K_n is proportional to the width-to-length ratio (W/L) of the transistor, it is clear that R_{MOS} is inversely proportional to W and, hence, by choosing a narrow width transistor, a large R_{MOS} can be realized. In the differential configuration of Figure 5.24b, the nonlinear term in R_{MOS} gets cancelled out and one can derive the differential resistance value as

$$R = \frac{1}{\mu_n C_{ox}(V_C - V_{Thn})}\left(\frac{L}{W}\right) = R_{sh}\left(\frac{L}{W}\right), \text{ where } K_n = \frac{1}{2}\mu_n C_{ox}\left(\frac{W}{L}\right) \tag{5.82}$$

With $\mu_n C_{ox} = 25$ μA V^{-2} and $V_C - V_{Thn} = 1$ V, a value $R_{sh} = 40$ kΩ can be realized.

5.10.2
Capacitors in IC Technology

Capacitor structures available in monolithic form include pn junction, MOS, and polysilicon capacitors. Silicon dioxide or silicon nitride sandwiched between two layers of doped (n- or p-type) silicon forms the basis of capacitors in a typical IC technology. These, however, need extra masking steps. More naturally occurring capacitors are formed between semiconductor junctions, for example, in BJT process technology. In any case, typical values are very small (viz., 0.05–0.5 pF/mil^2) (Grebene, 1984) and values larger than 10 pF are not practical because of the large substrate area consumed.

5.10.2.1 Junction Capacitors

Application of a reverse bias across a semiconductor junction results in a depletion layer that is devoid of mobile carriers. The situation closely resembles two parallel plates separated by a dielectric of width equal to the width of the depletion area. The area of the parallel plate is the total junction area. A typical analytical expression for the capacitance per unit area takes the form $C = C_{xo}/(1 + V/V_t)^n$, where C_{xo} is the zero bias capacitance (per unit area), V_t is the built-in junction potential, and the exponent n is dependent on the profile of the junction transition. This changes between 1/2 and 1/3, as the transition contour changes from a step junction to a linearly graded junction. As in a semiconductor resistor, a capacitor built using a semiconductor process is far from ideal. Thus, it will have few parasitic components arising out of the properties of the semiconductor layers. In a BJT process, three capacitors are available (see Figure 5.25), and C_{BC} is mostly used for practical purposes. C_{CS} is essentially a parasitic capacitance and is not conveniently measurable. C_{EB} can give relatively high value (0.001 pF/μm^2) of capacitance but has a small breakdown voltage. C_{BC} has a higher break down voltage, but a smaller value for capacitance (2.3 $\times$ 10^{-4} pF/μm^2) compared with C_{EB} (Grebene, 1984).

5.10.2.2 MOS Capacitors

An MOS capacitor is formed between the n+ diffusion (deposited while creating the channel region) and an aluminum (or polysilicon) layer with a layer of SiO$_2$ deposited (500–1000 Å) in between them. Sometimes, the silicon nitride layer is used as dielectric because of slightly higher dielectric constant (4–9) than silicon dioxide (2.7–4.2). A typical cross section and the electrical model are shown in Figure 5.26 (Grebene, 1984).

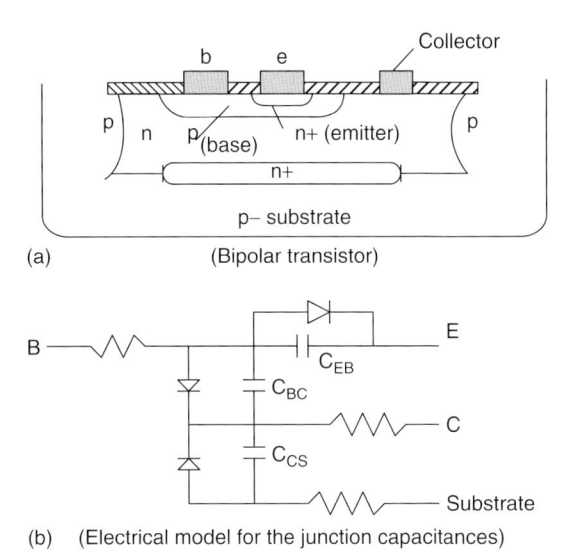

Figure 5.25 (a) Cross section of a BJT transistor and (b) electrical circuit model showing the junction capacitors.

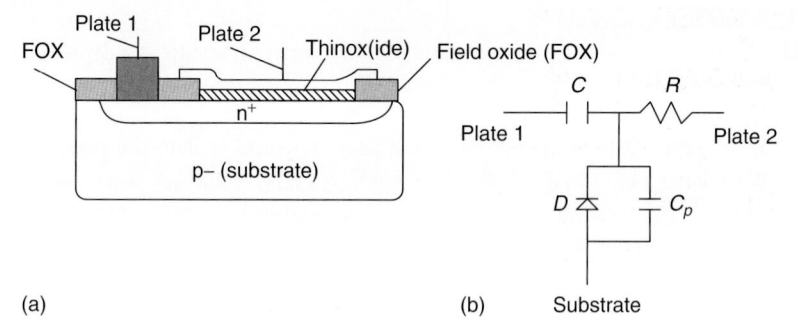

Figure 5.26 Basic MOS capacitor: (a) cross section and (b) electrical circuit model.

It may be noticed that, while C is the desired capacitance, it is invariably associated with parasitic components like R, C_p, and a diode D.

5.10.2.3 Polysilicon Capacitor

Polysilicon capacitors are conveniently available in metal-oxide semiconductor field effect transistor (MOSFET) technology, where the gate of the MOSFET transistor is made of polysilicon material. Figure 5.27a shows a typical structure of a polysilicon capacitor, where a thin oxide is deposited on top of a polysilicon layer and serves as an insulating layer between the top-plate metal layer and the bottom-plate polysilicon layer. The polysilicon region is isolated from the substrate by a thick oxide layer that forms a parasitic parallel-plate capacitance between the polysilicon layer and the substrate. The equivalent circuit model, which reflects this, is shown in Figure 5.27b.

5.10.3
Inductors

Interest in monolithic inductors has been on the increase with the availability of submicron CMOS IC technology. Initially, the role of active-RC filters was to replace the use of inductances. The signal processing case was for low frequencies

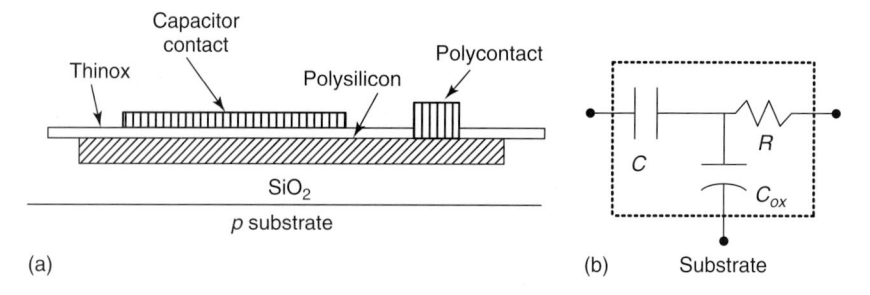

Figure 5.27 MOS polysilicon capacitor: (a) cross section and (b) electrical circuit model.

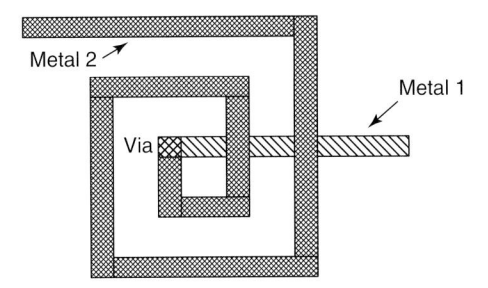

Figure 5.28 Typical layout of a spiral inductor in CMOS IC technology.

(300 Hz to 3.4 kHz) and the inductors usable at these frequencies were too bulky. As the feature size of the transistors in the IC technology shrunk, the possibilities of processing high-frequency signals on a monolithic substrate came into play. At these frequencies (300 MHz and more), the inductors needed are small in value (nano henries) and can be easily laid out using the metal lines available in a modern IC technological process. Both rectangular and circular spiral patterns have been used for signal processing at frequencies of 900 MHz and more (Nguyen and Meyer, 1990). A typical pattern for a rectangular spiral monolithic inductor is shown in Figure 5.28 (Chen, 1995). Laying two such independent patterns in an interleaved way can produce a transformer on a monolithic substrate. Like a monolithic capacitor, the use of a monolithic inductor is primarily limited by the substrate area that it will occupy. Practical limits are expected to be around a value of 10 nH.

5.10.4
Active Building Blocks

5.10.4.1 Operational Amplifier (OA)

OA is the most important and most widely used building block in active filter design (Geiger, Allen, and Strader, 1990; Johns and Martin, 1997; Baker, Li, and Boyce, 1998). The desired characteristics of an OA, for the design of a filter, are very high input impedance, high open-loop gain, very wide gain-bandwidth product value, and very low output impedance. Other important characteristics are low offset voltage, high common-mode rejection ratio (CMRR) and power supply rejection ratio (PSRR), high slew rate, low settling time, and large dynamic range. Several excellent designs of OA have been reported in the literature. Many high-quality OAs are available as commercial products. Examples are AD648, HA 2600, LM 101, and μA 741 chips. One typical realization is shown in Figure 5.29. This implementation uses a 5-μm CMOS process; the transistor dimensions are shown in Table 5.11 (Toumazou, Lidgey, and Haigh, 1990). This OA uses a bias current of 25 mA with ±5 V supply. The open-loop gain is about 73 dB and the GB value is 2.2 MHz. The first corner frequency (for one-pole model) is at 400 Hz.

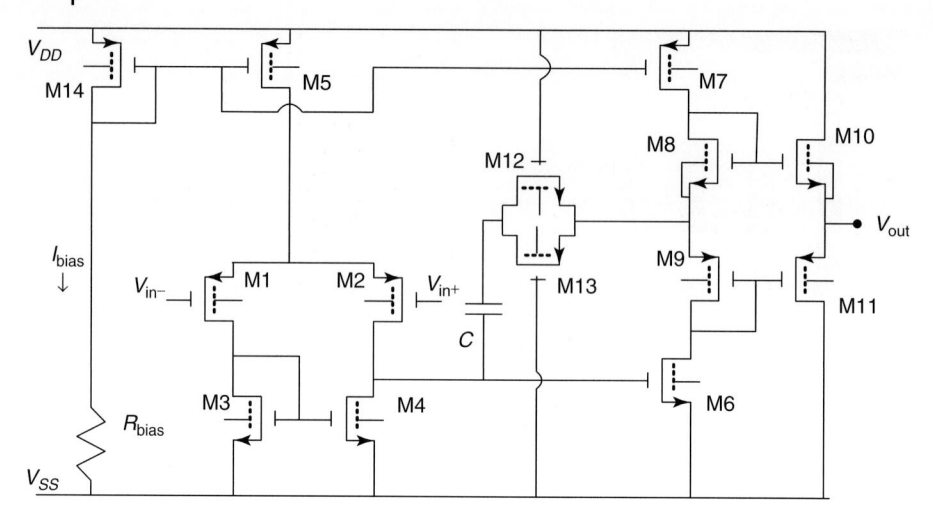

Figure 5.29 A typical realization of an OA (adapted from Toumazou et al., 1990).

Table 5.11 Geometric dimensions of the OA shown in Figure 5.29.

Transistor	M1	M2	M3	M4	M5	M6	M7	M8	M9	M10	M11	M12	M13	M14
W (μm)	120	120	50	50	150	95	150	135	207.5	135	402.5	7.5	7.5	150
L (μm)	7	7	10	10	10	10	10	5	5	5	5	15	5	10

In Fig. 5.29, transistor M14 works like a diode providing the bias current I_{bias} for M5 and M7, M8, M9, M6 chain. The PMOS transistors M1 and M2 form the input differential stages. The current mirror load M3 and M4 convert the differential signal to a single-ended output. M6 provides additional gain. M8 and M9 provide for biasing M10 and M11 for class AB operation. The output (V_{out}) is taken from the source terminals of M10 and M11, thereby facilitating a low output impedance condition. The transistors M12 and M13 and the capacitance C form a series RC feedback path for frequency compensation and stability.

5.10.4.2 Operational Transconductance Amplifier (OTA)

The OTA has gained popularity as a building block for monolithic solution for implementation of analog filters (Geiger and Sanchez-Sinencio, 1985; Mohan, 2002). In OTA-based implementation, the resistance can be replaced by the OTA itself, as illustrated in Table 5.9. An OTA closely follows the structure of an OA at its input, but the low output impedance stage is replaced by a high impedance output stage. Large open circuit voltage gain is easily available, but low output impedance is not obtainable. Low output impedance is not essential if the amplifier is driving a high impedance load, such as a capacitor. Such is the case, for example, in

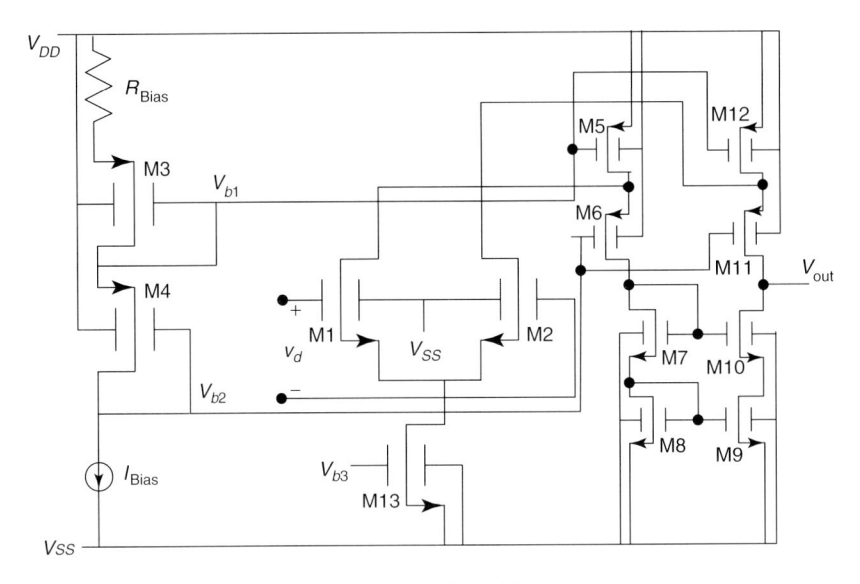

5.10 Technological Implementation Considerations

Figure 5.30 A typical realization of an OTA (adapted from Geiger et al., 1990).

switched capacitor filters. Several OTA chips are available as commercial product. The LM13700 chip is one such example. The pertinent transfer function of an OTA is the small signal (ac) short circuit transconductance. Figure 5.30 shows a typical OTA network where the popular folded-cascade architecture has been used (Geiger, Allen, and Strader, 1990). The output stage consists of a cascode current mirror whose output impedance is very high (100 MΩ and more). Table 5.12 presents the geometrical dimensions and DC bias currents in the various transistors. The transistors M1 and M2 comprise the input differential stage. The loads for these amplifiers are the common gate stages arranged by transistors M6 and M11. Since the common gate stage has low input impedance, the differential voltage gain is compromised, but a large transconductance gain is achieved. The signal currents in M6 and M11 are passed via the cascode mirror transistors M7–M10. The differential-output current at the drain nodes of M10 and M11 transistors produces a large voltage gain across the high output impedance available at this node.

Table 5.12 Geometric dimensions for the OTA shown in Figure 5.30.

Transistor	M1	M2	M3	M4	M5	M6	M7	M8	M9	M10	M11	M12	M13
W (μm)	15	15	30	30	70	70	15	15	15	15	70	70	30
L (μm)	5	5	5	5	5	5	5	5	5	5	5	5	5
I_{DC} (μA)	5	5	10	10	10	10	5	5	5	5	5	5	10

OTAs with dual and multiple outputs are also available and have been used in filter design (Deliyanis, Sun, and Fidler, 1999; Tu et al., 2007; Wang, Zhou, and Li, 2008; Hwang et al., 2008)

5.10.4.3 Transconductance Amplifiers (TCAs)

TCAs have become popular in the last decade for realizing high-frequency wideband continuous-time filters in monolithic form. The function of a TCA is similar to that of an OTA, but the TCA does not necessarily follow the architecture of an OTA. The number of transistors employed to implement a TCA is usually much smaller than that for an OTA. The transconductance factor in a TCA is also usually smaller than that available in an OTA. The TCAs provide a low-voltage, low-power solution toward high-frequency wideband signal processing operation. Figure 5.31 presents a TCA, which uses eight MOS transistors. Assuming that all the transistors are operating in saturation and using the square-law model for the transistors, the expression of the output current is given by (Raut, 1992)

$$I_{out} = \mu_P C_{oxP} \left(\frac{W}{L}\right)_P (V_{C_1} - V_{C_2})(V_1 - V_2) \tag{5.83}$$

where μ_P, C_{oxP}, and $(W/L)_P$ are same for all the PMOS transistors. The form of Eq. (5.83) shows that I_{out} depends linearly upon both $(V_{C_1} - V_{C_2})$ and $(V_1 - V_2)$ and thus the transconductor can be used for both linear and nonlinear signal processing. Applications of the TCAs for a variety of signal processing, including filtering, have been reported in the literature (Raut, 1993; Raut and Daoud, 1993). It should also be mentioned that some new elements such as the current differencing transconductance amplifier (CDTA) and the multioutput TCA, have been introduced recently, and their use in the realization of current-mode filters has been discussed (Biolek, Hancioglu, and Keskin, 2008; Prasad, Bhaskar, and Singh, 2009; Tangsrirat, Tanjaroen, and Pukkalamm, 2009).

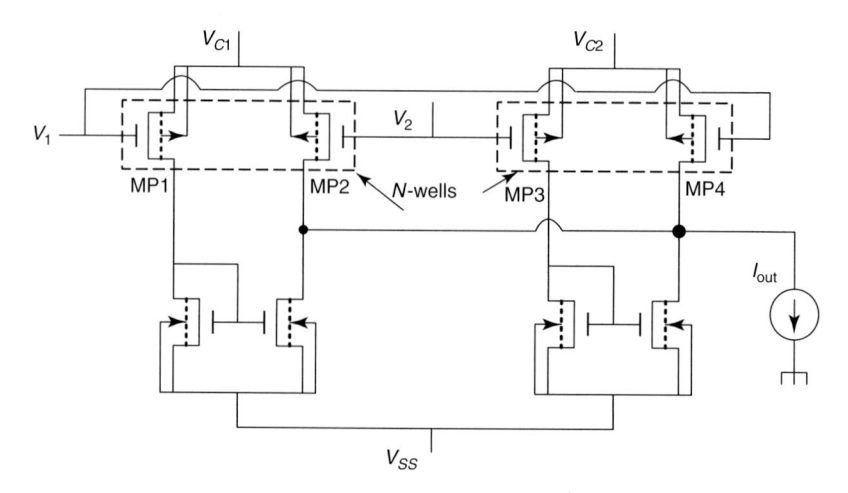

Figure 5.31 A typical realization of a TCA (adapted from Raut, © IEEE, 1992).

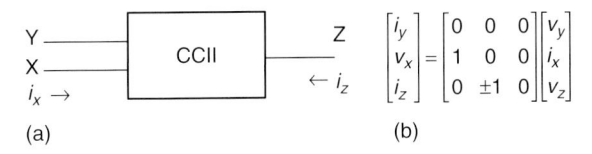

5.10 Technological Implementation Considerations

(a) Symbol: Y, X ($i_x \rightarrow$) inputs to CCII block; Z output ($\leftarrow i_z$).

(b) $$\begin{bmatrix} i_y \\ v_x \\ i_z \end{bmatrix} = \begin{bmatrix} 0 & 0 & 0 \\ 1 & 0 & 0 \\ 0 & \pm1 & 0 \end{bmatrix} \begin{bmatrix} v_y \\ i_x \\ v_z \end{bmatrix}$$

Figure 5.32 Current conveyor type II. (a) Symbol representing CCII. (b) Input-output relationship in matrix form.

5.10.4.4 Current Conveyor (CC)

CCs have become popular in the past decade for implementing current-mode filters. A CC (Sedra and Smith, 1970) maintains a virtual short circuit at its input nodes as in an OA. But one of the input nodes of a CC is a low-impedance node. A signal current injected at this node can be conveyed to the output terminal with a proportionality factor $\pm k$, where k is a constant and is usually unity. Symbolically, a CC is represented as shown in Figure 5.32a, and the output–input relationship is given by a matrix, as shown in Figure 5.32b.

The BJT IC-technology-based AD844 (Analog Devices) has been used by many researchers to illustrate CC-based filter designs (Sedra and Smith, 2004; Fabre and Alami, 1995; Abuelma'atti and Shabra, 1966; Mohan, 2002). Good CMOS-based CCs have also been reported in the literature (Elwan and Soliman, 1996). A dual-output CC, implemented in a 3-µm CMOS technology, is shown in Figure 5.33 (Minaei, Kuntman, and Cicekoglu, 2000). The capacitors C are used to stabilize the system. The signal currents at the gates of M4 and M6 are transported to terminals Z_1 and Z_2. The output currents at Z_1 and Z_2 are of opposite phases; therefore, this

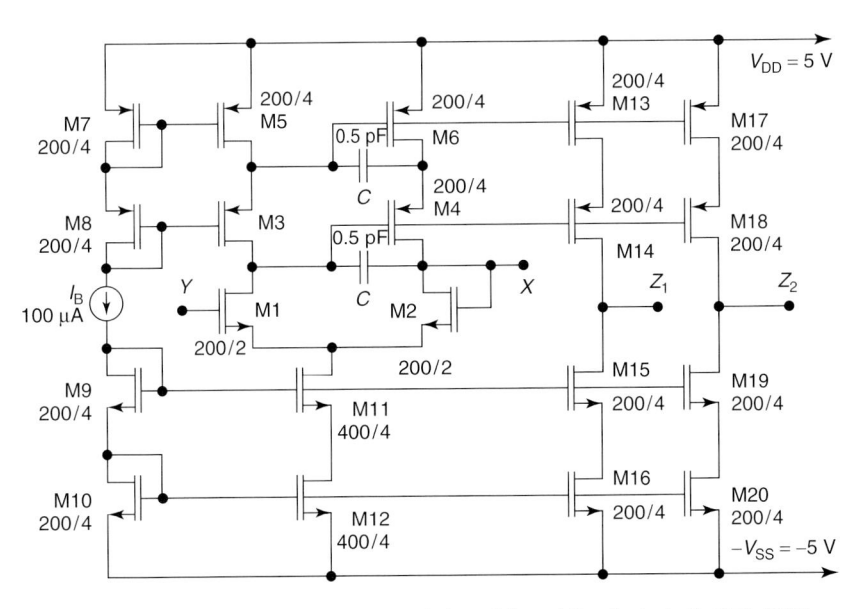

Figure 5.33 A dual-output current conveyor (adapted from Minaei, et al., © IEEE, 2000).

network can realize the CC function with either polarity. Applications of the CC in realizing current-mode filters are discussed in Chapter 8.

It should be mentioned that a number of other circuit elements, such as the current feedback OA, differential current-mode OA, dual-output current OA, differential difference current conveyors (DDCCII), and fully difference current conveyors (FDCCII), have been introduced and their applications in the current-mode filter design are given (Mohan, 2002; Altun and Kuntman, 2008; Cheng and Wang, 1997; Toumazou, Lidgey, and Haigh, 1990; Jiraseree-amornkun and Surakampotorn, 2008; Chang, Soliman, and Swamy, 2007; Soliman, 1996, 1997; Elwan and Soliman, 1996; Chen, 2009a, 2009b; Chiu and Horng, 2007).

Practice Problems

5.1 (a) Design a Sallen–Key biquad to realize the voltage transfer function, $H(s) = \frac{H_0 s^2}{s^2 + 0.01s + 1}$. Let $C_1 = C_3 = 1$ F and $R_2 = R_4$. (b) Using RC:CR transformation, get the corresponding LP filter.

5.2 (a) Design a Sallen–Key (SK) band-pass biquad with a pole frequency of 1000 Hz, a pole Q of 10, and a peak gain of 4. Use equal element values and $0.01\,\mu$F capacitors. (b) If you use the RC:CR transformation, what kind of a filter do you obtain? Draw the circuit diagram of the resulting filter and find the various element values.

5.3 (a) Design an SK BP filter with the voltage transfer function, $\frac{600s}{s^2 + 600s + 3 \times 10^8}$. (b) If you use the RC:CR transformation, what kind of filter do you obtain? Draw the circuit diagram of the filter so obtained and find the various element values.

5.4 Consider Figure 5.3. If Y_1, Y_2, and Y_3 correspond to admittances of resistors, and Y_4 and Y_5 to that of capacitors, what kind of filter characteristic does it realize?

5.5 Assume that $R_1 = R_2 = R_3 = 1\,\Omega$ and $C_4 = 10$ F. Determine the values of C_5 and K such that the denominator of the voltage transfer function in Problem 5.4 above has the form $D(s) = s^2 + 0.2s + 1$.

5.6 Design a second-order Bessel–Thomson (BT) filter using SK architecture. The filter has a delay of 1 ms at DC. Choose equal-C and equal-R design strategy. Let $C = 1$ F.

5.7 Use the infinite-gain multiple feedback single-amplifier biquad (IGMFSAB) network to design a band-pass filter with $f_p = 10$ kHz and $Q_p = 10$. Assume equal-capacitor values of 1 nF each.

5.8 Consider the band-pass filter network of Figure P5.8. It combines both positive and negative feedback around an ideal OA. Find an expression for the voltage transfer function. Then design the element values to meet the specifications $f_p = 10$ kHz, $Q_p = 10$, and each of the capacitors is 0.1 μF. Limit the resistance value spread to within 100.

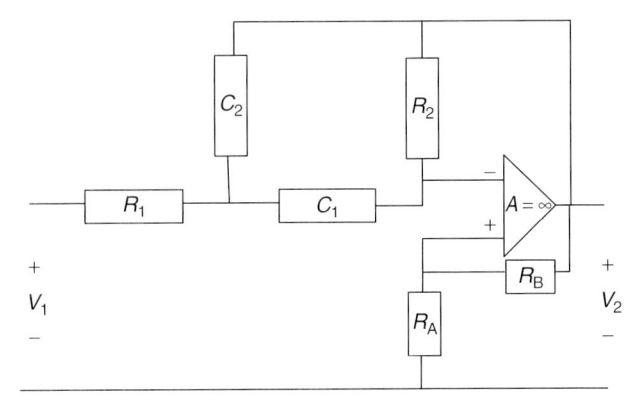

Figure P5.8

5.9 Design a KHN low-pass biquad such that $\omega_p = 1$ rad s^{-1} and $Q_p = 20$. Use equal capacitors of 1 F each and $R_1 = R_2 = 1\,\Omega$. What is the gain of this filter at DC?

5.10 Design a Tow–Thomas band-pass biquad with $\omega_p = 1$ rad s^{-1}, $Q_p = 15$, and a peak gain of 5. Use equal capacitors of 1 F each.

5.11 Design a Fleischer–Tow high-pass biquad with $\omega_p = 1$, $Q_p = 8$, and a flat gain of 2.

5.12 Design a Fleischer–Tow band-reject biquad with $\omega_p = 1$, $Q = 12$, and $H(0) = H(\infty) = 1$.

5.13 Figure P5.13 is known as the *Bainter band-reject biquad* (Bainter, 1975). Derive the transfer function V_o/V_s.
Choosing $C_1 = C_2 = 1$ F, derive the set of design equations involving the resistances $R_1, R_2, \ldots, R_6$ to realize the transfer function, $\frac{V_o}{V_s} = \frac{s^2 + \omega_z^2}{s^2 + \left(\frac{\omega_p}{Q_p}\right)s + \omega_p^2}$.

5.14 Derive an expression for $S_K^{Q_p}$ for the case where the denominator polynomial of a filter transfer function has the form $D(s) = s^2 + \left(\frac{\omega_p}{Q_p}\right)s + \omega_p^2 = (s+1)^2 - K\alpha s$, where α is not a function of K.

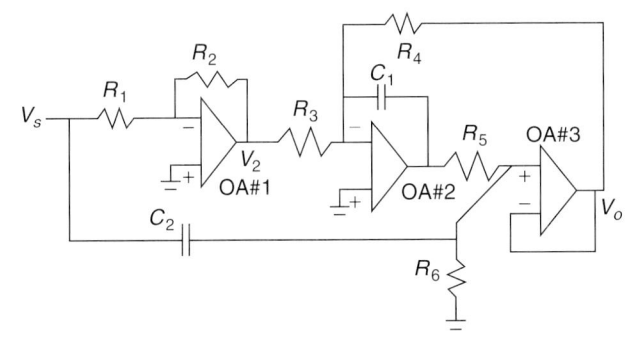

Figure P5.13

5.15 In Problem 5.9, find the expressions for the Q-sensitivity with respect to the capacitors.

5.16 In Problem 5.13, find the expressions for Q-sensitivity with respect to the resistors.

5.17 Use a Sallen and Key low-pass filter structure to realize the transfer function, $\frac{V_2(s)}{V_1(s)} = \frac{H_0}{s^2 0.5714s + 1}$ with equal-resistance, equal-capacitor assumption. (a) Find the design values for R, C, and K assuming that the gain K is implemented with an ideal OA. What will be the value of H_0 that will be realized? (b) If the OA has a frequency-dependent gain with normalized $GB_n = \omega_t/\omega_p = 7.5$ and the gain model $A(s) = \omega_t/s$, find the network function that will be actually realized. What will be the resulting values of ω_p and Q_p? What will be the numerical value of H_0 that is realized? (c) Verify your calculation results with SPICE (or similar) simulation. For the nonideal OA, you may use the model shown in Figure P5.17.

5.18 Repeat parts (a)–(c) of Problems 5.17 for an SK band-pass filter design with the transfer function:

$$\frac{V_2(s)}{V_1(s)} = \frac{Hs}{s^2 + 0.5714s + 1}.$$

Use equal-R, equal-C design strategy.

5.19 Design a band-pass filter for which the pole frequency is 10^5 rad s^{-1} with a bandwidth of 10^3 rad s^{-1}. Scale the network so that all the element values are in a practical range.

(In many of the following problems you may need to use a cascade connection of second- and first-order systems. Use your discretion.)

5.20 A low-pass filter has the following specifications: $f_c = 10$ kHz, $f_a = 60$ kHz, $A_p = 1$ dB peak-to-peak ripple, and $A_a = 50$ dB. Design an active-RC filter to satisfy the specifications.

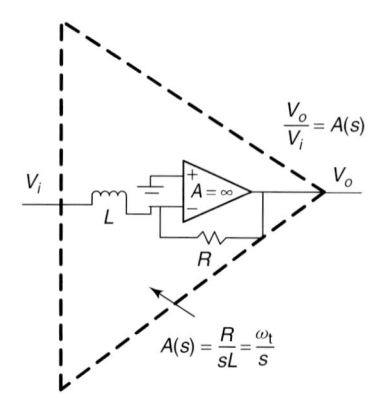

Figure P5.17

5.21 A high-pass filter has the specifications: passband from 10^4 rad s^{-1} to infinity. Passband peak-to-peak ripple less than 2 dB. For $\omega \leq 2000$ rad s^{-1}, the loss must be greater than 50 dB. Design an active-RC filter to satisfy the specifications.

5.22 An equiripple band-pass filter is required to satisfy the specifications: (a) The passband extends from $\omega = 1000$ to 4000 rad s^{-1}. The peak-to-peak ripple in the passband does not exceed 0.5 dB. (b) The magnitude characteristic is at least 30 dB down at $\omega = 12 \times 10^3$ rad s^{-1} from its peak value in the passband. (c) Design an active-RC filter to satisfy the specifications.

5.23 A band-reject filter has the specifications. (a) The stopband extends from 10 to 100 kHz. (b) The magnitude characteristic is at least 30 dB down from its peak value at 20 kHz. (c) The peak-to-peak ripple in the passband does not exceed 1 dB. Design an active-RC filter to satisfy the above specifications. Use state-variable biquad section(s).

5.24 Using the state-variable biquad, realize a second-order all-pass function having pole and zero frequency $\omega_p = \omega_z = 5 \times 10^4$ rad s^{-1}, and $Q_p = Q_z = 10$.

5.25 Figure P5.25 represents a second-order state-variable filter. Provide an implementation of the filter using two capacitors and three OAs, and several resistors. The OAs provide the summing operation while the capacitors produce the integrating function. The signal gains are shown by the letters $a, b, c, \ldots$ by the side of the line branches. Derive an expression for the voltage transfer function V_2/V_1.

5.26 Design a normalized Chebyshev second-order low-pass filter with $A_p = 2$ dB and a DC gain of 2. Use (a) multiple feedback (infinite-gain OA) structure and (b) state-variable structure.

5.27 The circuit of Figure P5.27 can produce (Friend, Harris, and Hilberman, 1975) a general biquadratic voltage transfer function V_2/V_1. Show a design to obtain $\frac{V_2}{V_1} = \frac{s^2-5s+10}{s^2+5s+10}$, which is an all-pass filter function.

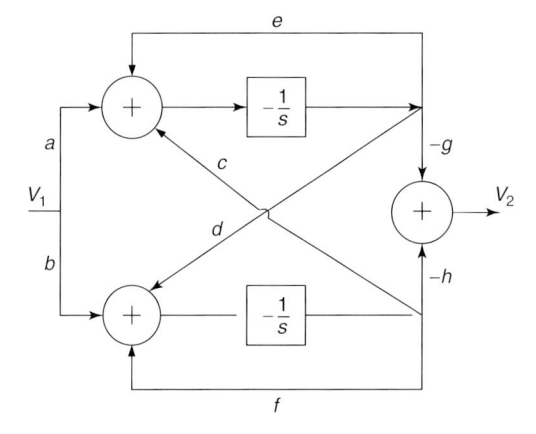

Figure P5.25

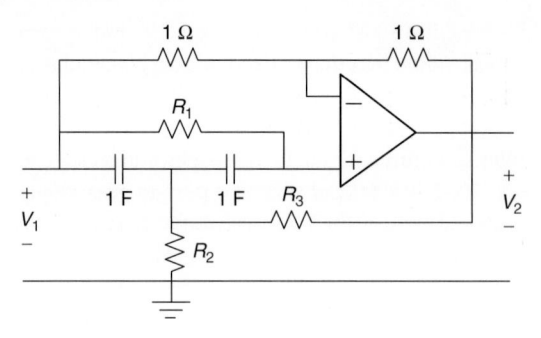

Figure P5.27

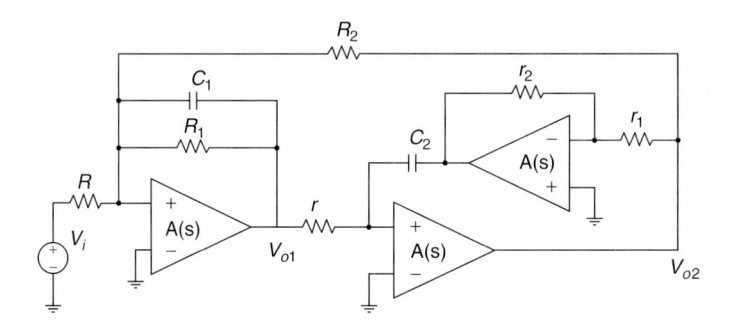

Figure P5.28

5.28 The A & M filter (Ackerberg and Mossberg, 1974) which can produce both low-pass and band-pass filter responses depending upon the choice of the output signal node is shown in Figure P5.28. Show that the band-pass voltage transfer function is given by

$$\frac{V_{o1}}{V_i} = -\frac{s/RC_1}{s^2 + s\frac{1}{R_1 C_1} + \frac{r_1}{C_1 C_2 R_2 rr_2}}.$$

5.29 Consider the Filter in Figure P5.28, where the OA has a frequency-dependant gain function A. Under this condition, show that the low-pass voltage transfer function is given by

$$\frac{V_{o2}}{V_i} = -\frac{\frac{1}{rR}}{\frac{1}{R_2 r} + \left(\frac{\frac{1}{R}+\frac{1}{R_2}}{A} + \left(1+\frac{1}{A}\right)\left(\frac{1}{R_1}+sC_1\right)\right)\left(\frac{\frac{sC_2}{r_1}}{\frac{1}{r_2}+\frac{1/r_2+1/r_1}{A}} + \frac{\frac{1}{r}+sC_2}{A}\right)}.$$

5.30 A fifth-order normalized low-pass filter can be realized by a cascade of first- and second-order functions as shown below:

$$H(s) = \frac{H_1}{s+1} \frac{H_2}{s^2 + 1.61803s + 1} \frac{H_3}{s^2 + 0.61803s + 1}$$

A possible design for the system is proposed in Figure P5.30 (Huelsman, 1993). Prove its adequacy or provide an alternative design.

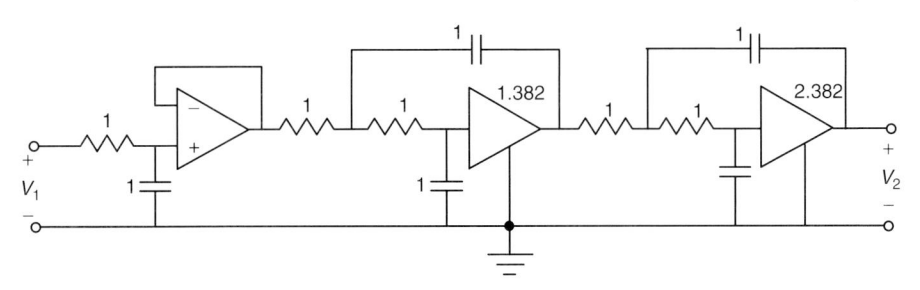

Figure P5.30

5.31 Figure P5.31 shows the Deliyannis band-pass filter circuit (Deliyanis, 1968; Friend, 1970) which uses both positive and negative feedback around the OA. Derive an expression for the voltage transfer function $V_o(s)/V_i(s)$.

5.32 Rederive the voltage transfer function for the system in Figure P5.31, assuming a nonideal $A(s) = \frac{\omega_t}{s}$.

5.33 For the following second-order filter specifications, determine the design parameters (C and g_m values) that need to be implemented for the g_m–C filters. Assuming that the g_m is generated from a simple differential pair in BJT technology (Figure P5.33) with ±5 V power supply system, calculate the DC power consumption that will be required in each case. (a) An HP filter with $\omega_p = 2000$ rad s^{-1}, $Q_p = 5$, $H_0 = 3$ (gain at infinite frequency). (b) A BP filter with $\omega_p = 5000$ rad s^{-1}, $Q_p = 10$, $H_0 = 10$ (gain at resonant frequency). (c) An AP filter with $\omega_p = 300$ rad s^{-1}, $Q_p = 3$, $H_0 = 1$. (d) An LP notch filter with $\omega_p = 2000$ rad s^{-1}, $Q_p = 5$, $H_0 = 1$, $\omega_z = 3000$ rad s^{-1}. (e) An HP notch filter with $\omega_p = 300$ rad s^{-1}, $Q_p = 5$, $H_0 = 1$, $\omega_z = 150$ rad s^{-1}.

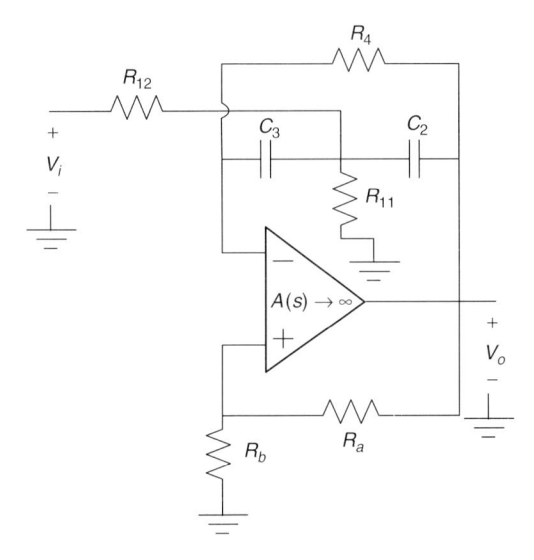

Figure 5.31

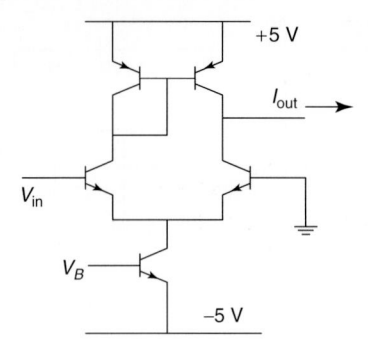

Figure P5.33

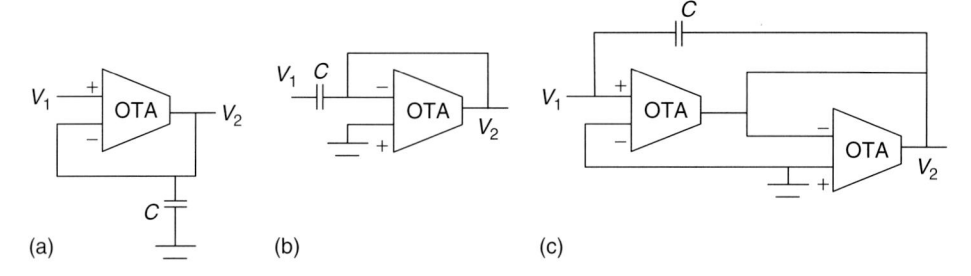

Figure P5.34

5.34 Determine the voltage transfer functions for each of the OTA-based networks in Figures P5.34a–5.34c. Note that the g_m could be positive or negative depending upon the input node polarity.

5.35 For the circuit of Figure P5.35, find the transfer function V_o/V_i.

5.36 For the biquad shown in Figure P.5.36, find the transfer functions $\frac{V_{o1}}{V_i}$, $\frac{V_{o2}}{V_i}$, and $\frac{V_{o3}}{V_i}$. Also, calculate the sensitivities of the pole frequency and pole Q w.r.t. the various components. (Note: Observe that the total sum of the sensitivities over all the elements for the pole frequency as well as the pole Q is zero. This is true for any OTA-C filter (Swamy, Bhusan, and Thulasiraman, 1972)

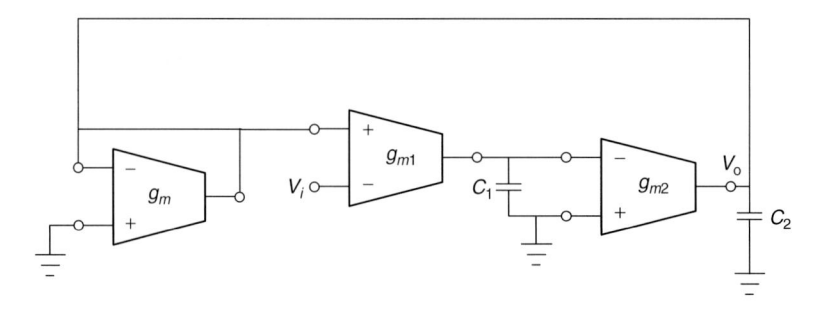

Figure P5.35 A dual-output current conveyor (adapted from Minaei, et al., © IEEE, 2000).

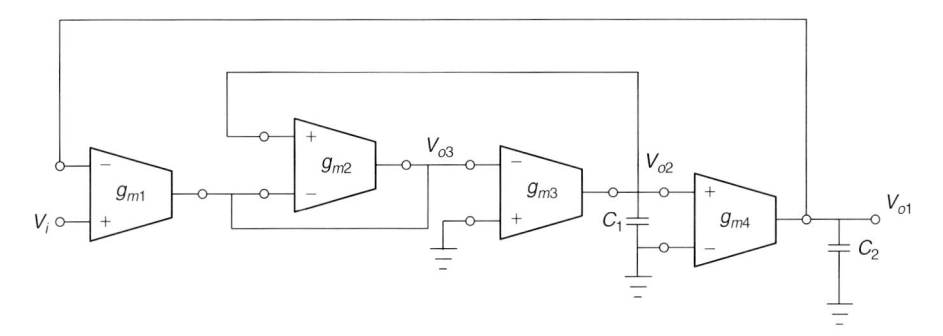

Figure P5.36

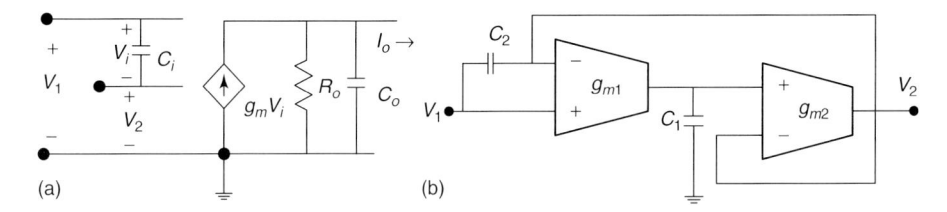

Figure P5.37

5.37 Figure P5.37 shows an approximate first-order equivalent model for a practical OTA operating as a single-ended device. At low frequencies, the parasitic capacitances can be ignored. Figure P5.37b shows a second-order filter network built from two OTA devices. Find the voltage transfer function V_2/V_1, at low frequencies, assuming (a) ideal OTA and (b) practical OTA.

5.38 A $g_m - C$ integrator must be designed for a unity-gain frequency of 9 MHz. The available transconductor is known to have the parameters $g_m = 250$ µS, $C_i = 0.05$ pF, and $C_o = 0.19$ pF (see Figure P5.37a). The DC gain must be at least 70 dB. (a) Determine the required load capacitor. (b) Determine the minimum value of output resistance R_o the OTA must have. (c) Verify your calculations using a network simulation tool (such as SPICE).

5.39 A very large inductor has a series loss resistor of 250 Ω. This resistance can be cancelled by a simulated negative resistance. Figure P5.39 shows a possible strategy to implement a negative resistance with an OTA. What will be the value of g_m needed? Verify your design with a network simulator.

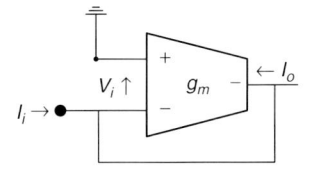

Figure P5.39

5.40 Design and test (by simulation) a first-order low-pass filter with a transconductance amplifier with $g_m = 250\,\mu S$ so that the following specifications are met: dc gain = 70 dB and unity-gain frequency = 10 MHz.

5.41 Using a cascade of first- and second-order g_m–C filter sections, design and test (by simulation) the following: (a) A third-order low-pass filter with maximally flat magnitude characteristic, having $A_p = 3$ dB and DC gain of 30. (b) A Chebyshev low-pass filter for the specifications: $A_p = 0.3$ dB, $f_c = 10$ MHz, $A_a = 22$ dB, and $f_a = 25$ MHz. The DC gain is unity. (c) A seventh-order Chebyshev high-pass filter with $A_p = 1$ dB, $f_p = 15$ MHz, and $H(\infty) = 4.5$ dB. (d) A fifth-order band-pass filter with a Chebyshev magnitude response having $A_p = 0.1$ dB, $f_o = 850$ kHz, 0.1 dB bandwidth of 70 kHz, and a center-of-band gain of 25 dB.

5.42 Realize the fifth-order low-pass filter function given by

$$H(s) = H_1(s)H_2(s)H_3(s)$$
$$= \frac{k(s^2 + 29.2^2)(s^2 + 43.2^2)}{(s + 16.9)(s^2 + 19.4s + 20.01^2)(s^2 + 4.72s + 22.52^2)}$$

as a cascade of three g_m–C filters. Test your design with a network simulator.

6
Switched-Capacitor Filters

In this chapter, we introduce the topic of second-order SC filters. Interest in SC networks revived in the mid-1970s, because of the potential to implement low-frequency (voice band) analog filters using a monolithic IC technology (i.e., CMOS) (Fried, 1972; Temes and Mitra, 1973; Hodges, Gray, and Brodersen, 1978; Gray and Hodges, 1979; Gregorian and Nicholson, 1979). The possibility of realizing a resistance, by switching a capacitance to and fro between two nodes was, however, known even at the time of Sir J. C. Maxwell. In connection with monolithic realization, the issue that appeared very attractive was that the values of the resistors and capacitors are inversely related; therefore, a resistance of high value could be generated by using an on-chip capacitor of small value, which is attractive for monolithic fabrication since it occupies a small substrate area.

In the following, we first deal with the equivalence between a resistance and an SC. The notion of discrete time (i.e., sampled-data) operation is introduced and several possible transform relations between the frequency in sampled-data time domain and continuous-time frequency domain are explored (Allen and Sanchez-Sinencio, 1984). The case of bilinear transformation (BLT) is addressed in more detail. The situation of parasitic capacitances associated with physical switches (made from MOS transistors) and capacitors is considered and the technique of parasitic-insensitive (PI) operation is introduced (Ghausi and Laker, 1981). A simple method of analysis of SC networks (Kurth and Moschytz, 1979; Vandewalle, DeMan, and Rabey, 1981; Laker, 1979; Hokënek and Moschytz, 1980; Raut and Bhattacharyya, 1984a), using discrete-time and z-transformed equations, is presented. This is followed by suggesting techniques for analyzing SC networks using common network analysis tools (for example, SPICE) (Nelin, 1983). The design of a second-order SC filter, using standard structures introduced by several researchers in this area (Ghausi and Laker, 1981; Fleischer and Laker, 1979; Mohan, Ramachandran, and Swamy, 1982, 1995; Raut, Bhattacharyya, and Faruque, 1992), is presented next. The potential of realizing high-frequency SC filters with unity-gain voltage amplifiers as the active building blocks (Raut, 1984; Fan et al., 1980; Malawka and Ghausi, 1980; Fettweis, 1979a, 1979b) is then discussed. Some second-order filter structures based on unity-gain amplifiers (UGAs) are presented.

Modern Analog Filter Analysis and Design: A Practical Approach. Rabin Raut and M. N. S. Swamy
Copyright © 2010 WILEY-VCH Verlag GmbH & Co. KGaA, Weinheim
ISBN: 978-3-527-40766-8

6.1
Switched C and R Equivalence

Consider Figure 6.1, which shows a capacitor C grounded at one end, and is switched between nodes 1 and 2 by two nonoverlapping clock signals ϕ_1 and ϕ_2. Each of the clock signals has a period of $T(=1/f_s)$, and they are so interleaved in time that they do not overlap with each other. That is, ϕ_2 does not become ON (i.e., high) until ϕ_1 is completely zero (OFF) and vice versa. When ϕ_1 is ON, C is connected to a voltage source of value v_1 and thus acquires a charge of Cv_1. When ϕ_2 becomes ON, the capacitor C is connected to the voltage v_2, thereby acquiring a charge Cv_2. Before ϕ_1 becomes ON again, that is, in the interval T, the capacitor C has transferred a charge of amount $C(v_1 - v_2) = \Delta Q$. Thus, the rate of charge transfer between nodes 1 and 2 is $\Delta Q/T$. In the limit when $T \to 0$ and $\Delta Q \to 0$, the charge transfer rate dq/dt implies a flow of current $i = dq/dt$ between the nodes 1 and 2. According to Ohm's law, if there is a resistance R between nodes 1 and 2, this current would also be equal to $(v_1 - v_2)/R$. Thus, in the case of a very high clock frequency ($f_s = 1/T \to \infty$) and an infinitesimal amount of charge transfer ($\Delta Q \to 0$), one can assume

$$C(v_2 - v_1)/T = (v_2 - v_1)/R$$

or

$$R = T/C = 1/(Cf_s) \tag{6.1}$$

The above equation is the famous equation for the equivalence of a switched C and a resistance R. It is apparent that with a small value of C (few picofarads) and a reasonable value of f_s (say, tens of kHz), a resistance value of tens of megaohms can be achieved. This is very desirable for implementing low-frequency filters on a monolithic substrate using only a small value of capacitance. The SC network just discussed is known as a toggle-switched-capacitor (TSC) network. Table 6.1 presents few other SC networks with the associated RC equivalent relations (Allen and Sanchez-Sinencio, 1984).

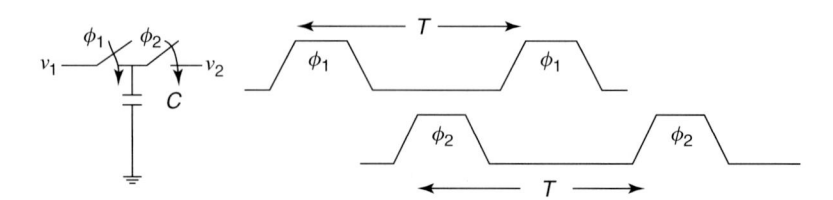

Figure 6.1 Toggle-switched capacitor (TSC) with a biphase clock signal.

Table 6.1 A few SC networks with associated RC equivalences.

Designated name	SC network	RC equivalence relation
A Series-switched C		$R = T/C$
B Series-shunt switched C		$R = T/(C_1 + C_2)$
C Toggle-switched C		$R = T/2C$
D Differential toggle-switched C		$R = T/4C$

6.2 Discrete-Time and Frequency Domain Characterization

The development above has been carried out in an approximate way, assuming that the clock frequency is very high compared to the signal frequencies. In reality, this is not true and, hence, it is necessary to use accurate timing information to account for the transfer of charge between two nodes by an SC connected between these nodes. The pertinent system of equations then becomes discrete-time equations, instead of a continuous-time one. The associated frequency domain characterization is termed *discrete or sampled-data frequency domain characterization*. Instead of the conventional continuous frequency variable $s = j\omega$, we have to use the sampled-data frequency variable $z = \exp(j\Omega T)$, where $F(z)$ is the z-transform of $f(nT)$ and Ω is the sampled-data analog of the continuous-time-domain angular frequency ω. Depending upon the way the SC network is operated, distinct relationships between the s and the z variables arise. This, thus, leads to several possible transformation relationships between the continuous domain frequency variable and the sampled-data domain frequency variable. Several of these are illustrated in the section to follow, by considering the SC equivalent to a continuous-time-domain RC integrator.

6.2.1
SC Integrators: $s \leftrightarrow z$ Transformations

Consider Figure 6.2. In Figure 6.2a, we show an active-RC integrator using an OA; in Figure 6.2b, a corresponding SC integrator with the resistance replaced by a TSC resistor is shown. For the RC integrator, the voltage transfer function is unique and is given by

$$T(s) = \frac{V_o(s)}{V_i(s)} = -\frac{1}{sRC} \tag{6.2}$$

The clock signal (Figure 6.2c) ϕ_1 goes ON at instants of time $(n-1)T, nT, \ldots$, and the clock signal ϕ_2 goes ON at instants of $(n-3/2)T, (n-1/2)T, \ldots$, so that each clock signal has a period of T. They do not overlap each other. For sample data operation, the sampling instants are considered to be coincident with the instants $(n-3/2)T, (n-1)T, (n-1/2)T, \ldots$, and so on. It has to be understood that, during the ON period of each clock phase, the network has a specific topology (i.e., interconnection pattern) and the topology changes as the clock phase alternates between ϕ_1 and ϕ_2. Thus, Figure 6.3a shows the topology of the network when clock

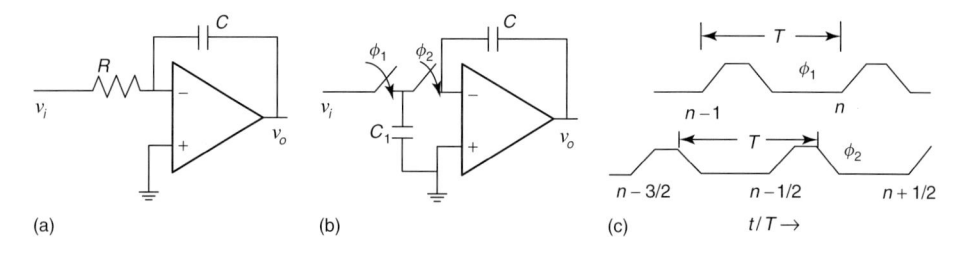

Figure 6.2 (a) Active-RC integrator, (b) TSC integrator, and (c) nonoverlapping clock signals.

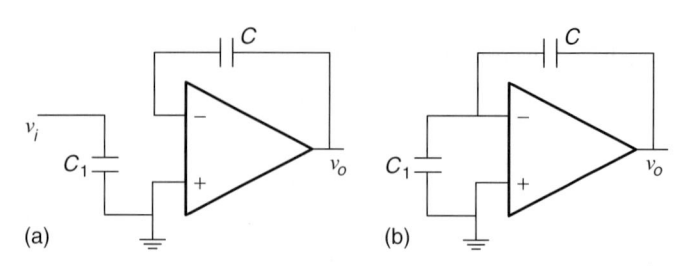

Figure 6.3 Configuration of the TSC integrator when (a) clock signal ϕ_1 is ON and (b) clock signal ϕ_2 is ON.

ϕ_1 is ON, say at $t/T = (n-1)$, that is, $t = (n-1)T$. At this time, C_1 has a voltage $v_i(n-1)$ across it, while C has a voltage $v_o(n-1)$ across it. Note that, for simplicity, we are using (n), $(n-1)$, ..., to imply the time instants nT, $(n-1)T$, ..., and so on. The voltage $v_o(n-1)$ is, however, the sampled and held version of voltage $v_o(n-3/2)$, since it was at this instant of time (i.e., $t = (n-3/2)T$) that C happened to be connected to C_1 and was subjected to the charge redistribution operation with C_1. Therefore, $v_o(n-3/2)T$ is like an initial value that is held until $t = (n-1)T$. Similarly, at $t = (n-1/2)T$, when ϕ_2 goes ON, the network topology is as shown in Figure 6.3b. The capacitor C_1, being across the input terminals of the OA, is virtually shorted and hence the charge across it becomes zero. The charge on C is now $Cv_o(n-1/2)$. At the instant $t = nT$, the scenario at $t = (n-1)T$ repeats, because the time interval is the same as the periodic interval T of the clock sequence. Thus, over one periodic interval of time T, the following changes occur in the charges held on the capacitors C and C_1.

For C,

$$\Delta Q_C = Q_C(n-1/2) - Q_C(n-1) = Cv_o(n-1/2) - Cv_o(n-1) \quad (6.3a)$$

For C_1,

$$\Delta Q_{C_1} = Q_{C_1}(n-1/2) - Q_{C_1}(n-1) = 0 - C_1 v_i(n-1) \quad (6.3b)$$

According to the principle of charge conservation, $\Delta Q_C = \Delta Q_{C_1}$. Then, from Eqs. (6.3a) and (6.3b), we can write

$$Cv_o(n-1/2) - Cv_o(n-1) = 0 - C_1 v_i(n-1) \quad (6.4)$$

The above is an equation in the discrete-time domain. To move to the discrete-frequency domain, we have to apply the z-transform to both sides of the equation. We must also remember that we are dealing here with the z-transforms for two distinct sampling sequences. One is due to the clock ϕ_1 and the other is due to the clock ϕ_2 – both having an identical sampling period T. Assigning the superscript (1) for the z-transform corresponding to ϕ_1 and the superscript (2) for the z-transform corresponding to ϕ_2, we write, from Eq. (6.4):

$$C[V_o(z)]^{(2)} z^{-1/2} - C[V_o(z)]^{(1)} z^{-1} = -C_1 [V_i(z)]^{(1)} z^{-1} \quad (6.5)$$

In the above, we have used the convention that $z[v(nT)] = V(z)$, where $z[.]$ implies "the z-transform of" $[.]$. Since, however, we have discussed that $v_o(n-1) = v_o(n-3/2)$, by taking the z-transform, we can write

$$[V_o(z)]^{(1)} z^{-1} = [V_o(z)]^{(2)} z^{-3/2}, \text{ that is, } [V_o(z)]^{(1)} = z^{-1/2} [V_o(z)]^{(2)} \quad (6.6)$$

Equation (6.6) is a special relationship between $[V_o(z)]^{(1)}$ and $[V_o(z)]^{(2)}$ that holds because of the sample-and-hold relation that is satisfied by the voltage $v_o(nT)$. This kind of relationship may not necessarily hold for V_i, unless we ascribe a similar sample-and-hold condition for the signal $v_i(nT)$.

Figures 6.2 and 6.3 indicate that the output v_o changes only during the phase ϕ_2, that is, when C is switched ON to the inverting input terminal of the OA. Therefore, we shall maintain $V_o(z)^{(2)}$ as the significant output variable. Then, we rewrite Eq. (6.5), after using the relation in Eq. (6.6):

$$CV_o(z)^{(2)}z^{-1/2} - CV_o(z)^{(2)}z^{-3/2} = -C_1 V_i(z)^{(1)}z^{-1}$$

Dividing out by $z^{-1/2}$, we get

$$CV_o(z)^{(2)} - CV_o(z)^{(2)}z^{-1} = -C_1 V_i(z)^{(1)}z^{-1/2} \tag{6.7}$$

Now taking the ratio, we get the z-domain transfer function for (1,2) phase combination (ignoring the argument z for simplicity):

$$\frac{V_o^{(2)}}{V_i^{(1)}} = H^{(1,2)} = -\frac{C_1}{C}\frac{z^{-1/2}}{1-z^{-1}} \tag{6.8a}$$

An alternative transfer function could be defined by considering the (1,1) phase group, that is,

$$\frac{V_o^{(1)}}{V_i^{(1)}} = H^{(1,1)} = -\frac{C_1}{C}\frac{z^{-1}}{1-z^{-1}} \tag{6.8b}$$

An interesting matter is immediately apparent, that is, we have been able to define two distinct z-domain transfer functions for the SC integrator network, while there is only one unique transfer function for the associated RC integrator, as given by Eq. (6.2). Using the equivalence $R \rightarrow T/C_1$ for the TSC-R equivalence in Eq. (6.8a), we get

$$H^{(1,2)} = -\frac{T}{RC}\frac{z^{-1/2}}{1-z^{-1}} \tag{6.9}$$

We can draw an interesting correspondence between the continuous-time-domain frequency variable s and the sampled-data domain frequency variable z by comparing Eq. (6.2) with Eq. (6.9). Thus,

$$T(s) \leftrightarrow H^{(1,2)}$$

leads to

$$\frac{1}{sRC} \leftrightarrow \frac{T}{RC}\frac{z^{-1/2}}{1-z^{-1}}$$

That is,

$$s \leftrightarrow \frac{1}{T}\frac{1-z^{-1}}{z^{-\frac{1}{2}}}$$

or

$$s \leftrightarrow \frac{1}{T}\left(z^{\frac{1}{2}} - z^{-\frac{1}{2}}\right) \tag{6.10}$$

The above represents a popular $s \leftrightarrow z$ transformation known as lossless digital integrator (LDI) transformation. One can appreciate the implication of this transformation by conceptualizing that, if an integrator-based RC-active filter is converted to an SC filter with the RC integrators (as in Figure 6.2a) replaced by TSC-C integrators (as in Figure 6.2b), then the transfer function of the SC filter in z-domain can be written down from the transfer function of the RC-active filter simply by replacing "s" by the function of "z," as given by the $s \leftrightarrow z$ transformation relation Eq. (6.10). It must also be understood that the output of each SC integrator in the filter must be sampled and transferred to the following stage in the filter, during the clock phase ϕ_2, so that the transformation in Eq. (6.10) is applicable. If, on the other hand, the output of each SC integrator in the filter is sampled during clock phase ϕ_1, the pertinent transformation will be $s \leftrightarrow \frac{1}{T}(z-1)$. One can derive this transformation relation by comparing the transfer function of Eq. (6.2) with the z-domain transfer function given by Eq. (6.8b). Two other important $s \leftrightarrow z$ transformations can be derived by considering an alternative implementation of a resistance by an SC network, while replacing an RC integrator with an SC integrator. Table 5.2 presents the various $s \leftrightarrow z$ transformations. For more details regarding the various transformations, one may refer to Mohan, Ramachandran, and Swamy (1982, 1995) and Allen and Sanchez-Sinencio (1984).

6.2.2
Frequency Domain Characteristics of Sampled-Data Transfer Function

By now, it is clear that a given continuous-time transfer function $H(s)$, when implemented using SC network, produces a transfer function $H(z)$ in the discrete-time domain. Corresponding to a given $H(s)$, there could be more than one possible $H(z)$ because of the presence of more than one clock signal in the SC network. For frequency domain characterization of $H(z)$, we use the relation $z = \exp(j\Omega T)$, where Ω is the sampled-data angular frequency. This is identical with the real physical frequency used to test and simulate a sampled-data system. Since $\exp(j\Omega T)$ is a periodic function, it is obvious that $H(z)$ will be a periodic function. This is an important distinction between the continuous-time frequency domain transfer function $H(s)$ and the sampled-data (i.e., discrete-time) frequency domain transfer function $H(z)$. One may visualize the situation as though the given $H(s)$, when implemented as an SC filter function, causes $H(s)$ to be modulated in frequency domain (multiplied in time domain) by the clock frequency $f_s = 1/T$ of the clock signals. Thus, corresponding to an LP $|H(j\omega)|$, we get a periodic train for $|H(e^{j\Omega T})|$, which will consist of lobes of $|H(j\omega)|$ around zero frequency (i.e., DC) and around $\pm 2\pi f_s, \pm 4\pi f_s, \ldots, \pm 2n\pi f_s$. This is illustrated in Figure 6.4, where $f_s = 1/T$.

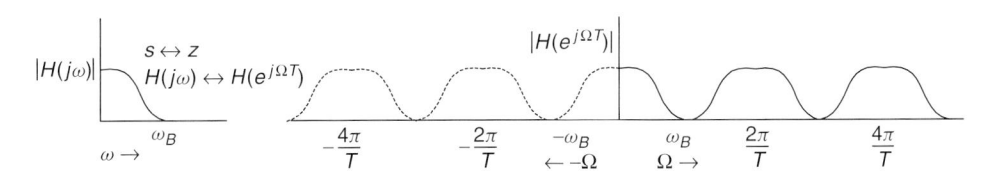

Figure 6.4 Sampling operation on $|H(j\omega)|$ and the sampled spectrum $|H(e^{j\Omega T})|$.

From the above, it is clear that in order for the adjacent response lobes in $|H(e^{j\Omega T})|$ not to overlap (i.e., alias) with each other, one must maintain $\omega_s = 2\pi f_s \geq 2\omega_B$, where ω_B is the significant bandwidth in the continuous-time-domain transfer function $H(s)$. This observation is consistent with the Nyquist rate sampling, as used in communication systems. In practice, a value of $\omega_s \geq 8\omega_B$ is employed. When f_s is fixed by other system constraints, the bandwidth of the incoming signal is restricted to a value less than $f_s/2$ by an antialiasing filter. This filter could be a simple passive-RC section. The band-limited signal is then subjected to the SC filter system. In cases where a continuous-time response is required, the SC filter output is smoothened by a continuous-time smoothing or reconstruction filter. The reconstruction filter must reject all the lobes in $|H(e^{j\Omega T})|$ beyond the frequency of ω_B. The system diagram is illustrated in Figure 6.5a. Since the antialiasing and reconstruction filters are simple RC networks, one can consider Figure 6.5b to get an idea of the frequency response characteristics of such filters. The passband of these filters must cover frequencies up to ω_B. The transition band of the filter will depend upon the separation of ω_s and ω_B, specifically on $\omega_s - \omega_B$. As Figure 6.5b indicates, one must have the TB $(\geq \omega_s - \omega_B) - \omega_B = \omega_s - 2\omega_B$ to avoid aliasing and to obtain a faithful reconstruction of the signal processed by the SC filter system.

Apart from the periodic characteristic of $H(z)$, there is another interesting characteristic of the sampled-data frequency variable $z = \exp(j\Omega T)$. As the frequency varies over the range $-\alpha \leq \Omega \leq \alpha$, z moves along the periphery of a circle of unit radius. Considering the complex frequency variable $s = \sigma + j\omega$ and $z = \exp(sT)$, the left-hand side (LHS) of the s-plane (i.e., $\sigma < 0$) is mapped into the interior of the circle of unit radius in the z-plane (i.e., $|z| < 1$) and the RHS of the s-plane (i.e., $\sigma > 0$) is mapped to the exterior of the unit circle in the z-plane (i.e., $|z| > 1$). The $j\omega$ axis in the s-domain is mapped on to the periphery of the unit circle in the z-domain (i.e., $|z| = 1$).

The various $s \leftrightarrow z$ transformations shown in Table 6.2 lead to similar characteristics related to $H(z)$ as discussed above. But the accuracy with which the $H(z)$ associated with a specific $H(s)$ matches the given $H(s)$ and varies with the type of $s \leftrightarrow z$ transformations chosen. The reason is the exponential relationship between the continuous-time and the sampled-data frequency variables, that is, s and z. Usually, an optimization algorithm may be required to improve the matching between a given analog $H(s)$ and the associated sampled-data $H(z)$, which results after applying a specific $s \leftrightarrow z$ transformation. Synthesizing the $H(z)$ corresponding to

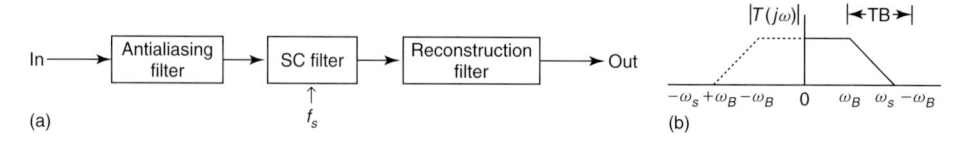

Figure 6.5 (a) Use of switched-capacitor filters in a practical system. (b) Frequency response characteristics of the antialiasing and reconstruction filters.

Table 6.2 Some well-known $s \leftrightarrow z$ transformations.

Transformation	Various integrators
$s \leftrightarrow \dfrac{1}{T}(z^{1/2} - z^{-1/2})$	Lossless digital integrator (LDI)
$s \leftrightarrow \dfrac{1}{T}(z - 1)$	Forward difference integrator (FDI)
$s \leftrightarrow \dfrac{1}{T}(1 - z^{-1})$	Backward difference integrator (BDI)
$s \leftrightarrow \dfrac{2}{T}\dfrac{1 - z^{-1}}{1 + z^{-1}}$	Bilinear integrator (BLI)

Table 6.3 Various steps in the design of an SC filter.

Given $A_p, A_a, f_p,$ and f_a
↓
Synthesize $H(s)$
↓
Apply $s \leftrightarrow z$ transformation ("prewarp" if possible) – see Section 6.3
↓
Derive the corresponding $H(z)$
↓
Implement $H(z)$ using SC and OA

an SC filter needs the application of one of the $s \leftrightarrow z$ transformation relations to a known $H(s)$, which is derived from the basic specifications for the filter, as already discussed in Chapter 3. A flow chart regarding this procedure is shown in Table 6.3.

6.3
Bilinear $s \leftrightarrow z$ Transformation

Of the several transformations, the BLT

$$s \leftrightarrow \frac{2}{T}\frac{1 - z^{-1}}{1 + z^{-1}} \tag{6.11}$$

is very popular because the entire analog frequency range of zero to infinity is transformed to a range of zero to π/T in the sampled-data frequency. This can be verified by putting $z = \exp(j\Omega T)$ in Eq. (6.11) to get

$$\omega = \frac{2}{T}\tan\frac{\Omega T}{2} \tag{6.12}$$

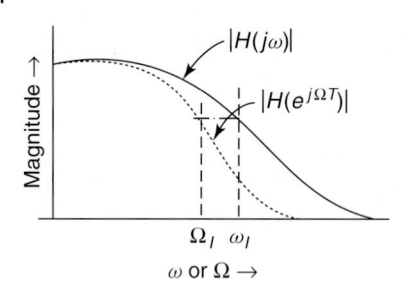

Figure 6.6 Effect of warping due to bilinear transformation on $H(s)$.

Thus, when $\omega = 0, \Omega = 0$, but when $\omega = \infty$, one gets $\tan \frac{\Omega T}{2} = \infty$; hence, $\Omega = \pi/T$. This also implies that the response $H(z)$ will be somewhat squeezed in the sampled-data frequency domain, if $H(z)$ were derived from $H(s)$ by the application of the BLT. This is known as the *warping* effect of the BLT. Thus, a given frequency ω_I in the analog domain is squeezed to Ω_I in the sampled-data domain with $\Omega_I < \omega_I$. Actually, $H(z)|_{z=\exp(j\Omega_I T)}$ becomes equal to $H(s)|_{s=j\omega_I}$. This is easily illustrated by Figure 6.6.

This warping effect can be compensated by what is known as *prewarping* the frequencies of the analog filter. Thus, if ω_I is a significant frequency (for example, the passband edge frequency) in $H(s)$, one can rewrite $H(s)$ with ω_I replaced by the prewarped frequency $\hat{\omega}_I$, given by

$$\hat{\omega}_I = \frac{2}{T} \tan\left(\frac{\omega_I T}{2}\right) \tag{6.13}$$

The corresponding transfer function may be labeled as $H(\hat{s})$ and is known as the *prewarped analog transfer function*. After application of the BLT to this new transfer function $H(\hat{s})$, one gets $H(z)$. The frequency Ω_I in $H(z)$ will then, by virtue of the BLT,

$$\Omega_I = \frac{2}{T} \tan^{-1}\left(\frac{\hat{\omega}_I T}{2}\right) = \frac{2}{T} \tan^{-1}\left(\frac{T}{2}\frac{2}{T}\tan\left[\frac{\omega_I T}{2}\right]\right) = \omega_I \tag{6.14}$$

which is same as the original frequency of interest. Thus, by prewarping, the effect of distortion due to the BLT is eliminated at the frequency ω_I. If there is more than one critical frequency in $H(s)$ (for example, the passband edge frequency ω_c and the stopband edge frequency ω_a), then the process of prewarping is applied to all these frequencies before obtaining the prewarped analog transfer function $H(\hat{s})$. Table 6.4 presents the set of biquadratic transfer function expressions in the sampled-data domain associated with the corresponding analog transfer function expressions (Raut, 1984), when the BLT is used.

Example 6.1. Consider the second-order BP analog transfer function given by $H(s) = \dfrac{2027.9s}{s^2 + 641.28s + 1.0528 \times 10^8}$. Obtain the corresponding sampled-data transfer function $H(z)$ using the BLT. The clock frequency is 8 kHz.

From the given $H(s)$, one can calculate the following:

$$\omega_p = \sqrt{1.0528 \times 10^8} = 1.026 \times 10^4, \quad \omega_p/Q_p = 641.28,$$

$$Q_p = 16, \quad H_{BP} = 2027.9/641.28 = 3.162 \rightarrow 10 \text{ dB}$$

Table 6.4 Sampled-data filter functions associated with standard biquadratic filter functions when the BLT is used.

Filter type	Analog transfer function $H(s) = \dfrac{N(s)}{F_1(s)}$	Sample-data transfer function $H(z) = h_D \dfrac{1 + a_{1N}z^{-1} + a_{2N}z^{-2}}{1 - a_{1D}z^{-1} + a_{2D}z^{-2}}$		
		h_D	a_{1N}	a_{2N}
LP	$H_{LP}\omega_p^2/F_1$	$H_{LP}\hat{\omega}_p^2/F_2$	2	1
HP	$H_{HP}s^2/F_1$	$H_{HP}a^2/F_2$	-2	1
BP	$H_{BP}(\omega_p/Q_p)s/F_1$	$H_{BP}(\hat{\omega}_p/Q_p)a/F_2$	0	-1
AP	$H_{AP}[F_1 - 2(\omega_p/Q_p)s]/F_1$	$H_{AP}a_{2D}$	$-a_{1D}/a_{2D}$	$1/a_{2D}$
Notch	$H_N(s^2 + \omega_n^2)/F_1$	$H_N(a^2 + \hat{\omega}_n^2)/F_2$	$-2(a^2 - \hat{\omega}_n^2)/(a^2 + \hat{\omega}_n^2)$	1

Note: $a = 2/T$, $\hat{\omega}_p = a\tan(\omega_p/a)$, $\hat{\omega}_n = a\tan(\omega_n/a)$.
$F_1 = s^2 + (\omega_p/Q_p)s + \omega_p^2$, $F_2 = a^2 + (\hat{\omega}_p/Q_p)a + \hat{\omega}_p^2$.
$a_{1D} = 2(a^2 - \hat{\omega}_p^2)/F_2$, $a_{2D} = [a^2 - (\hat{\omega}_p/Q_p)a + \hat{\omega}_p^2]/F_2$.

Since $f_s = 8$ kHz, $T = 125$ μs. Hence,

$$\hat{\omega}_p = 2f_s \tan(\omega_p/2f_s) = 1.1944 \times 10^4, \quad \hat{\omega}_p/Q_p = 746.488$$

Thus, the prewarped analog transfer function will be

$$H(\hat{s}) = \frac{H_{BP}(\hat{\omega}_p/Q_p)\hat{s}}{\hat{s}^2 + (\hat{\omega}_p/Q_p)\hat{s} + \hat{\omega}_p^2} = \frac{3.162 \times 746.488\hat{s}}{\hat{s}^2 + 746.488\hat{s} + 1.42659 \times 10^8}$$

One can now apply the BL $s \leftrightarrow z$ transformation to derive the various coefficients associated with the sampled-data transfer function $H(z)$. Thus,

$$h_D = 9.192 \times 10^{-2}, \quad a_{1D} = 0.5521, \quad a_{2D} = 0.9418$$

Therefore,

$$H(z) = \frac{9.192 \times 10^{-2}(1 - z^{-2})}{1 - 0.5521z^{-1} + 0.9418z^{-2}} \tag{6.15}$$

It may be remarked that one could skip the intermediate step of writing down $H(\hat{s})$ and get to $H(z)$ directly by applying the relations given in Table 6.4.

To calculate the magnitude response, one has to substitute $z = \exp(j\Omega T)$ and assign different values to the frequency variable Ω. For the problem on hand, $T = 125$ μs. The curves in Figure 6.7a depict the magnitude responses corresponding to $H(j\omega)$ and $H(e^{j\Omega T})$, as obtained by the MATLAB program. The listing of the program is given in Program 6.1. The differences between $|H(j\omega)|$ (solid line) and $|H(e^{j\Omega T})|$ (dotted line) at frequencies other than ω_p is due to the fact that the prewarping has been used at only one frequency, that is, at ω_p. If one would rather choose the -3 dB frequencies of the response function and apply prewarping to these two frequencies, the derived $H(e^{j\Omega T})$ would show a better matching with the given $H(j\omega)$.

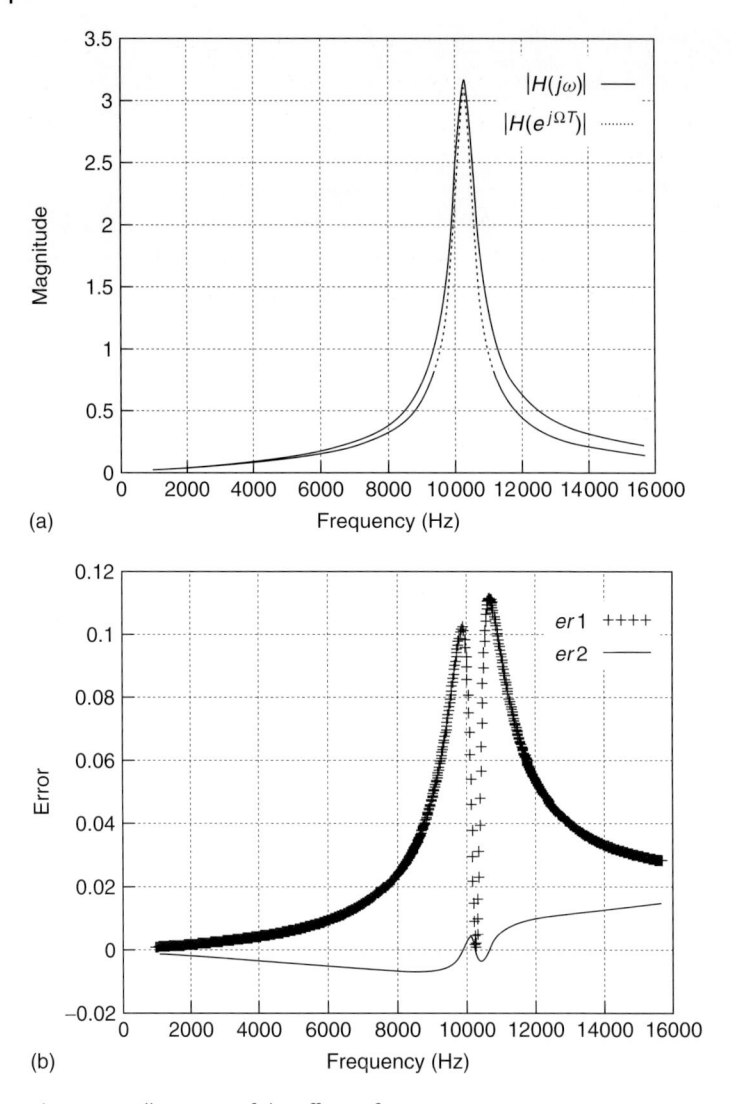

Figure 6.7 Illustration of the effects of prewarping:
(a) applied to the center frequency and (b) applied to the two-band edge frequencies of a band-pass filter.

To accomplish the above task, one first finds out that the bandwidth of the BP filter is $\omega_p/Q_p = 641.28$ rad s^{-1}. If ω_h and ω_l are the higher and lower -3 dB frequencies of the BP filter, then $\omega_h - \omega_l =$ bandwidth $= 641.28$. The geometric symmetry of a BP filter dictates that $\omega_h \times \omega_l = \omega_p^2 = 1.0528 \times 10^8$. These give (solving the quadratic equation), $\omega_h = 1.059 \times 10^4$ and $\omega_l = 9.945 \times 10^3$. We can now apply prewarping to these two frequencies. We then get $\hat{\omega}_h = 1.246 \times 10^4, \hat{\omega}_l = 1.146 \times 10^4$, thereby giving the prewarped bandwidth $\hat{\omega}_h - \hat{\omega}_l = 9.995 \times 10^2$. Remembering that the prewarped resonant frequency is now given by $\hat{\omega}_p^2 = \hat{\omega}_h \times \hat{\omega}_l = 1.428 \times 10^8$, the prewarped transfer function will become $H'(\hat{s}) = \frac{3161\hat{s}}{\hat{s}^2 + 999.5\hat{s} + 1.428 \times 10^8}$. On application of the BLT, the sampled-data transfer function will be

$$H'(z) = \frac{0.1219(1 - z^{-2})}{1 - 0.5455 z^{-1} + 0.9229 z^{-2}} \tag{6.16}$$

The curves in Figure 6.7b show the errors $er1 = |H(j\omega)| - |H(e^{j\Omega T})|$ (the curve with +++) and $er2 = |H(j\omega)| - |H'(e^{j\Omega T})|$ (the solid line curve) as a function of frequency. The error in $|H'(e^{j\Omega T})|$ relative to $|H(j\omega)|$ is less than that with $|H(e^{j\Omega T})|$, as expected.

Program 6.1 MATLAB code for the magnitude responses of $H(j\omega)$ and $H(e^{j\Omega T})$, and the error magnitudes

```
%sampled-data filter
Fs=8e3;
% analog filter response
a=[1 641.28 1.0528e8];
b=[0 2027.9 0];
f=linspace(150,2500,901);
w=2*pi*f;
h=freqs(b,a,w);
mag1=abs(h);
b=[9.192e-2 0 -9.192e-2];
a=[1 -.552 0.9418];
f=linspace(150,2500,901);
w=2*pi*f;
h=freqz(b,a,f,Fs);
mag2=abs(h);
b=[.1219 0 -.1219];
a=[1 -.5455 0.9229];
f=linspace(150,2500,901);
w=2*pi*f;
h=freqz(b,a,f,Fs);
mag3=abs(h);
er1=(mag1-mag2)/3.316;
er2=(mag1-mag3)/3.316;
%plot(w,mag1,'w-',w,mag2,'w.')
plot(w,er1,'w+',w,er2,'w-')
grid
xlabel('Freq.(Hz)-->');
%ylabel('Magnitude |H(s)|--, |H(z)|..');
ylabel('Error er1++, er2--');
end
```

6.4 Parasitic-Insensitive Structures

It has already been mentioned that SC filters received considerable attention because of the potential for fabrication as an IC. In this respect, however, a problem arises because of the presence of unwanted capacitances that are invariably present in an IC structure. Virtually, any two distinct semiconductor layers will have a capacitance associated with the junction between them. This is true both for the MOS transistors that are used as switches in an SC network, as well as for the MOS capacitors that have additional layers of semiconductors as part of the IC technological process flow. One can consider Figures 6.8a and 6.8b which depict the typical interlayer capacitances that arise in a MOS transistor and in a double-poly MOS capacitor (Schaumann, Ghausi, and Laker, 1990) respectively.

For the MOS transistor in Figure 6.8a, the bulk (substrate) is usually connected to a fixed DC potential so that this node can be considered to be signal ground (AC ground). The capacitors C_{bd1}, C_{bc1}, and C_{bs1} can be lumped as parasitic capacitances C_{SW} at the drain and source ends as shown on the RHS of Figure 6.8a. The parasitic capacitances (C_{gdo}, C_{gc}, C_{gso}, and C_{gb}) from the gate to drain, to channel, to source, and to bulk nodes are responsible for clock feedthrough effects. Similarly, in Figure 6.8b, the actual design capacitance C_R between the two poly layers is attended by parasitic capacitances such as C_j, C_m from the top plate, and C_b from the bottom plate. Because of closer proximity to signal ground (i.e., fixed DC potential), C_b has a higher value than C_j or C_m. The equivalent circuit on the RHS

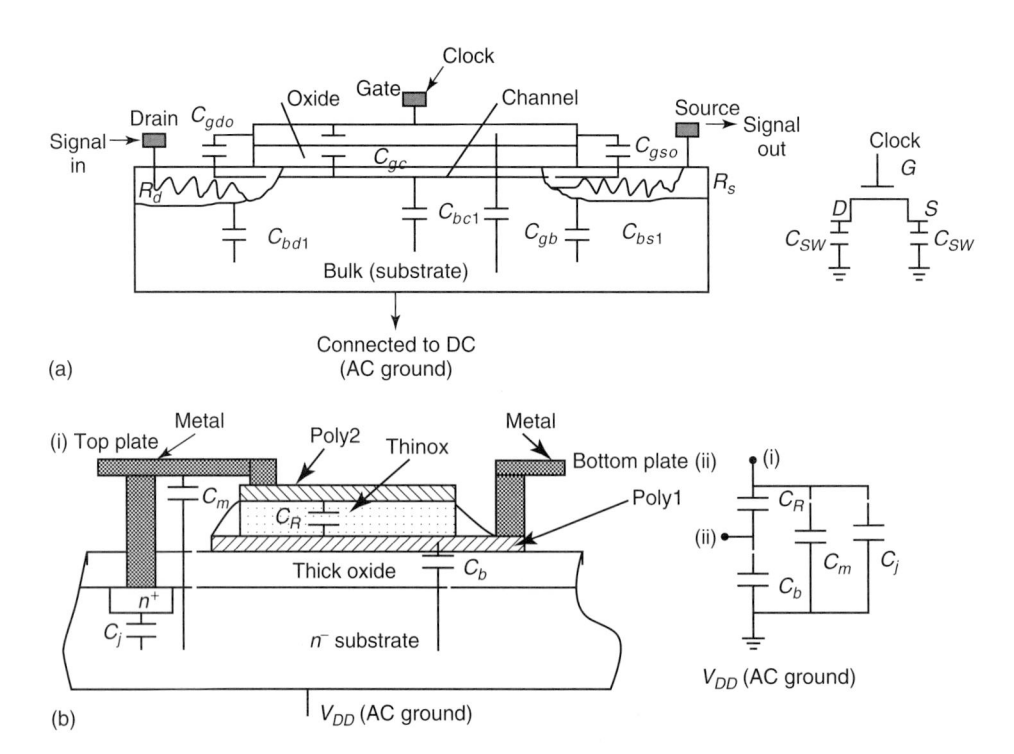

Figure 6.8 Parasitic capacitances in (a) an MOS transistor and (b) an MOS capacitor.

6.4 Parasitic-Insensitive Structures

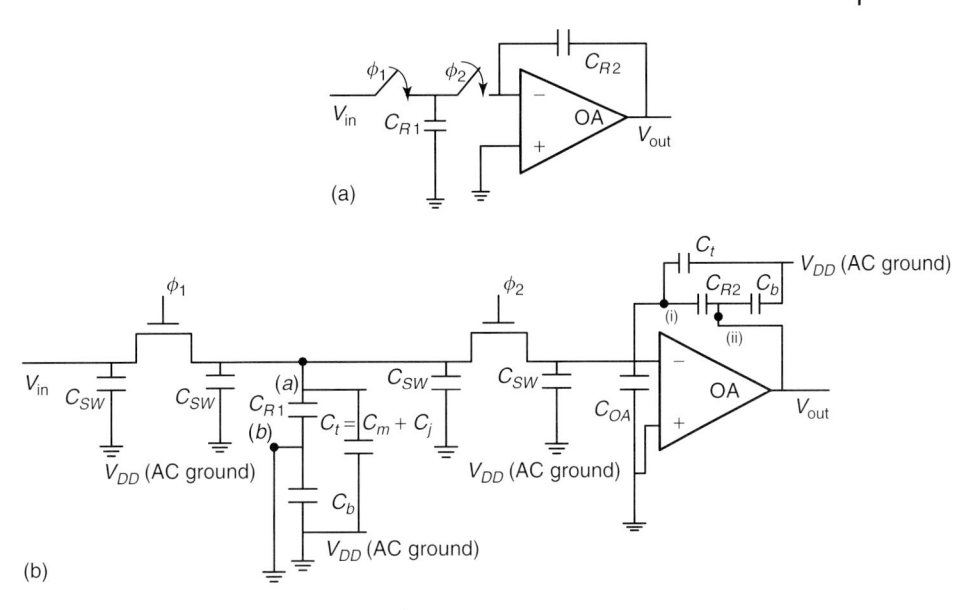

Figure 6.9 (a) A TSC integrator ignoring the parasitic capacitances. (b) The TSC integrator including the parasitic capacitances of the MOS switches and MOS capacitors.

of Figure 6.8b presents the situation for a design capacitance C_R in a typical MOS (or CMOS) IC technology.

If we incorporate the above information in a TSC integrator (see Figure 6.9a when no parasitic capacitances are assumed to be present), where a TSC of value C_{R1} is used to replace the resistance and C_{R2} is the integrating capacitor, the schematic including the parasitic capacitances will appear as shown in Figure 6.9b. In the schematic of Figure 6.9b, the switches are replaced with MOS transistors driven by appropriate clock signals (i.e., ϕ_1 and ϕ_2). Out of the several parasitic capacitances, those connected to a voltage source node (i.e., V_{in}, V_{out}) and those at the virtual ground input node of the OA can be ignored for signal processing since they either instantly hold the signal voltages or are at zero signal potential. The parasitic capacitances that are neither at a voltage source node nor at the virtual ground node contribute to the charge conservation equation (CCE) during the switching phases ϕ_1 and ϕ_2. A more simplified schematic obtained by considering only the important parasitic capacitances is shown in Figure 6.10. It is seen that the switch parasitic capacitances C_{SW} and the top-plate capacitance C_t of the MOS capacitance C_{R1} should be taken into consideration when writing the discrete-time equations such as Eqs. (6.3) and (6.4). This does not pose any theoretical problem, but the practical problem is the uncertainty about the exact values of the parasitic capacitances. The values are dependent upon the actual value of the design capacitance as well as on the layout of the transistors and the MOS capacitors. To save the space on the substrate, the value of C_R (see Figure 6.8b) should be limited to within 10 pF

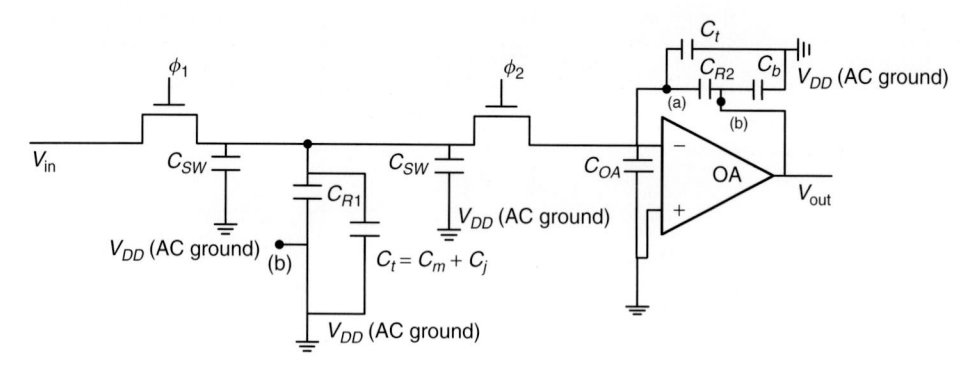

Figure 6.10 Simplified schematic of the TSC integrator of Figure 6.9b, showing only the important parasitic capacitances.

and an uncertain value of the parasitic capacitance, which could in some cases be as large as 80% of the design capacitance, makes accurate theoretical calculation rather difficult. This difficulty is overcome by the use of PI–SC networks that are described below (Gregorian and Nicholson, 1979; Mohan, Ramachandran, and Swamy 1982; Schaumann, Ghausi, and Laker, 1990).

6.4.1
Parasitic-Insensitive-SC Integrators

In PI-SC integrators, the parasitic capacitance is discharged to signal ground by the appropriate clock signal, before the design capacitance (i.e., which is replacing the resistance) acquires a new charge or shares its charge with another capacitor. Thus, while the parasitic capacitance is present physically, it does not contribute toward the CCE.

In the following, the switching symbol (wherever used) implies the presence of an MOS transistor operated by (i.e., driven at the gate terminal) a clock signal $\phi_i (i = 1, 2)$.

6.4.1.1 Lossless Integrators
Figure 6.11a shows a PI-SC integrator, known as the *inverting PI-SC integrator*. Prior to clock phase ϕ_1, the two plates of C_1 are discharged to signal ground, thereby removing all the charges on the parasitic capacitances C_{p1} and C_{p2}. Figure 6.11b shows a PI-SC integrator known as the *noninverting PI-SC integrator*.

It may be seen that, in Figure 6.11b, the left-side plate of C_2 is charged during ϕ_2, while the right-side plate and the parasitic C_{p2} are discharged to ground. During ϕ_1, a reverse operation takes place with C_2, since the left-side plate (and the parasitic C_{p1}) is now grounded, while the right-side plate gets connected to the input of the OA. Because of this signal reversal, accomplished by reversing the plates of C_2, the system produces a noninverting mode of operation.

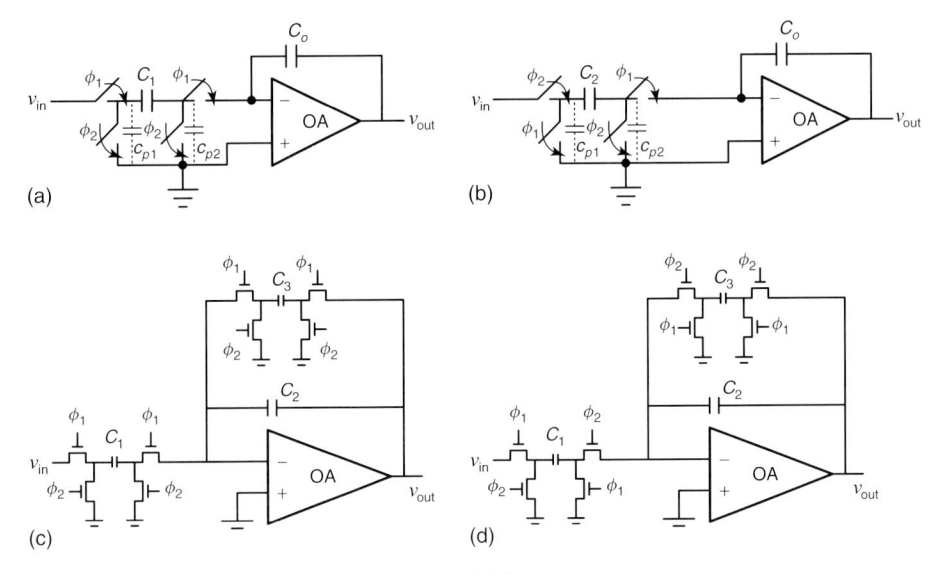

Figure 6.11 Parasitic-insensitive (PI) integrators: (a) lossless inverting, (b) lossless noninverting, (c) lossy inverting, and (d) lossy noninverting.

6.4.1.2 Lossy Integrators

Figures 6.11c and 6.11d show two lossy integrator configurations using PI-switching schemes. Basically, a lossy integrator is formed by connecting a PI-SC resistance across the integrating capacitor C_2. Both lossless and lossy integrators are important for active filter realizations. The inverting and noninverting lossy integrator structures are depicted in Figures 6.11c and 6.11d, respectively. The capacitor C_3 affords the lossy integrator operation.

In the following, we discuss the technique of analysis of PI-SC integrators in a modular way. The analysis can be used to derive the transfer functions of the PI-SC integrators, and can be easily extended to more complex SC networks involving PI features.

6.5 Analysis of SC Networks Using PI-SC Integrators

6.5.1 Lossless and Lossy Integrators

6.5.1.1 Inverting Lossless Integrator

Consider Figure 6.12a, which shows a PI-SC integrator with an input signal v_{in}. The path for v_{in} is the inverting mode of operation. The signal is switched to the OA input only during the ON phase of the clock ϕ_1. When the clock signals are either ON or OFF, the respective network topologies are linear time invariant; hence, the

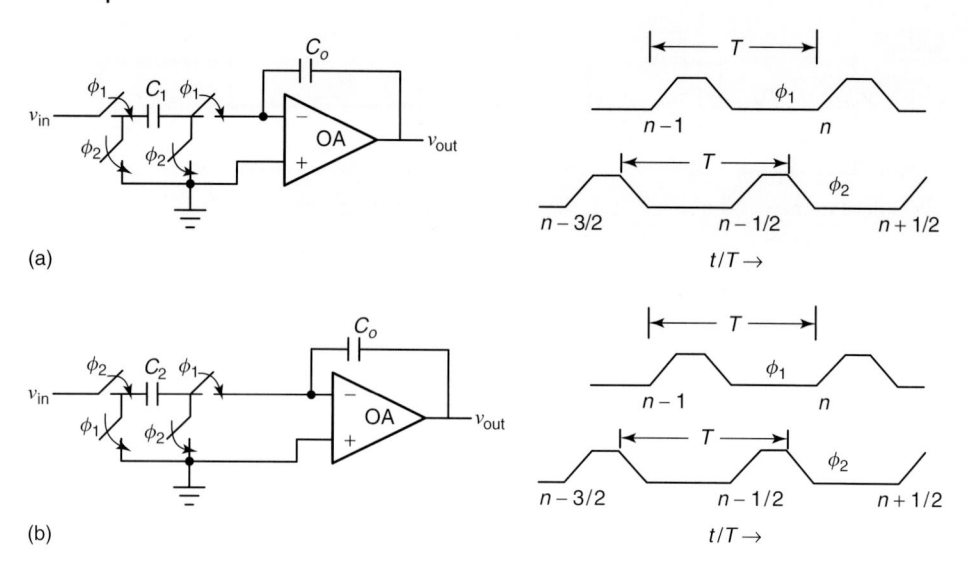

Figure 6.12 (a) Inverting lossless PI-integrator and (b) non-inverting lossless PI-integrator, along with the clock signals.

principle of superposition holds. Considering the discrete-time equation for the signals v_{in} and v_{out}, the CCE gives

$$0 - C_1 v_{in}(n-1) = C_o v_{out}(n-1) - C_o v_{out}(n-3/2)$$

Using the condition $v_{out}(n-3/2) = v_{out}(n-2)$, taking z-transform and simplifying, one arrives at

$$V_{out}^{(1)} = -\frac{C_1}{C_o} \frac{1}{1-z^{-1}} V_{in}^{(1)} \tag{6.17}$$

where $V_{out}^{(1)}$ and $V_{in}^{(1)}$ are the z-transforms of $v_{out}(n)$ and $v_{in}(n)$ respectively due to the clock signal ϕ_1. One may note the appearance of the negative sign in front of (C_1/C_o) on the RHS of Eq. (6.17) because of the inverting integration operation. Since v_{in} is processed through C_1 without any intermediate delay, and is connected to the inverting input of the OA, a sign inversion is inevitable. The term $(1-z^{-1})$ in the denominator of the RHS of Eq. (6.17) appears because of integration by sample-and-hold operation over one full clock period T. Because of the sample-and-hold operation, we can further derive

$$V_{out}^{(2)} = z^{-\frac{1}{2}} V_{out}^{(1)} \tag{6.18}$$

clock ϕ_2 being nominally one-half period delayed relative to clock ϕ_1. It should be noted that since C_o receives a new packet of charge each time the clock ϕ_1 comes ON, the signal v_{out} is available in reality during phase 1. Hence, we use $V_{out}^{(1)}$ as the primary variable in Eq. (6.17).

If the clock signals in Figure 6.12a are interchanged, we can similarly obtain

$$V_{out}^{(2)} = -\frac{C_1}{C_o} \frac{1}{1-z^{-1}} V_{in}^{(2)} \tag{6.19}$$

and
$$V_{\text{out}}^{(1)} = z^{-\frac{1}{2}} V_{\text{out}}^{(2)} \tag{6.20}$$

6.5.1.2 Noninverting Lossless Integrator

Consider Figure 6.12b, with an input signal v_{in}. Because of the reversal of the plates of C_2 between clock phases, the CCE becomes

$$0 - (-C_2 v_{\text{in}}(n - 3/2)) = C_o \left[v_{\text{out}}(n-1) - v_{\text{out}}(n - 3/2) \right]$$

But $v_{\text{out}}(n - 3/2)$ is the same as $v_{\text{out}}(n-2)$ sampled to and held on C_o. In the z-transform domain, this implies $z^{-\frac{3}{2}} V_{\text{out}}^{(2)} = z^{-2} V_{\text{out}}^{(1)}$, where $V_{\text{out}}^{(1)}$ is the z-transform of $v_{\text{out}}(n)$ for the clock signal ϕ_1, and $V_{\text{out}}^{(2)}$ is the z-transform of $v_{\text{out}}(n)$ owing to clock signal ϕ_2. Taking z-transform on both sides and relating $v_{\text{out}}(n - 3/2)$ with $v_{\text{out}}(n-2)$ as mentioned above, we get

$$C_2 z^{-3/2} V_{\text{in}}^{(2)} = C_o z^{-1} V_{\text{out}}^{(1)} - C_o z^{-3/2} V_{\text{out}}^{(2)} = C_o z^{-1} V_{\text{out}}^{(1)} - C_o z^{-2} V_{\text{out}}^{(1)}$$

where $V_{\text{in}}^{(2)}$ is the z-transform of $v_{\text{in}}(n)$ due to the clock signal ϕ_2. Dividing both sides by z^{-1} and simplifying, one gets

$$V_{\text{out}}^{(1)} = \frac{C_2}{C_o} \frac{z^{-1/2}}{1 - z^{-1}} V_{\text{in}}^{(2)} \tag{6.21}$$

One may note that since the signal v_{in} in this case is processed for noninverting integration, there is no negative sign in front of (C_2/C_o) on the RHS of Eq. (6.21). Further, since v_{in} is held on C_2 during the ON period of ϕ_2 and switched on to the input of the OA during the ON period of ϕ_1, that is, an interval of $T/2$ later, the term $z^{-1/2}$ appears in the numerator of the RHS of Eq. (6.21). This signifies one-half period delay in the signal path. The denominator $(1 - z^{-1})$ in Eq. (6.21) is a characteristic of the integration and appears in view of the sample-and-hold operation over a full period T by the integrating capacitor C_o. Because of the time relationship between ϕ_1, and ϕ_2, we can write as before,

$$V_{\text{out}}^{(2)} = z^{-\frac{1}{2}} V_{\text{out}}^{(1)} \tag{6.22}$$

If the clock signals in Figure 6.12b are interchanged, we can similarly obtain

$$V_{\text{out}}^{(2)} = \frac{C_2}{C_o} \frac{z^{-\frac{1}{2}}}{1 - z^{-1}} V_{\text{in}}^{(1)} \tag{6.23}$$

and

$$V_{\text{out}}^{(1)} = z^{-\frac{1}{2}} V_{\text{out}}^{(2)} \tag{6.24}$$

6.5.1.3 Inverting and Noninverting Lossless Integration Combined

Consider Figure 6.13a, where signal v_1 is processed for inverting integration, while signal v_2 is processed for noninverting integration. Equations (6.17) and (6.21) were derived assuming the presence of only one signal at a time. When both v_1 and v_2

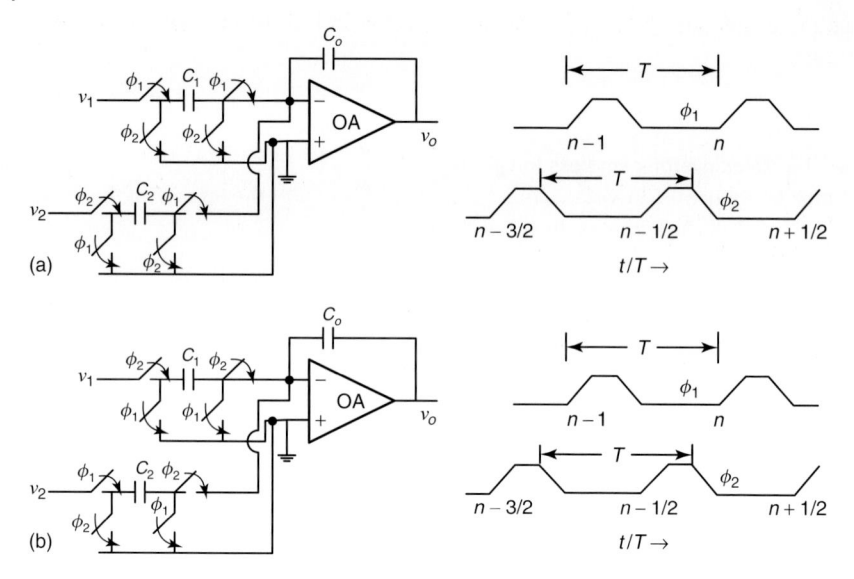

Figure 6.13 Parasitic-insensitive (PI) inverting and noninverting integrators combined: OA input connected to the signal paths during (a) clock phase ϕ_1 and (b) clock phase ϕ_2.

are operating simultaneously, one can invoke the superposition principle of linear time-invariant networks and write

$$V_o^{(1)} = \frac{C_2}{C_o} \frac{z^{-1/2}}{1-z^{-1}} V_2^{(2)} - \frac{C_1}{C_o} \frac{1}{1-z^{-1}} V_1^{(1)} \quad (6.25)$$

In the above, $V_x^{(y)}$ ($x = 1, 2$ and $y = 1, 2$) represent the z-transformed variables pertaining to the signals v_x ($x = 1, 2$) due to the two clock signals ϕ_1 and ϕ_2. Equation (6.25) represents the z-transformed output signal, when two signals v_1 and v_2 are fed to the OA input. Notice that, although the two signals are processed by the clock signals in different ways, they are connected to the input of the OA at the same time (i.e., clock phase ϕ_1). Since the signals are switched on to the OA input during clock ϕ_1, the significant output signal V_o is labeled with (1) as the superscript.

If the clock phases in the paths for v_1 and v_2 are switched around (see Figure 6.13b) with the OA input being connected to the signal paths during the clock phase ϕ_2, we could similarly derive

$$V_o^{(2)} = \frac{C_2}{C_o} \frac{z^{-1/2}}{1-z^{-1}} V_2^{(1)} - \frac{C_1}{C_o} \frac{1}{1-z^{-1}} V_1^{(2)} \quad (6.26)$$

Note that, since now the signals are switched to the input of the OA during the clock signal ϕ_2, the significant signal at the output is recognized as $V_o^{(2)}$. The reader is advised to practice writing down similar equations with other combinations of clock phasing and signal positioning. It is obvious that the

6.5.1.4 Lossy PI-SC Integrator

Referring to Figure 6.11c, which depicts an inverting lossy integrator, we can visualize that the signals v_{in} and v_{out} are switched to the OA input via the PI-SC elements C_1, and C_3. We can thus use Eq. (6.17), once for $V_{in}^{(1)}$, and then for $V_{out}^{(1)}$, followed by superposition. Hence, we get

$$V_{out}^{(1)} = -\frac{C_1}{C_2}\frac{1}{1-z^{-1}}V_{in}^{(1)} - \frac{C_3}{C_2}\frac{1}{1-z^{-1}}V_{out}^{(1)} \tag{6.27}$$

The above equation leads to the transfer function

$$\frac{V_{out}^{(1)}}{V_{in}^{(1)}} = \frac{-C_1/(C_2+C_3)}{1-C_2 z^{-1}/(C_2+C_3)} \tag{6.28}$$

The above is an *inverting* lossy integrator transfer function.

Now, for the circuit of Figure 6.11d, note that v_{in} is sampled in during ϕ_1 but applied to the OA input during ϕ_2, with a phase reversal across the capacitor C_1. Therefore, we can use Eq. (6.23) for $V_{in}^{(1)}$. Similarly, we can use Eq. (6.19) for $V_{out}^{(2)}$. Using superposition thereafter, we get

$$V_{out}^{(2)} = \frac{C_1}{C_2}\frac{z^{-\frac{1}{2}}}{1-z^{-1}}V_{in}^{(1)} - \frac{C_3}{C_2}\frac{1}{1-z^{-1}}V_{out}^{(2)} \tag{6.29}$$

The above leads to the transfer function of the noninverting lossy integrator

$$\frac{V_{out}^{(2)}}{V_{in}^{(1)}} = \frac{C_1 z^{-1/2}/(C_2+C_3)}{1-C_2 z^{-1}/(C_2+C_3)} \tag{6.30}$$

6.5.2 Application of the Analysis Technique to a PI-SC Integrator-Based Second-Order Filter

Consider the Tow–Thomas (TT) biquad structure of Chapter 5 (Figure 5.16) with $R_4 = R/k$, $R_3 = R_2 = R$, $R_1 = QR$, and $r_1 = r_2 = r$. This leads to LP and BP filter realizations using the RC-active filter configuration. The third OA in the TT network serves to provide a gain of -1. This function can be easily obtained in the SC version by employing a phase reversed SC network, as used in the noninverting PI-SC integrator. This technique saves one OA in the SC version, which is presented in Figure 6.14. Our objective is to derive the z-domain transfer function $V_{o2}^{(2)}/V_i^{(1)}$, using the analysis technique discussed above.

In order to apply the analysis technique presented in Section 6.5.1, one needs to observe that, around the amplifier OA1, the (kC, C_1) pair forms an inverting integrator for the signal v_i, the (C/Q, C_1) pair forms an inverting integrator for v_{o1}, and the (C, C_1) pair forms a noninverting integrator for v_{o2}. Similarly, for the amplifier OA2, the pair (C, C_2) forms an inverting integrator for the signal v_{o1}.

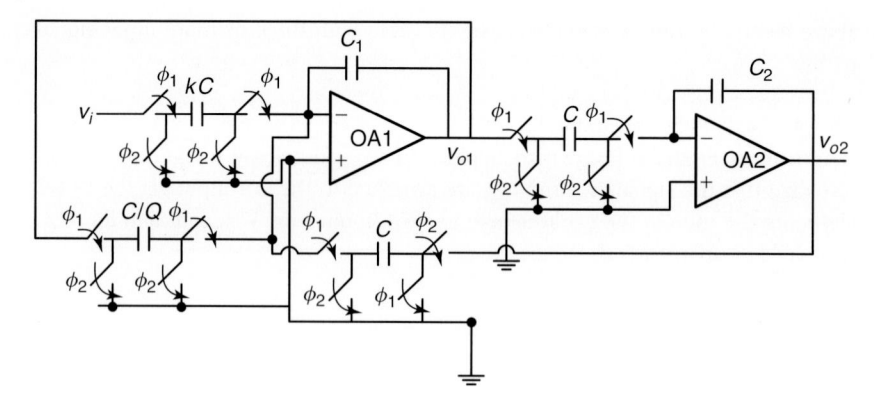

Figure 6.14 A second-order switched-capacitor filter using PI integrators.

With the above observations, one can write the following:

$$V_{o1}^{(1)} = \frac{-C/Q}{C_1} \frac{1}{1-z^{-1}} V_{o1}^{(1)} - \frac{kC}{C_1} \frac{1}{1-z^{-1}} V_{in}^{(1)} + \frac{C}{C_1} \frac{z^{-1/2}}{1-z^{-1}} V_{o2}^{(2)}$$

$$V_{o1}^{(1)} = \frac{-C/Q}{C_1} \frac{1}{1-z^{-1}} V_{o1}^{(1)} - \frac{kC}{C_1} \frac{1}{1-z^{-1}} V_{in}^{(1)} + \frac{C}{C_1} \frac{z^{-1/2}}{1-z^{-1}} V_{o2}^{(2)}$$

On rearranging and substituting $A = C/QC_1$, $B = kC/C_1$, and $D = C/C_1$ in the above equation, we get

$$V_{o1}^{(1)}\left(1 + \frac{A}{1-z^{-1}}\right) = -\frac{B}{1-z^{-1}} V_{in}^{(1)} + D\frac{z^{-1/2}}{1-z^{-1}} V_{o2}^{(2)} \qquad (6.31)$$

For the subnetwork around OA2, we can similarly write

$$V_{o2}^{(1)} = -\frac{C}{C_2} \frac{1}{1-z^{-1}} V_{o1}^{(1)} \qquad (6.32)$$

Writing $E = C/C_2$, and noting that $v_{o2}(n-1/2) = v_{o2}(n-1)$, in view of the sample-and-hold operation of C_2, we get

$$V_{o1}^{(1)} = -\frac{1-z^{-1}}{E} V_{o2}^{(1)} \qquad (6.33a)$$

and

$$V_{o2}^{(2)} = z^{-\frac{1}{2}} V_{o2}^{(1)} \qquad (6.33b)$$

Substituting Eq. (6.33a,b) in Eq. (6.31), we have

$$\frac{V_{o2}^{(1)}}{V_{in}^{(1)}} = \frac{BE}{1+A-(A-DE+2)z^{-1}+z^{-2}} \qquad (6.34)$$

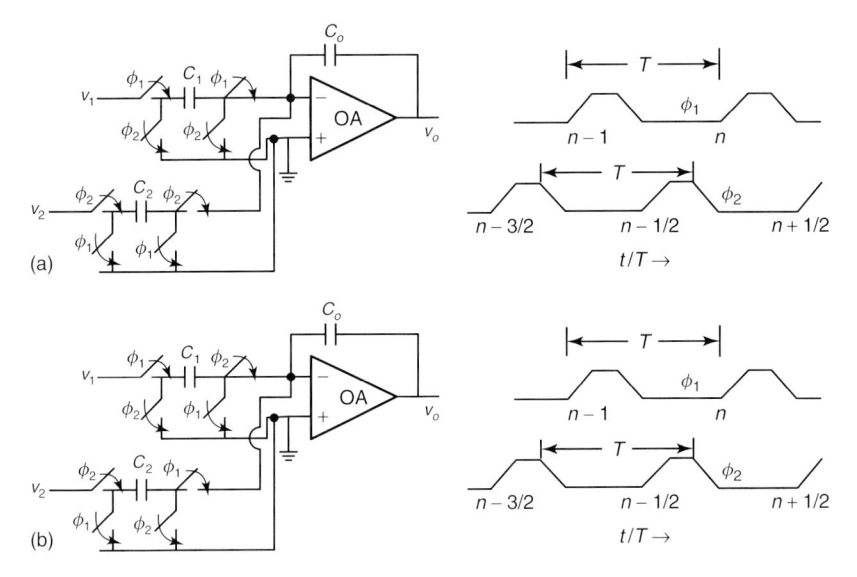

Figure 6.15 (a) PI inverting integrator having signals switched to the OA input during both the clock phases (ϕ_1 and ϕ_2). (b) PI noninverting integrator having signals switched to the OA input during both the clock phases (ϕ_1 and ϕ_2).

6.5.3
Signals Switched to the Input of the OA during Both Phases of the Clock Signal

Consider Figure 6.15a, which represents this case for two input signals v_1 and v_2. One should note that under the present case the condition $v_o(n - 3/2) = v_o(n - 2)$ cannot be applied any more. One can write the equations:

$$-C_1 v_1(n-1) = C_o v_o(n-1) - C_o v_o(n - 3/2)$$

and

$$-C_2 v_2(n - 1/2) = C_o v_o(n - 1/2) - C_o v_o(n - 1)$$

On taking the z-transforms and simplifying the above equations, we get

$$C_o V_{out}^{(1)} - z^{-1/2} C_o V_{out}^{(2)} = -C_1 V_1^{(1)} \tag{6.35a}$$

and

$$-z^{-1/2} C_o V_{out}^{(1)} + C_o V_{out}^{(2)} = -C_2 V_2^{(2)} \tag{6.35b}$$

One can put the above in the following compact form:

$$\begin{bmatrix} C_o & -z^{-1/2} C_o \\ -z^{-1/2} C_o & C_o \end{bmatrix} \begin{bmatrix} V_{out}^{(1)} \\ V_{out}^{(2)} \end{bmatrix} = - \begin{bmatrix} C_1 V_1^{(1)} \\ C_2 V_2^{(2)} \end{bmatrix}$$

$$\tag{6.36}$$

In Figure 6.15a, the SC networks are configured to produce an inverting integration operation. If these are reconfigured to produce a noninverting integration operation (see Figure 6.15b), one could similarly derive

$$\begin{bmatrix} -z^{-1/2}C_o & C_o \\ C_o & -z^{-1/2}C_o \end{bmatrix} \begin{bmatrix} V_{out}^{(1)} \\ V_{out}^{(2)} \end{bmatrix} = z^{-1/2} \begin{bmatrix} C_1 V_1^{(1)} \\ C_2 V_2^{(2)} \end{bmatrix} \quad (6.37)$$

6.6
Analysis of SC Networks Using Network Simulation Tools

When the network on hand is large and complex, hand analysis becomes very laborious and prone to errors (Nelin, 1983). The analytical effort becomes even more complex for SC networks because of the presence of the clock phases, ϕ_1 and ϕ_2. If one models the MOS switches as voltage-controlled switches, it will be possible to simulate a given SC network in the time domain, using the well-known simulation programs such as SPICE, PSpice, HSPICE, and so on. For frequency domain analysis, however, this approach may be too time consuming since measurements have to be taken at many frequencies. Assuming the switches as ideal, however, leads to a simple method of modeling an SC network using standard components in the simulation routine. These components are ideal delay lines and voltage amplifiers. The procedure is discussed below.

If one reviews the various z-domain expressions discussed in Section 6.4, one can conclude that these expressions have the general form: $V_{out} = kV_i + Az^{-1/2}V_{o1} + Bz^{-1}V_{o2} + \ldots$, where k, A, and B are constants, V_i, V_{o1}, and V_{o2} are the z-transformed voltage signals, and $z^{-1/2}, z^{-1}, \ldots$, imply delays by half-a-period, one full period, and so on. In simulation programs, such as SPICE, a voltage signal can be modeled as it is, the constant coefficients can be modeled as amplification (or attenuation) factors, and the delays can be modeled by transmission lines with a fixed delay. The simulation of the operation of a filter can then be carried out in the same way as simulating a system of equations in a state-variable filter using multipliers, adders, and integrators. The technique for the SC filter is illustrated in the following.

6.6.1
Use of VCVS and Transmission Line for Simulating an SC Filter

Consider the case of a PI inverting integrator transfer function $\dfrac{V_{out}^{(1)}}{V_i^{(1)}} = -\dfrac{C_1}{C_o}\dfrac{1}{1-z^{-1}}$.

Writing the constant $A = C_1/C_o$, we can write $V_{out}^{(1)} = -AV_i^{(1)} + z^{-1}V_{out}^{(1)}$. This is a simple feedback equation and can be modeled easily by the block diagram shown in Figure 6.16. This block diagram can be emulated as in Figure 6.17 using ideal voltage amplifiers (i.e., VCVS) and ideal transmission lines, each with a delay of one-half period of the clock signal.

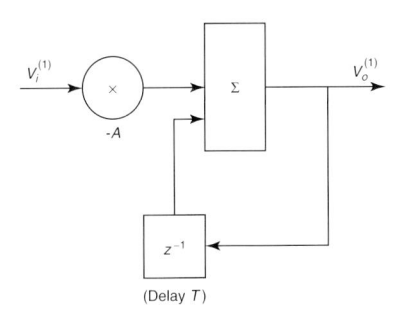

Figure 6.16 Block diagram realizing $V_o^{(1)} = -AV_i^{(1)} + z^{-1}V_o^{(1)}$ using basic delay, add, and multiply operations.

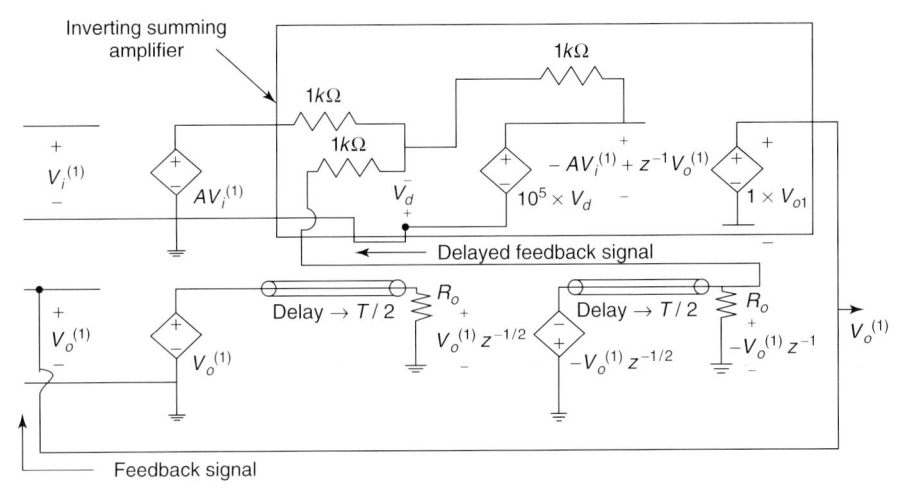

Figure 6.17 Implementing the delay, add, and multiply operations in Figure 6.16, using standard circuit elements in a SPICE program.

A systematic procedure to use this simulation technique will involve the following steps.

1) For the given SC filter, identify the SC subnetworks around each OA and write down the expressions relating the output voltage variable to the input voltage variable and the feedback voltage variables, in the form

$$V_{\text{out}} = kV_i + Az^{-1/2}V_{o1} + Bz^{-1}V_{o2} + \ldots$$

In case of feedback, one of the signals $V_{o1}, V_{o2}, \ldots$, will be same as V_o.

2) Use a half-period delay unit as the basic delay block and implement this with a lossless transmission line having a delay of $T/2$ (T = full period) isolated by a unit gain voltage amplifier buffer (i.e., an ideal VCVS of gain 1 or -1). See, Figure 6.17, for example.

3) Implement ideal VCVS networks with gain values corresponding to the coefficients in the equation derived in Step 1 above (i.e., $k, A, B, \ldots$).

4) Connect the delay blocks and the VCVS blocks in conformity with the expression derived in Step 1 above.

5) Repeat Steps 1–4 for the subnetworks around each OA in the SC filter. Ensure that the gain blocks have proper polarities with the gain so as to satisfy the system of equations correctly.
6) Write down the netlist file (or draw the schematic) for the entire network obtained after Step 5.
7) Run the simulation in accordance with the guidelines of the simulation program (such as, PSpice) that is available.

Example 6.2. Consider the TT biquad shown in Figure 6.18a. With the values shown, the filter will have a center frequency of $\omega_o = 1000$ rad/s and a Q_p of 5. An SC implementation of the filter using a clock frequency of $f_s = 10$ kHz is shown in Figure 6.18b. The netlist for simulating the response of the SC filter using the Pspice program is given in Program 6.2. The simulated frequency response is

Program 6.2 PSpice code for simulation of the response of the SC filter of Figure 6.18b

```
**SC filter simulation - TT-BIQUAD, ideal VCVS, switch
.subckt del 1 2 4
***clock period T=0.1ms
e1 3 0 1 2 1
t1 3 0 4 0 zo=50 td=.05m
rt 4 0 50
.ends del
vs 1 0 ac 1
e2 2 0 1 0 .1
r1 2 3 1k
r2 15 3 1k
r3 14 3 1k
r4 10 3 1k
e3 15 0 4 0 .02
e4 14 0 12 0 -1
e5 10 0 9 0 -.1
e6 16 0 9 0 -1
x1 4 0 11 del
x2 11 0 12 del
x3 7 0 8 del
x4 8 0 9 del
r5 3 4 1k
e7 4 0 0 3 1e4
ro1 4 0 1
e8 5 0 4 0 0.1
r6 5 6 1k
r7 16 6 1k
r8 6 7 1k
e9 7 0 0 6 1e4
ac dec 51 10 1k
print ac vm(4) vp(4)
probe
end
```

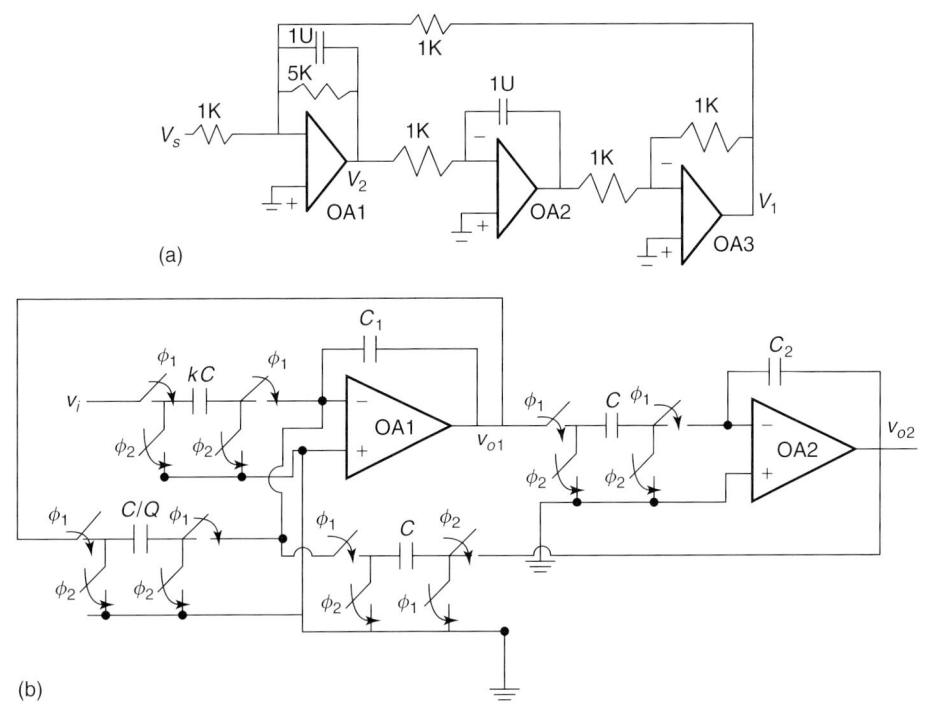

Figure 6.18 (a) Tow–Thomas active-RC second-order filter and (b) a corresponding SC filter using PI switched-capacitor integrators.

shown in Figure 6.19a. The time domain response at frequencies of 160, 100, and 200 Hz are shown in Figures 6.19b to 6.19d.

6.7
Design of SC Biquadratic Filters

In the ground-breaking era of SC filtering technique, the existing RC-active filters were converted to active SC filters simply by using capacitors equivalent to the resistors in the RC structures in accordance with the ideal relation $C = 1/Rf_s$, where f_s is the sampling clock frequency. However, with the knowledge of PI-SC structures and the requirement of using z-transform techniques for accurate prediction of the behavior of the SC filters, the design techniques leaned more toward realizing a given z-domain transfer function. Thus, instead of designing an SC filter equivalent to a parent active-RC filter, PI-SC networks and OAs were interconnected in special ways to realize standard second-order z-domain transfer functions (i.e., $H(z)$) such as those shown in Table 6.5. We now consider a popular structure proposed by Fleischer and Laker (1979).

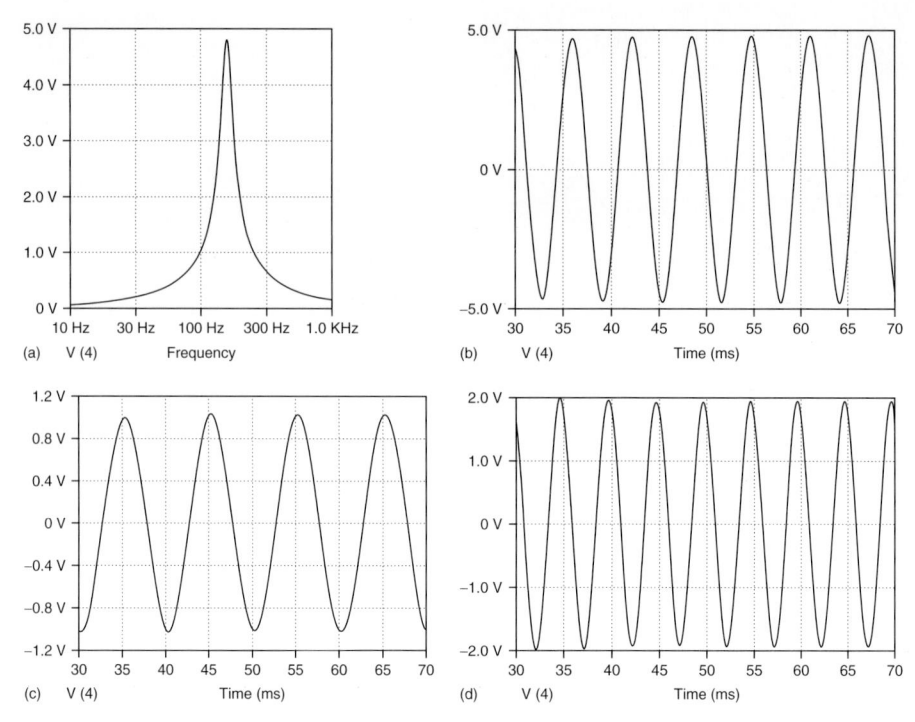

Figure 6.19 Response of the filter of Figure 6.18b using the technique presented in Section 6.6: (a) frequency-domain response, and time-domain response at (b) $f = 160$ Hz, (c) $f = 100$ Hz, and (d) $f = 200$ Hz, respectively.

6.7.1
Fleischer–Laker Biquad

The biquad in its most general form is shown in Figure 6.20. The switches are shown as MOS transistors operated by the two-phase clock signals ϕ_1 and ϕ_2 (Fleischer and Laker, 1979; Ghausi and Laker, 1981).

On carefully examining Figure 6.20, one can see that the forward paths (i.e., $V_i \to V_1 \to V_2$) in the network comprise the capacitors H, G, L, and D around OA1 and A, I, J, K, and B around OA2. The feedback paths (i.e., backwards from V_2) comprise the capacitors F around OA2 and C and E around OA1. The integrating capacitors are D and B. One can also figure out that the capacitor pair (H, D) provides noninverting integration, the capacitor pair (G, D) provides inverting integration, and the pairs (L, D), (E, D), and (K, B) provide simple voltage magnification (or attenuation). One can similarly determine the functions of the capacitors A, I, J, B, and F. Taking into consideration the above facts, one can apply the method of analysis discussed under Section 6.4 to derive the following second-order z-domain transfer functions:

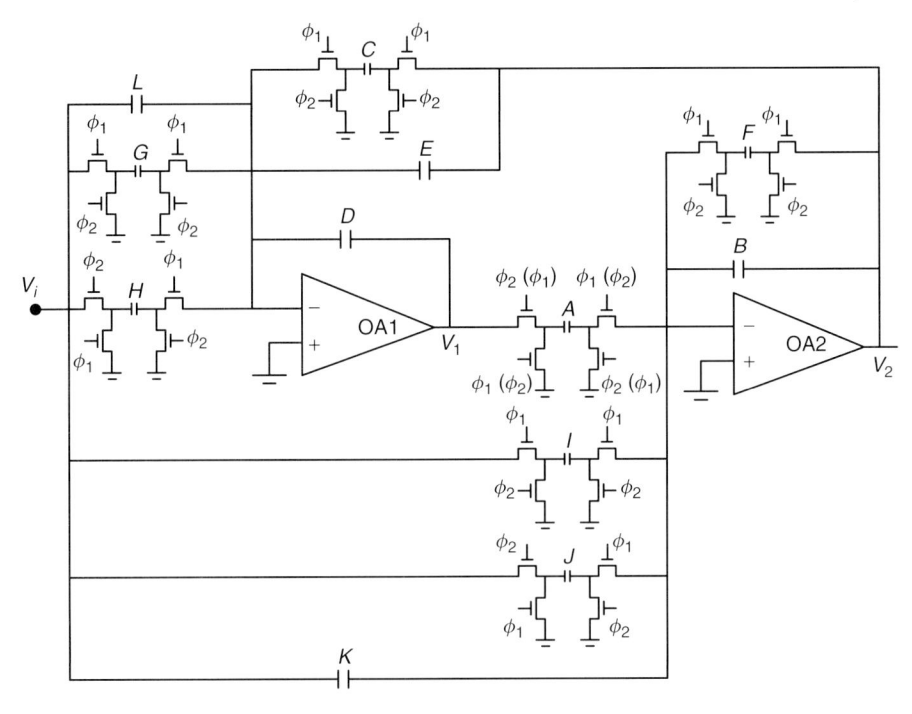

Figure 6.20 Fleischer–Laker SC second-order filter structure.

$$H_{2,1}(z) = \frac{V_1}{V_i}$$

$$= \frac{\left[zC + E(z-1)\right]\left[zI - J + K(z-1)\right] - \left[zF + B(z-1)\right]\left[zG - H + L(z-1)\right]}{A\left[zC + E(z-1)\right] + D(z-1)\left[zF + B(z-1)\right]}$$

(6.38)

$$H_{2,2}(z) = \frac{V_2}{V_i} = -\frac{A\left[zG - H + (z-1)L\right] + D(z-1)\left[zI - J + (z-1)K\right]}{A\left[zC + E(z-1)\right] + D(z-1)\left[zF + B(z-1)\right]}$$

(6.39)

Considerable simplification can be obtained by assuming the absence of the capacitors L, K, and F. In that case, one can get from Eq. (6.39),

$$H_{2,2}(z) = -\frac{DI + \left[AG - D(I + J)\right]z^{-1} + (DJ - AH)z^{-2}}{DB + \left[A(C + E) - 2DB\right]z^{-1} + (DB - AE)z^{-2}}$$

(6.40)

It may be remarked that in an SC network transfer function, the various coefficients (such as h_D and a_{1D}) are dependent upon the ratio of the capacitors in the network. Thus, one may normalize all the capacitors in terms of the integrating capacitors B and D by setting $B = D = 1$. On substituting this in Eq. (6.40), it is seen that the capacitor A remains as a free parameter, which can be chosen arbitrarily. Therefore, choosing $A = 1$, one can get the simplified form for $H_{2,2}(z)$ as

$$H_{2,2}(z) = -\frac{I + [G - (I + J)]z^{-1} + (J - H)z^{-2}}{1 + [(C + E) - 2]z^{-1} + (1 - E)z^{-2}} \quad (6.41)$$

The example below illustrates the use of the above expressions toward designing a second-order SC filter.

Example 6.3. Consider the realization of the second-order BP filter of Example 6.1. The sampled-data transfer function $H(z) = \frac{9.192 \times 10^{-2}(1-z^{-2})}{1-0.5521z^{-1}+0.9418z^{-2}}$ given by Eq. (6.15) is to be realized using the SC network of Figure 6.20. The simplified network function given by Eq. (6.41) can be used.

One can reorganize Eq. (6.41) in the form:

$$H_{2,2}(z) = -I\frac{1 + [G/I - (1 + J/I)]z^{-1} + (J/I - H/I)z^{-2}}{1 + [(C + E) - 2]z^{-1} + (1 - E)z^{-2}}$$

By comparing this with the specified $H(z)$, one can write the design equations:

$$I = 9.192 \times 10^{-2}$$
$$G/I - (1 + J/I) = 0$$
$$J/I - H/I = -1$$
$$C + E - 2 = -0.5521$$
$$1 - E = 0.9418$$
$$A = B = D = 1 \text{ each} \quad (6.42)$$

From the above set, one can immediately get $E = 0.0582$ and $C = 1.3897$. Further $G = I + J$ and $J - H = -I$. Hence, we have three unknowns, namely, G, J, and H with only two equations, and therefore we can choose $H = 1$. Then $J = 0.908$ and $G = 1$. Therefore, the design capacitances are now

$$A = B = D = 1, I = 0.09192, E = 0.0582, C = 1.3897, G = 1, H = 1, \text{ and } J = 0.908$$

We have to now focus on the practicality of the design values obtained above. A given IC technology has a limit on the minimum valued capacitor that can be reliably produced. If this value is, say 1 pF, the minimum design capacitance value must be equal or higher than 1 pF. In the above design, the capacitance $E = 0.0582$ is the capacitance of minimum value. Therefore, we need to use the principle of impedance scaling by bringing E to the value of 1 pF. Thus, the scale factor is $1/0.0582$ pF $= 17.182$ pF. Multiplying all the capacitors by this factor, the practical set of design capacitances now become

$$A = B = D = G = H = 17.182 \text{ pF}, \quad C = 23.878 \text{ pF},$$
$$E = 1 \text{ pF}, \quad I = 1.579 \text{ pF}, \quad \text{and} \quad J = 15.601 \text{ pF}$$

Note that there were nine capacitors (eight capacitance ratios) to be designed, but only five distinct design equations – see Eq. (6.42). Hence, we could make four free choices. These four choices could be conveniently used to satisfy further specifications in the design, such as minimum total capacitance, minimizing the capacitance spread, equalizing the dynamic range of the two stages, and so on.

6.7.2
Dynamic Range Equalization Technique

Dynamic range equalization implies that the gain levels of the two stages have to be so adjusted that the maximum output levels of V_1 and V_2 are equalized while realizing the specified overall gain level, for example, $|V_2/V_i|$ in the case of the biquad of Figure 6.20 (Ghausi and Laker, 1981). If T_2 is the maximum value of the gain $|V_2/V_i|$ over the frequency range of interest, while $T_1(\neq T_2)$ is the maximum value of the gain $|V_1/V_i|$ and if T_2 is the desired gain, then T_1 should be scaled by a factor $\mu = T_2/T_1$, that is, we need to set $T_1 = T_2/\mu$. If the gains are calculated in decibels scale, then $\mu =$ antilog $[(T_2 - T_1)/20]$. For the structure of Figure 6.20, if T_1 is to be scaled by μ without altering T_2, the only recourse left is to scale the capacitors that are connected to the output of stage 1 (i.e., to the node of V_1) by the same scaling factor. These capacitors are D and A. Thus, if the capacitor D is scaled to D/μ, then the gain T_1 will be increased by μ (since the voltage gain is proportional to $1/D$). If at the same time the capacitor A is scaled to A/μ, the gain $|V_2/V_1|$ will be reduced by μ (since for the second stage the voltage gain is proportional to A), thereby keeping the overall gain T_2 unchanged.

After the capacitors are adjusted for equalized dynamic range, impedance (or admittance) scaling is used, if necessary, to bring the capacitor of minimum value to be equal to the value allowed by the IC technological process. In order that the dynamic range adjustment does not get altered, impedance scaling is performed on the group of capacitors that are incident on the same input node of each of the active devices (i.e., OAs). Thus, in the case of Figure 6.20 (excluding K, L, F capacitors), these group of capacitors will be

- Group incident on OA1: (C, D, E, G, H)
- Group incident on OA2: (A, B, I, J).

It may be remarked that the detailed procedure for dynamic range adjustment is very dependent upon the case on hand, that is, it is specific to the network being designed. The discussion above provides some general guidelines. Detailed numerical calculations will be needed followed by careful considerations of the system level specifications to carry out the actual dynamic range adjustment procedure for a specific design case on hand.

6.8
Modular Approach toward Implementation of Second-Order Filters

Modularity is very attractive when implementing a system in an IC technology (Raut, Bhattacharyya, and Faruque, 1992). This makes the artwork for mask production more efficient and also facilitates the estimation of the substrate space

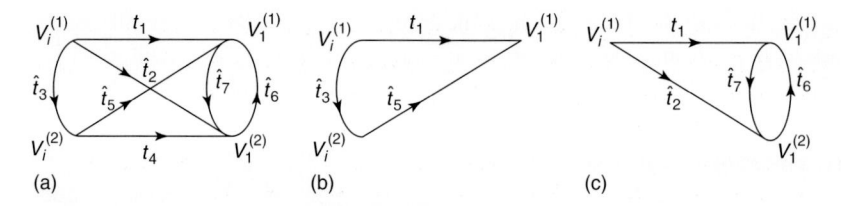

Figure 6.21 Signal flow graphs for a first-order sampled-data transfer function with biphase clocking: (a) most-connected graph, (b) and (c) possible subgraphs of (a).

that will be required for the implementation. In terms of hardware components, modularity implies presence of a known number of components associated with each active device and a repetition of the same interconnection pattern around the active devices in the path of signal processing. If one puts an additional constraint of only local connectivity, the resulting structure could be viewed as a systolic array architecture (SAA), which is considered to be very efficient for implementation of digital signal processing systems (Kung, 1982; Kuo, 1967). In the following, we present a modular approach for realizing a second-order SC filter transfer function in the z-domain. The method uses the technique of signal flow graphs (SFGs) and relies on the following observations.

1) The basic functions that are required in sampled-data filtering (SC filtering) are weighted addition, delay, and feedback. These are the same as those required in a digital filter. In an analog filtering case, the required operations are weighted addition, integration (or differentiation), and feedback.
2) Since $z = \exp(j\Omega T)$, z^{-k} implies a delay of kT or by k periods, since T is the sampling period. Thus $z^{-1/2}$ implies a half-period delay and z^{-1} implies a full-period delay.
3) The basic function to be carried out in SC filtering is to relate the z-transform of a voltage signal to the z-transforms of other voltage signals through a relationship of the form, $V_j = aV_i + bz^{-1/2} V_k + cz^{-1} V_m$, that is, a weighted sum of delayed and direct versions of the voltage signals. In case of a feedback, it is possible to have $V_k = V_j$, and so on.

The requirement in three of the above can be very easily appreciated by considering the SFGs shown in Figures 6.21a to 6.21c.

In Figure 6.21a, the input voltage source V_i, with phases (1) and (2) corresponding to the clock signals ϕ_1 and ϕ_2, is shown. $V_1^{(1)}$ and $V_1^{(2)}$ are the voltages at the output of the first OA at the clock phases ϕ_1 and ϕ_2. A direct gain path between the input voltage and the output voltage is shown by the branch $t_i (i = 1, 2, \ldots)$ without a hat. This represents a delay-free term like a capacitance ratio C_i/C_j in the SC network. A delay-free edge occurs between voltage nodes operated by the same clock phase (i.e., ϕ_1 or ϕ_2). A delayed gain path between V_i and V_1 is shown as $\hat{t}_k$ ($k = 1, 2, \ldots$), where the hat implies a half-period delay, that is, $\hat{t}_k = z^{-1/2} t_k$, where t_k is a term like C_k/C_j in the SC network. Such a path will exist, for example, in the case of a noninverting PI-SC integrator, where phase-reversal switching is used at the input capacitor. It must be noted that only forward paths exist from the input source node, that is, there cannot be any feedback path between V_1 and V_i.

Since in an SC integrator, the possibility of sample-and-hold operation exists, the delayed gain edges $\hat{t}_6$ and $\hat{t}_7$ are inserted between $V_1^{(1)}$ and $V_1^{(2)}$ to accommodate this scenario. Since these two edges join the voltage variables at the same node (i.e., output of the first OA), they represent simply terms like $z^{-1/2}$, that is, a delay by half a period. Since no switched path can precede the input source, only one delay edge (i.e., $\hat{t}_3$) is inserted between $V_i^{(1)}$ and $V_i^{(2)}$ to take care of the possibility of a sample-and-hold operation at the input voltage source. The direction of this edge is arbitrary. Figure 6.21a is the most complete (i.e., most connected) SFG for a first-order transfer function in the z-domain. This can be verified by the application of Mason's gains formula (Kuo, 1967) to derive

$$\frac{V_1^{(1)}}{V_i^{(1)}} = \frac{\left[t_1 + (t_2 t_6 + t_3 t_4 t_6 + t_3 t_5)\, z^{-1}\right]}{\left[1 - t_6 t_7 z^{-1}\right]} \tag{6.43}$$

We can easily see that the elimination of t_6 and/or t_7 will reduce the above transfer function to a first-order function. This could be used as a building block to generate, by cascading, a higher-order polynomial function of z^{-1}. This provides an example of realizing a finite impulse response (FIR) system. On examining the numerator of Eq. (6.43), one can observe that there is redundancy in the coefficient of the z^{-1} term. Thus, one or more edges could be removed and still a first-order transfer function in the sample data variable z can be realized. Removal of such edges will generate subgraphs such as the ones shown in Figures 6.21b and 6.21c. Each of these subgraphs will also produce a first-order transfer function in the form $(a + bz^{-1})/(c + dz^{-1})$. In terms of hardware implementation, removal of the edges amount to eliminating one or more capacitances in the SC network and will be preferred for a cost-effective realization.

The first-order SFG can be repeated and two such sections can be cascaded to generate a second-order transfer function in the z-domain. This is a modular approach toward implementation of a higher-order transfer function since repeated use is made of a module like the first order SFG. Since local (i.e., adjacent nodes) connectivity is allowed, both feed-forward and feedback gain paths are allowed between the voltage nodes V_1 and V_2. In an SC filter network, V_2 will be the voltage at the output of the second OA. The general second-order SFG is depicted in Figure 6.22. One can apply Mason's gain formula to derive the second-order transfer function $\frac{V_2^{(1)}(z)}{V_i^{(1)}(z)} = \frac{N(z)}{D(z)}$, where the expressions for $D(z)$ and $N(z)$ are given by

$$N(z) = \left[t_1 t_8 + (t_1 t_9 t_{15} + t_1 t_7 t_{13} t_{15} + t_1 t_7 t_{11} + t_2 t_3 t_{11} + t_2 t_4 t_8 + t_2 t_3 t_{13} t_{15}\right.$$
$$\left. + t_5 t_6 t_8 + t_5 t_{11} + t_5 t_{13} t_{15})\, z^{-1} + (t_2 t_4 t_9 t_{15} + t_2 t_4 t_7 t_{13} t_{15}\right.$$
$$\left. + t_2 t_4 t_7 t_{11} + t_2 t_3 t_6 t_9 t_{15} + t_5 t_6 t_9 t_{15})\, z^{-2}\right] \tag{6.44}$$

6 Switched-Capacitor Filters

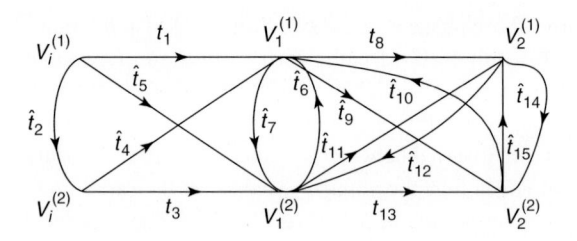

Figure 6.22 Most-connected signal flow graph for a second-order transfer function.

$$D(z) = \big[1 - \big(t_6 t_7 + t_{14} t_{15} + t_9 t_{10} + t_{11} t_{12} + t_6 t_8 t_{12} + t_7 t_{10} t_{13} + t_{12} t_{13} t_{15} + t_8 t_{10} t_{14}$$
$$+ t_8 t_{10} t_{12} t_{13}\big) z^{-1} + \big(t_6 t_7 t_{14} t_{15} + t_9 t_{10} t_{11} t_{12} - t_7 t_{10} t_{11} t_{14} - t_6 t_9 t_{12} t_{15}\big) z^{-2}\big]$$
(6.45)

On examining the various terms associated with the coefficients of z^{-1} and z^{-2} in $D(z)$ and $N(z)$, one can clearly see that there are considerable number of extra terms. Some of these could be discarded and yet a second-order transfer function obtained. A modular implementation plan should try to maintain equal number of terms from both the halves of the second-order SFG and a cost-effective realization should try to minimize the number of such terms. On studying the distribution of the edges in the two parts (namely, V_i to V_1 and V_1 to V_2) of the SFG very carefully, one can isolate nine possible groups, which will satisfy the above criteria. These are shown in Table 6.6 under the column labeled SAA. If one allows additional feed-forward edges running between nodes V_i and V_2, the condition of local (i.e., adjacent nodes only) connectivity is relaxed. Such a structure can be named semi-SAA (SEMI-SAA). The corresponding second-order SFG is shown in Figure 6.23.

Under this condition, we get several extra terms in the numerator functions compared to that of $N(z)$ in Eq. (6.34). Denoting the new numerator as $\overline{N}(z)$, one

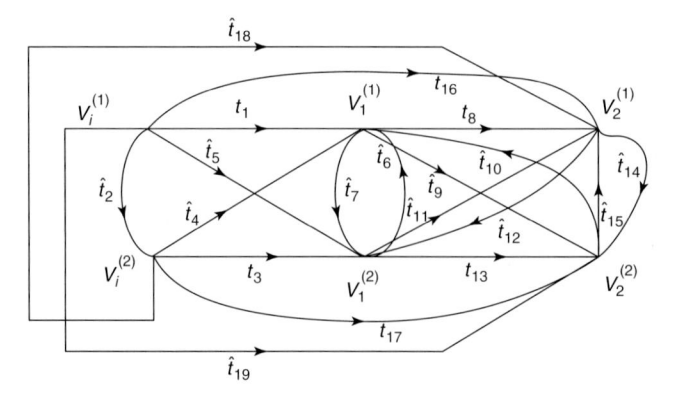

Figure 6.23 Signal flow graph with semisystolic architecture (i.e., including connections between nonadjacent nodes).

Table 6.5 Several possible edge groups for SAA and SEMI-SAA implementations.

Edge groups for SAA implementation			Edge groups for SEMI-SAA implementation		
Group #	Subgraph 1	Subgraph 2	Group #	Subgraph 1	Subgraph 2
1	1,5,6,7,10	8,9,14,15	10	1,2,5,6,7,10	8,9,14,15,17
2	1,5,6,7,10	8,9,13,14,15	11	1,2,4,6,7,10	8,13,14,15,19
3	1,5,6,7,10,11	8,9,14,15	12	2,5,6,7,10	8,9,14,15,16,17
4	1,2,3,5,6,7,10	8,9,13,14,15	13	2,4,6,7,10	8,13,14,15,16,19
5	1,5,6,10,11	8,9,12,14,15	14	1,2,5,6,7,10	8,9,14,15,19
6	1,2,4,6,7,10	8,9,13,14,15	15	1,2,3,6,7,10	8,9,14,15,18
7	1,2,4,6,7,10	8,13,14,15	16	2,5,6,7,10	8,13,14,15,16,19
8	1,2,3,6,7,10	8,9,13,14,15	17	2,3,6,7,10	8,9,14,15,16,18
9	1,2,3,6,7,10	8,9,14,15	18	1,2,4,6,7,10	8,13,14,15,18
			19	1,2,3,6,7,10	8,9,14,15,17
			20	2,4,6,7,10	8,13,14,15,16,18
			21	2,3,6,7,10	8,9,14,15,16,17
			22	1,2,4,6,7,10	8,13,14,15,17
			23	1,2,3,6,7,10	8,9,14,15,19
			24	2,4,6,7,10	8,13,14,15,16,17
			25	2,3,6,7,10	8,9,14,15,16,19

can derive

$$\overline{N}(z) = N(z) + (t_{15}t_{19} + t_8 t_{10} t_{19} + t_2 t_{15} t_{17} + t_2 t_8 t_{10} t_{17} + t_2 t_{18}) z^{-1} \quad (6.46)$$

Inclusion of these extra terms leads to additional structures that preserve the modularity feature with minimal number of elements in the SC network. These structures are listed under column SEM-SAA in Table 6.5. On the whole, we have 25 possible structures for implementing second-order SC filters. The architectures are very modular and each structure contains a minimum possible number of elements for practical implementation.

A lookup list of the SAA structures to implement the standard second-order filters is provided in Table 6.6. An example case is discussed below.

Table 6.6 Possible SAA edge groups for standard second-order filters.

Filter	SAA group # (see Table 6.5)
LP	2,3,4,5
HP	1,2,3,4,5
BP	4,6,7,8,9
AP	1,2,3,4,5
NOTCH	1,2,3,4,5

Example 6.4. Let us consider the case of implementation of the second-order BP filter with the following specifications.

Center-of-band gain: 10 dB
Center frequency: 1633 Hz
Pole Q: 16
Sampling clock frequency: 8 kHz.
Use bilinear $s \leftrightarrow z$ transformation for the SC filter implementation.

On analyzing the specifications, one can figure out that the analog domain transfer function of the filter is $H(s) = \frac{2027.9s}{s^2+641.28s+1.0528\times10^8}$. On applying the BL transformation to the two $-3\,\text{dB}$ frequencies, the prewarped analog transfer function becomes, $H(\hat{s}) = \frac{3161\hat{s}}{\hat{s}^2+999.5\hat{s}+1.428\times10^8}$. After applying the BL transformation, the z-domain transfer function becomes $H(z) = \frac{0.1219(1-z^{-2})}{1-0.5455z^{-1}+0.9229z^{-2}}$. We now consider realizing this transfer function by using the SAA structure 4 as suggested in Table 6.5 and listed in Table 6.6.

In SAA group 4, the edges to be considered are $t_1, t_2, t_3, t_5, t_6, t_7, t_{10}, t_8, t_9, t_{13}, t_{14}$, and t_{15}. At this point it will be good to assess the practical implications associated with the various terms that are to be designed. On considering the SFG, one can realize that the terms t_1, t_3, t_8, and t_{13} represent delay-free gains. For implementation using PI-SC integrators, we need to use inverting integrator networks. So, in reality, we need to design, say, $t_1 = -m_1, t_3 = -m_3, t_8 = -m_8$, and $t_{13} = -m_{13}$ as the capacitance ratios. The terms t_6 and t_7 are delay terms associated with voltage V_1, while t_{14} and t_{15} are delay terms associated with voltage V_2; the delay is by half a period. When we set any of these terms equal to unity, it implies a sample-and-hold operation by half a period. Thus, no extra capacitance is necessary to implement this case. When any of these terms is assigned a value not equal to unity, the practical realization will require a feedback path with half-a-period delay. This can be implemented with a noninverting integrator using PI-SC networks. All terms associated with half-period delay need to be implemented using PI noninverting SC integrator networks. Looking at the expressions for $N(z)$ and $D(z)$ in Eqs. (6.44) and (6.45) and comparing the coefficients of z^{-1} and z^{-2} with the coefficients of the standard form of $H(z)$ (see Table 6.4), we can write, for group 4

$$h_D = m_1 m_8$$
$$a_{1N} = \frac{1}{h_D}[-m_1 t_9 t_{15} + m_1 m_{13} t_7 t_{15} + t_2 m_3 m_{13} t_{15} - m_8 t_5 t_6 - m_{13} t_5 t_{15}]$$
$$a_{2N} = \frac{1}{h_D}[-t_2 m_3 t_6 t_9 t_{15} + t_5 t_6 t_9 t_{15}]$$
$$a_{1D} = t_6 t_7 + t_{14} t_{15} - t_7 t_{10} m_{13} - m_8 t_{10} t_{14}$$
$$a_{2D} = t_6 t_7 t_{14} t_{15} \tag{6.47}$$

There are only 5 equations, although 12 capacitance ratios have to be designed. Hence, many free choices are possible. As a first simplification, we shall attempt to put as many of the edges as possible to have the value of unity. This will enhance the accuracy for generating the masks for IC production and also lead toward a small value for the total capacitance area (Ghausi and Laker, 1981). Next, when grouped in a product form (i.e., $h_D = t_1 t_8, a_{2D} = t_6 t_7 t_{14} t_{15}$), we can make as many of them equal to one another as possible. This will be desirable for reliable mask generation and will also lead to the calculation of the filter coefficients more readily. On the basis of the discussion in the paragraph preceding Eq. (6.47), we choose $t_2 = 1, t_6 = t_7 = 1, t_1 = t_8 = -m_1, t_{14} = t_{15}$. We then get $t_1 = t_8 = -\sqrt{h_D} = -\sqrt{0.1219}, = -0.3491$, and $t_{14} = t_{15} = \sqrt{a_{2D}} = \sqrt{0.9229} = 0.9607$. The equations pertaining to a_{1N}, a_{2N}, and a_{1D} then lead to

$$-m_1 t_{14} t_9 + m_1 m_{13} t_{14} + m_3 m_{13} t_{14} - m_1 t_5 - m_{13} t_5 t_{14} = 0$$

$$-m_3 t_9 t_{14} + t_5 t_9 t_{14} = -h_D = -0.1219$$

$$1 + a_{2D} - m_{13} t_{10} - m_1 t_{10} t_{14} = a_{1D}$$

Clearly, the next stage of simplification will be to assume $t_5 = 0$. We then have

$$m_{13} = \frac{m_1 t_9}{m_1 + m_3}, \quad m_3 t_9 = \frac{h_D}{\sqrt{a_{2D}}}, \quad t_{10} = \frac{1 - a_{1D} + a_{2D}}{m_{13} + m_1 \sqrt{a_{2D}}}$$

We now have m_{13}, m_3, t_9, and t_{10} as unknowns, but only three equations. If we choose m_3, we get t_9, then we choose m_{13} to obtain t_{10}. By examining the relations above, it is clear that the terms are not related to one another in a linear manner. In such a case, it is advisable to look for a solution, which will minimize the sum of these terms (i.e., capacitance ratios). This will lead to savings in the substrate area for monolithic fabrication. By setting $\gamma_1 = h_D/\sqrt{a_{2D}}$, the sum of these capacitance ratio terms can be formulated as

$$S = m_3 + \gamma_1/m_3 + m_1 \gamma_1 / (m_3 m_1 + m_3^2) + \frac{1 - a_{1D} + a_{2D}}{m_1 \sqrt{a_{2D}} + m_1 \gamma_1 / (m_3 m_1 + m_3^2)} \tag{6.48}$$

It may be noted that $m_1 = -t_1 = 0.3491$, $\gamma_1 = 0.1269$, $1 - a_{1D} + a_{2D} = 1.3763$, and $a_{2D} = 0.9229$. A plot of S as a function of m_3 is shown in Figure 6.24. There is a minimum in S for $m_3 \approx 0.18$ and the minimum value is about 3.1. With this value for m_3, we then get, $t_9 = 0.705$, $m_{13} = 0.4652$, and $t_{10} = 1.7192$. The complete

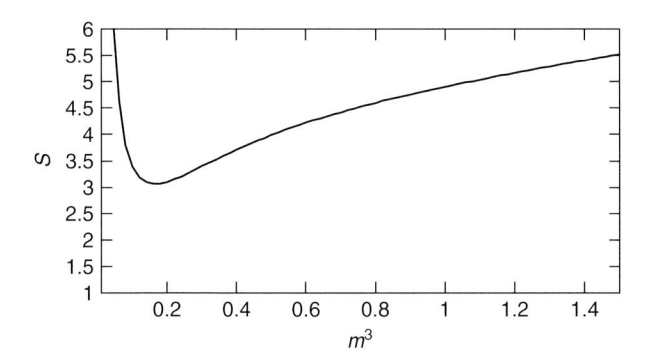

Figure 6.24 Total capacitance versus the parameter m leading to optimum m for minimum total capacitance design.

Figure 6.25 Signal flow graph associated with the solution of Example 6.4.

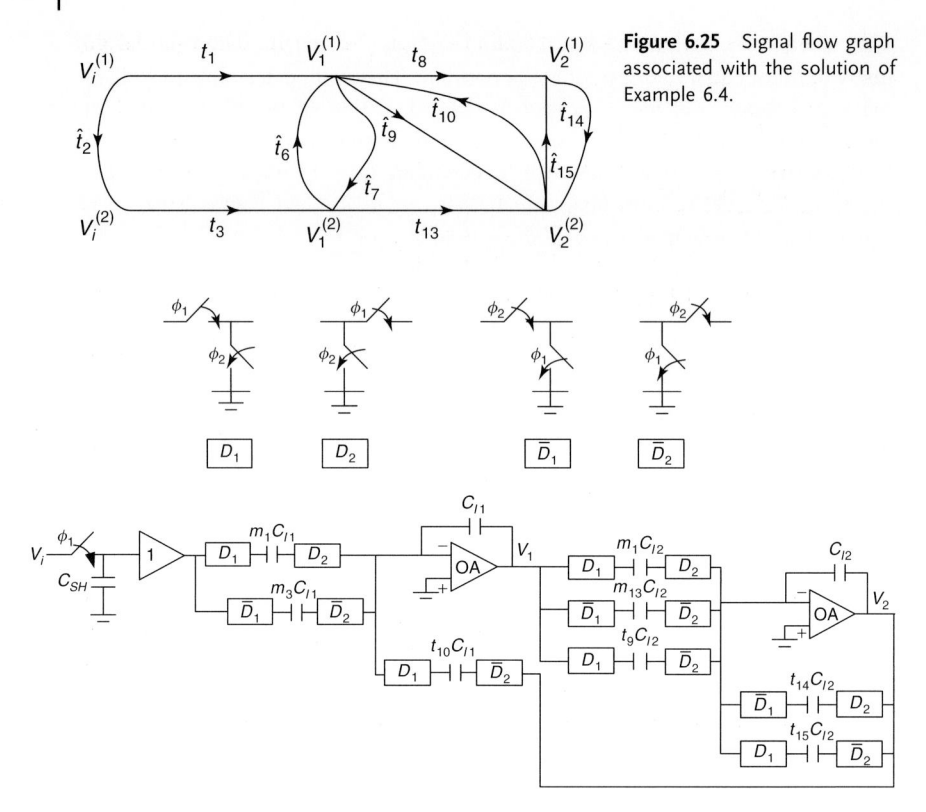

Figure 6.26 The biphase SC filter configuration corresponding to the graph in Figure 6.25.

solution is then

$$t_1 = -0.3491, t_2 = 1, t_3 = -0.18, t_5 = 0, t_6 = t_7 = 1, t_8 = -0.3491,$$
$$t_9 = 0.705, t_{10} = 1.7192, t_{13} = -0.4652, t_{14} = t_{15} = 0.9607$$

The SFG associated with this solution is shown in Figure 6.25.

Accordingly, the SC network involving only PI inverting and noninverting integrators are as shown in Figure 6.26. In the figure, $D1, D2, \overline{D}1$, and $\overline{D}2$ are the arrays of switches, as shown on the top part of Figure 6.26. To complete the design, one has to apply impedance scaling to bring the minimum capacitance equal to the level allowed by the IC technology on hand. Scaling to equalize the dynamic range can also be applied, if required.

Example 6.5. Consider the implementation of the design problem in Example 6.3 by using the SAA group 7 network.

In this case, the edges concerned are $t_1, t_2, t_4, t_6, t_7, t_{10}, t_8, t_{13}, t_{14}$, and t_{15}. Of these, t_1, t_8, and t_{13} are to be implemented using PI inverting integrators. Accordingly, we set $t_1 = -m_1, t_8 = -m_8$, and $t_{13} = -m_{13}$. By comparing with the coefficients of the standard sampled-data biquadratic function, we get the following design equations

$$t_1 t_8 = m_1 m_8 = h_D = 0.1219$$
$$m_1 m_{13} t_7 t_{15} - m_8 t_2 t_4 - m_8 t_2 t_4 t_6 = a_{1N} = 0$$
$$-\frac{t_2 t_4 t_7 m_{13} t_{15}}{h_D} = a_{2N} = -1$$
$$t_6 t_7 + t_{14} t_{15} - m_{13} t_7 t_{10} - m_8 t_{10} t_{14} = a_{1D} = 0.5455$$
$$t_6 t_7 t_{14} t_{15} = a_{2D} = 0.9229 \qquad (6.49)$$

We assume that $t_6 = t_7 = 1, t_{14} = t_{15}, t_1 = t_8,$ and $t_2 = 1$. This will lead to $t_1 = t_8 = -0.3491, t_4 = 0.2469, t_{14} = t_{15} = 0.9607,$ and $t_{10} = (1 - a_{1D} + a_{2D})/(m_{13} + \sqrt{h_D a_{2D}})$. The last relation provides an opportunity for minimizing the sum $t_{10} + m_{13}$. Using elementary calculus, we find the optimum value of m_{13} to be 0.838. Using this value for m_{13}, one gets $t_{10} = 1.1871$. Thus, the complete solution is

$$t_1 = -0.3491, t_2 = 1, t_4 = 0.2469, t_6 = t_7, t_{10} = 1.1871,$$
$$t_8 = -0.3491, t_{13} = -0.838, t_{14} = t_{15} = 0.9607$$

It may be interesting to compare the two designs regarding the total capacitance needed. This will be proportional to the sum $\sum_i |t_i|$, where i spans the design edges. For the design in Example 6.4, this sum is 5.689 units, while for the design in Example 6.5, the sum is 4.8916. Hence, the solution in Example 6.5 will require less area for implementation on the IC substrate.

6.9 SC Filter Realization Using Unity-Gain Amplifiers

An ideal unity-gain voltage amplifier has infinite input impedance, zero output impedance, and a voltage gain of unity between the input and output terminals (Fan et al., 1980; Malawka and Ghausi, 1980; Fettweis et al., 1980). Since the signal voltage difference between the input and output is zero, one can conceive of the presence of a virtual short circuit across the input and output nodes of a unity-gain voltage amplifier (UGA). In an OA-based SC filter, capacitors are charged to signal voltages in one clock phase and then are discharged at the next clock phase (in a system with two clock phases like ϕ_1 and ϕ_2) across the virtual short circuit at the input of the OA whose noninverting input terminal is grounded. At the same time, in the integrating capacitor across the input and output of the OA, which is connected to the same virtual ground node, a redistribution of the charge occurs, leading to the transfer of charge on to the integrating capacitor. The signal voltage on the integrating capacitor then attains a new value. This is how the signal is processed via an array of SCs and non-SCs in an SC network using an OA. Since a virtual ground exists across the input–output nodes of a UGA, similar signal processing should, in principle, be possible using an array of SCs and non-SCs, and a UGA.

The interest in implementing UGA-based SC filtering arose in view of the following reasons: (i) A UGA could be implemented with fewer transistors than those in an OA and thus would require less substrate area and less DC power, and (ii) since the voltage gain is unity, the resulting bandwidth available will be lot more than that available in an OA. This will make it possible for the implementation of SC filters at frequencies higher than that possible with OAs. In the following, we discuss a few important building blocks based on UGAs. These blocks can be suitably interconnected to produce a general z-domain transfer function of any arbitrary order and in particular, of second order.

6.9.1
Delay-and-add Blocks Using UGA

It must be clear by now that that the sampled-data transfer function $H(z)$ has the same form as the transfer function in a digital filter. The basic building blocks to implement a digital filter are delay, add, and multiply units. These are also the basic building blocks for realizing an SC filter transfer function. Figure 6.27a shows the type-1 delay-and-add block (DA1, in short) implemented using several capacitors and one UGA. When the clock phase 1 (i.e., ϕ_1) is ON, that is, $t = (n-1)T$, the input signal v_i is sampled and held on C_x. The capacitor C_{IN} is charged by the voltage difference $v_i(n-1) - v_x$. At the same time, the voltage held on C_{1N} from the previous interval is discharged across the UGA and fed to the input node marked X. In the next clock interval with half-a-period delay (clock phase ϕ_2 ON), the signal $v_i(n-1)$ held on C_x is connected to the input of the UGA and the voltage is stored on C_{1N}. Another half-a-period later, ϕ_1 comes ON and a new value $v_i(n)$ is sampled in on C_x. The CCE at this time will be

$$C_{ON}v_o(n) = C_{IN}[v_i(n) - v_o(n)] + C_{1N}v_i(n-1) \tag{6.50}$$

On taking z-transforms on both sides and rearranging, we get

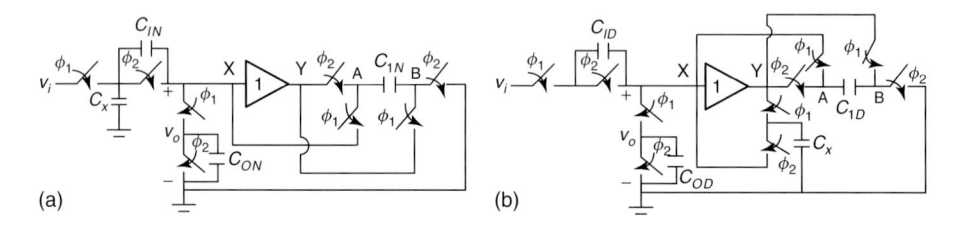

$$V_{out}(z) = \frac{C_{IN}}{C_{IN} + C_{ON}}[1 + \frac{C_{1N}}{C_{IN}}z^{-1}]V_i(z) \tag{6.51}$$

Figure 6.27 Unity-gain amplifier (UGA)-based SC configurations to produce (a) a first-order numerator $N(z)$ and (b) a first-order denominator $D(z)$.

Equation (6.51) represents a first-order transfer function in the z-domain with only a numerator term proportional to $1 + \frac{C_{1N}}{C_{IN}} z^{-1}$.

Considering Figure 6.27b, one can similarly formulate

$$C_{ID}[v_i(n) - v_o(n)] + C_{1D}v_o(n-1) = C_{OD}v_o(n) \qquad (6.52)$$

In this case, C_{1D} stores $v_o(n)$ instead of $v_i(n)$ as in Figure 6.27a. On taking z-transforms on both sides and simplifying, we get

$$V_{\text{out}}(z) = V_i(z)/D(z)$$

where

$$\frac{1}{D(z)} = \frac{C_{ID}}{C_{ID} + C_{OD}} \frac{1}{1 - \frac{C_{1D}}{C_{OD} + C_{ID}} z^{-1}} \qquad (6.53)$$

Equation (6.53) indicates that the arrangement in Figure 6.26b leads to a transfer function having only a denominator term, such as $D(z)$. It must be remarked that the sign associated with the z^{-1} term in Eqs. (6.41) and (6.43) can be reversed by interchanging the terminals of C_{1N} and C_{1D}. The network in Figure 6.27b is called *type 2 delay-and-add network* (DA2).

From the above, it is apparent that, by combining the DA1 and DA2 networks suitably, one can generate a rational transfer function of the form $N(z)/D(z)$, where both $N(z)$ and $D(z)$ are of degree z^{-1}.

6.9.2
Delay Network Using UGA

A delay network using an UGA, several switches and sample-and-hold capacitors is shown in Figure 6.28. It is easy to see that a signal sampled in phase 2 of the clock signal on to C_{x1} is delayed by two half intervals (i.e., a full period T) by the succeeding phases 1 and 2 of the clock signal and appears across the terminals XY. Thus, v_{xy} is a one-period delayed version of v_i. In the z-domain $V_{XY}(z) = z^{-1} V_i(z)$. The network in Figure 6.28 will be termed as a *D network* (D for delay).

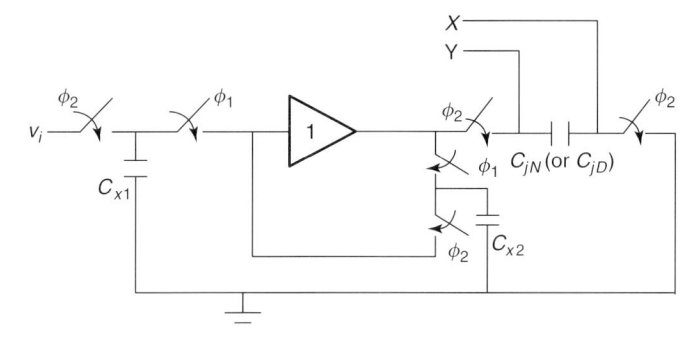

Figure 6.28 UGA-based delay network providing the delay z^{-1}.

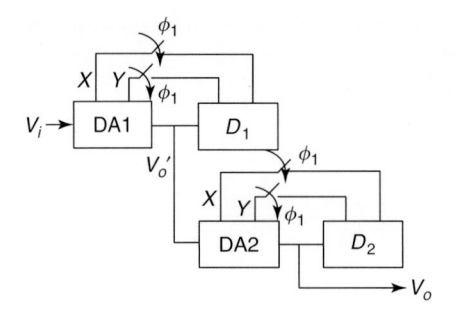

Figure 6.29 Block diagram realizing a second-order sampled-data transfer function using DA1, DA2, and D networks.

6.9.3
Second-Order Transfer Function Using DA1, DA2 and D Networks

Intuitively, it is clear that if one cascades a D network after a DA1 network, one will be able to produce a second-order numerator function of the form $N(z) = a_o + a_1 z^{-1} + a_2 z^{-2}$. Similarly, cascading one D network with a DA2 network, one can generate a denominator function of the form $1/D(z) = 1/[b_o + b_1 z^{-1} + b_2 z^{-2}]$. Figure 6.29 shows the interconnection that will be needed to produce a general second-order z-domain transfer function. In this diagram, the DA1, DA2, and D networks are shown as rectangular blocks, so that the clarity of the interconnection is not lost. One can derive (Raut, 1984)

$$H(z) = \frac{V_{out}(z)}{V_i(z)} = \frac{V_{out}(z)}{V'_{out}(z)} \frac{V'_{out}(z)}{V_i(z)} = \frac{1}{D(z)} N(z) = h_D \frac{1 + a_{1N} z^{-1} + a_{2N} z^{-2}}{1 - a_{1D} z^{-1} + a_{2D} z^{-2}}$$

(6.54)

In terms of the labeling of the components in DA1, DA2, and D networks, it can be shown that (see Table 6.4).

$$h_D = \frac{1}{(1 + C_{ON}/C_{IN})(1 + C_{OD}/C_{ID})}$$

$$a_{1N} = C_{1N}/C_{IN}, a_{2N} = C_{2N}/C_{IN}$$

$$a_{1D} = C_1 D/(C_{OD} + C_{ID}), a_{2D} = C_{2D}/(C_{OD} + C_{ID}) \quad (6.55)$$

For a specific design, the values of $h_{1D}, a_{1N}, a_{2N}, a_{1D}$, and a_{2D} will be known when the BLT is applied to the given second-order analog filter (i.e., continuous-time) transfer function (Table 6.4). It is then a simple matter to determine the various capacitances to satisfy the relations shown in Eq. (6.55). In essence, the design task in this case is much easier than in the case of OA-based implementation.

6.9.4
UGA-Based Filter with Reduced Number of Capacitances

The realizations proposed above require 7 capacitors for each order of the transfer function, and hence 14 capacitors are required to realize a second-order transfer function. Using an improved basic network can reduce the total number of capacitors to only 10. This improved basic block is known as composite

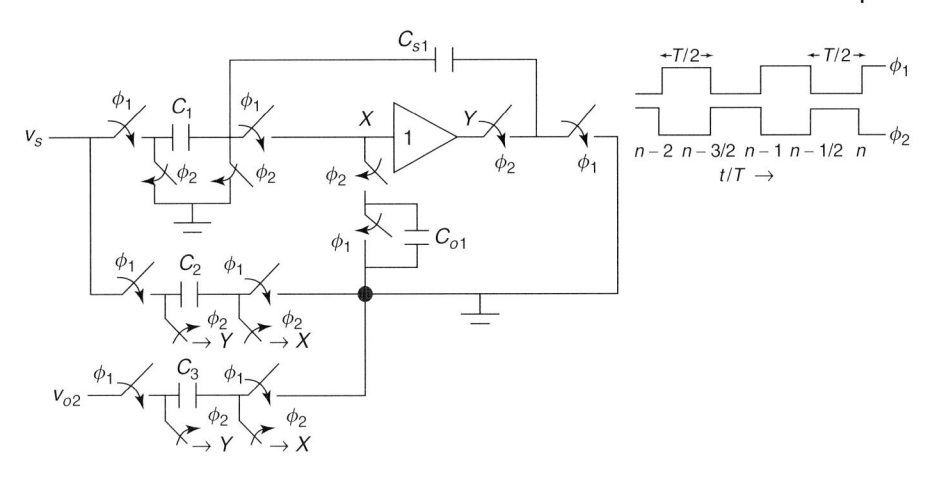

Figure 6.30 Composite delay-and-add (CDA) network to produce a UGA-based first-order transfer function.

delay-and-add network (*CDA*) and is presented in Figure 6.30. Considering the time instants at $t = (n - 1/2)T$ and $t = nT$, we can write down the following CCEs:

$$C_{o1}v_x(n - 1/2) + C_2 v_s(n - 1) + C_3 v_{o2}(n - 1) = 0 \tag{6.56}$$

$$C_1 \left[v_s(n) - v_x(n)\right] - C_1 \times 0 + C_{s1}\left[v_x(n) - v_x(n - 1/2)\right] = 0 \tag{6.57}$$

After eliminating $v_x(n - 1/2)$ from Eqs. (6.56) and (6.57), and taking z-transforms, one can arrive at the following equations:

$$V_{o1}^{(1)} = A_1 V_s^{(1)} + A_2 z^{-1} V_s^{(1)} + A_3 z^{-1} V_{o2}^{(1)}$$
$$A_1 = C_1/(C_1 + C_{s1}), \, A_2 = C_2 C_{s1}/[C_{o1}(C_1 + C_{s1})],$$
$$A_3 = C_3 C_{s1}/[C_{o1}(C_1 + C_{s1})] \tag{6.58}$$

In the above, we have assigned the superscript (1) to imply the clock phase ϕ_1 and we have ignored the explicit notation for the function of the variable z. Equation (6.55) contains the add, multiply and delay terms, which are basic for generating a general transfer function in the variable z. It may be noted that the signs of the coefficients A_2 and A_3 can be changed simply by interchanging the terminals of the associated capacitors (C_2 and C_3) in Figure 6.30.

Figure 6.31 presents the interconnection of two CDA networks to produce a general second-order transfer function in the z-domain. Omitting the argument z for simplicity, one can derive the following (Fan *et al.*, 1980).

$$V_{o1}^{(1)} = A_1 V_s^{(1)} + A_2 z^{-1} V_s^{(1)} + A_3 z^{-1} V_{o2}^{(1)} \tag{6.59}$$

$$V_{o2}^{(1)} = B_1 V_{o1}^{(1)} - B_2 z^{-1} V_{o1}^{(1)} + B_3 z^{-1} V_s^{(1)} \tag{6.60}$$

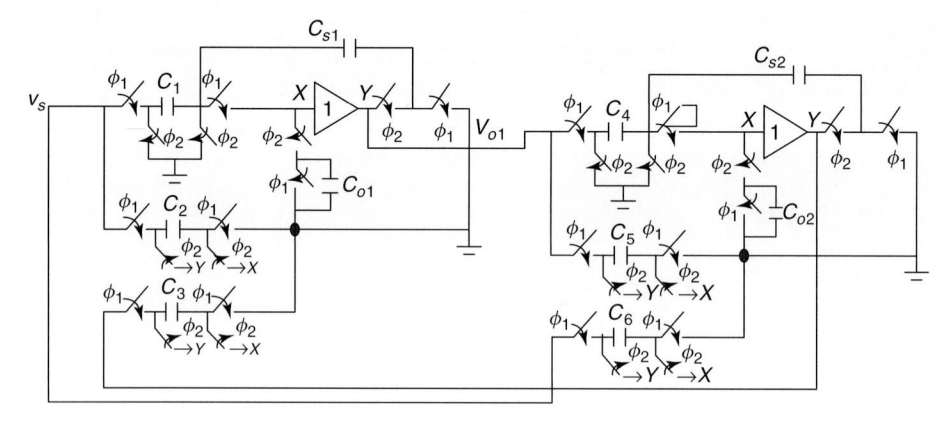

Figure 6.31 Use of two CDA networks to produce a second-order sampled-data transfer function.

The above two equations can be combined to yield the transfer function:

$$H(z) = \frac{V_{o1}^{(1)}}{V_s^{(1)}} = \frac{A_1 + A_2 z^{-1} + A_3 B_3 z^{-2}}{1 - B_1 A_3 z^{-1} + B_2 A_3 z^{-2}} \qquad (6.61)$$

In the above, the coefficients B_1, B_2, and B_3 are dependent upon the network capacitances as follows:

$$B_1 = C_4/(C_4 + C_{s2}),\ B_2 = C_5 C_{s2}/[C_{o2}(C_4 + C_{s2})],\ B_3 = C_6 C_{s2}/[C_{o2}(C_4 + C_{s2})] \qquad (6.62)$$

On comparing with the coefficients of the standard form of $H(z)$ (see Table 6.4), one can find that

$$A_1 = h_D,\ A_2 = h_D a_{1N},\ A_3 B_3 = h_D a_{2N},\ B_1 A_3 = a_{1D},\ \text{and}\ B_2 A_3 = a_{2D} \qquad (6.63)$$

The shortcoming of the UGA-based realization is that the SC networks used are not PI. The summing junction happens to be at the input of the UGA, that is, a high impedance node, and this node is not shorted to ground during any of the clock signal phases. Thus, the design capacitances should be chosen to be considerably higher than the highest possible value of the parasitic capacitances. In practice, optimization programs should be used for careful selection of the various design capacitances in the presence of top- and bottom-plate parasitic capacitances. Such designs are known as *parasitic tolerant designs* (Raut and Bhattacharyya, 1984b).

Practice Problems

6.1 A low-pass filter for an implantable device for medical applications has a bandwidth of 125 Hz. As the circuit has to be integrated, the largest capacitor that may be used is 15 pF. Approximately what are the resistor values needed to realize the required time constants in this low-frequency active-RC filter?

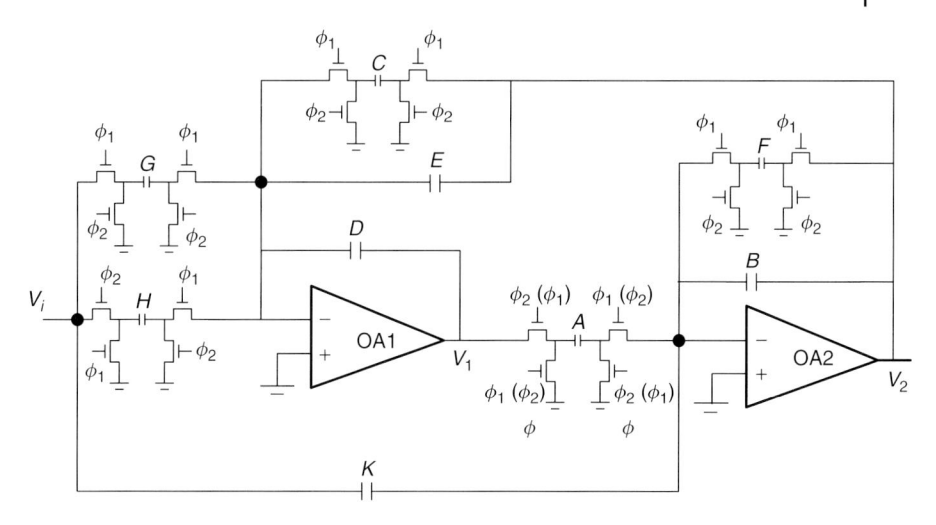

Figure P6.5

The filter should be realized as a switched-capacitor circuit, clocked at 100 times the filter's bandwidth. What is the size of the switched capacitors C_R needed to implement the resistors?

6.2 Perform prewarping for the transfer function $H(s) = \frac{s^2 + 1.4212 \times 10^5}{s^2 + 1004.2s + 6.9833 \times 10^5}$ and derive $H(z)$ using bilinear transform. Use the sampling frequencies of (a) 8 kHz and (b) 128 kHz. Discuss the response characteristics.

6.3 Use the SC filter structure of Figure 6.20 (Section 6.7.1) to realize the low-pass notch function $H(s) = \frac{0.891975 s^2 + 1.140926 \times 10^8}{s^2 + 356.047 s + 1.140926 \times 10^8}$. The sampling frequency is to be 128 kHz. You can ignore prewarping.

6.4 Synthesize the transfer function of a high-pass sampled-data (switched-capacitor/digital) filter that has (i) an equiripple passband for $f > 15$ kHz with $A_p = 0.5$ dB and (ii) a monotonic stopband for $f < 7.8$ kHz with $A_a = 35$ dB. You can use bilinear $s \leftrightarrow z$ transformation, and a clock frequency of 150 kHz. What is the attenuation of the synthesized filter function at $f = 7.5$ kHz?

6.5 For the SC filter network of Figure P6.5 (Ghausi and Laker, 1981), derive the expression for $\frac{V_2^{(1)}}{V_i^{(1)}}$. Assume that the capacitors $B = D = 1$ pF.

6.6 Derive the z-domain transfer functions given in Eq. (6.40) using the basic analysis technique presented in Section 6.5.

6.7 Using the approach suggested in Section 6.6, derive a simulation model for the transfer function in Eq. (6.40).

6.8 Using the structure of Figure 6.20 and a clock frequency of 64 kHz, design the filters in Problems 6.3–6.5. Apply the BL transformation and prewarping. Wherever possible, minimize the total sum of the capacitances. Discuss the response characteristics of the SC filters as compared to the corresponding active-RC filters satisfying the same set of specifications. Use necessary simulation tools and techniques.

6.9 Design the SK band-pass filter with the voltage transfer function as given in Problem 5.3 using the switched-capacitor technique. Use a clock frequency of 120 kHz.

6.10 Design the band-pass filter of Problem 5.8 using the switched-capacitor technique. Use a sampling frequency of 100 kHz.

6.11 Design a high-pass filter with $f_c = 1.2$ kHz, $Q_p = \sqrt{2}$, and high-frequency gain of 1 dB. Use a clock frequency of 256 kHz. The largest capacitor must be less than 20 pF.

6.12 Design a low-pass filter with $f_c = 3.4$ kHz, $Q_p = 1.3$ and dc gain of 0 dB. Use a clock frequency of 256 kHz. The largest capacitor must be less than 20 pF.

6.13 Use the modular approach of design given in Section 6.8 to derive suitable SC filter structures, which will satisfy the specifications in Problems 6.3–6.5 above. Apply the BL transformation and a clock frequency of 64 kHz. Wherever possible, minimize the total sum of the capacitances.

6.14 Design a parasitic-insensitive switched-capacitor version of the second-order Bessel–Thomson filter with voltage transfer function $T(s) = \frac{12 \times 10^6}{s^2 + 6 \times 10^3 s + 12 \times 10^6}$. Use bilinear $s \leftrightarrow z$ transformation. Use a switched-capacitor structure and a sampling frequency of your choice. Using a numerical program, compare the magnitude response of your designed filter with that of $T(s)$. Use three frequencies for this comparison, namely, $f = 0$, $f = f_s/8$, and $f = f_s/2$, where f_s is the sampling frequency you have chosen.

6.15 Use the UGA-based structures with DA1, DA2, and D networks to design the filters specified in Problems 6.3–6.5. Use a minimum capacitance value of 1 pF.

6.16 Repeat Problem 6.12 employing the UGA-based structure with the CDA network. Keep the minimum capacitance value of 1 pF. Compare the total capacitance in each of the designs with the corresponding design in Problem 6.12.

7
Higher-Order Active Filters

In Chapters 5 and 6, we have discussed the analysis and design aspects related to second-order active filters, covering both continuous-time (active-RC) and sampled-data (SC) filters. In Chapter 4, we introduced concepts related to synthesis of filters of any order using passive elements. In this chapter, we deal with realization of filters of order higher than 2 using active devices, resistances, and capacitances. The signal processing techniques will cover both the continuous-time and sampled-data domains. In the continuous-time domain, we discuss three approaches: (i) simulation of passive ladders using active-RC networks, (ii) cascading second-order filter sections with no feedback, and (iii) cascading second-order sections with special multiloop feedback or feed-forward techniques. In the sampled-data domain, we cover analogous approaches with SC networks.

It should be mentioned that the particular choice of realizing a higher-order filter is based on a number of factors such as sensitivity, simplicity of design, power dissipation, simplicity of tuning, dynamic range, noise, implementation in integrated form, and economic considerations.

7.1
Component Simulation Technique

It is common knowledge that filters with good frequency selectivity have to be of order higher than 2. Classically, LC ladder filters were used to accomplish this. It has been shown that a doubly terminated lossless LC ladder structure has minimum sensitivity to component variations in the frequency band of interest (Orchard, 1966). Thus, the performance of LC ladder filters is very reliable and stable. With the advent of electronic filters, it has become a common practice to replicate the operation of an LC ladder by means of active filter components to preserve the same low-sensitivity feature. One approach is to implement the operation of an inductance using active-RC components, and then replace each L in the LC ladder by the simulated inductance and expect that the overall filter will behave in the same manner as the prototype LC filter. This method is very useful in the case of HP filters, where the inductance is grounded; however, in the case of a filter like the LP filter, where the inductors are floating, this is not very suitable as it is

Modern Analog Filter Analysis and Design: A Practical Approach. Rabin Raut and M. N. S. Swamy
Copyright © 2010 WILEY-VCH Verlag GmbH & Co. KGaA, Weinheim
ISBN: 978-3-527-40766-8

difficult to simulate accurately a floating inductor. In such a case, a transformation is introduced to convert the inductors into resistors; but, in doing so, the capacitors get converted to "supercapacitors" or "frequency-dependent negative resistances" (FDNRs)," which are then replaced by active-RC circuits (Bruton, 1969, 1980). This approach of replacing inductors or FDNRs by active-RC circuits is known as the *component simulation technique*.

7.1.1
Inductance Simulation Using Positive Impedance Inverters or Gyrators

We recall that an impedance inverter is defined by its chain matrix (see Table 2.2) as

$$[a]_{II} = \begin{bmatrix} 0 & 1/G_1 \\ G_2 & 0 \end{bmatrix} \tag{7.1}$$

Further, it is called a *positive impedance inverter* or a *gyrator* if $(G_2/G_1) > 0$, and is represented in Figure 7.1. If, in addition, $G_1 = G_2 = G$, then it is called an *ideal gyrator*. Since most designs of a gyrator using active elements have a common ground between the input and the output, we consider only grounded gyrators for inductor simulation. Consider a grounded gyrator loaded by a capacitor C as shown in Figure 7.2.

Then, the DPI Z_{in} at port 1 is given by

$$Z_{in} = \frac{V_1}{I_1} = \frac{1}{G_1 G_2} \left(\frac{-I_2}{V_2} \right) = \frac{sC}{G_1 G_2} = s\frac{C}{G_1 G_2} = sL \tag{7.2}$$

Thus, a grounded inductor of value $L = C/(G_1 G_2)$ is simulated at port 1. Such inductors are useful in implementing LC ladder filters, where the inductors are grounded such as in the case of HP filters. For implementing a filter having floating inductors, such as in the case of an LP, BP, or BR filter, we need to simulate floating inductors. This can be achieved by using two gyrators with a grounded capacitor in between, as shown in Figure 7.3a.

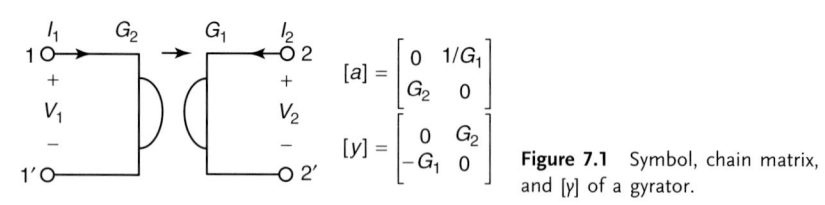

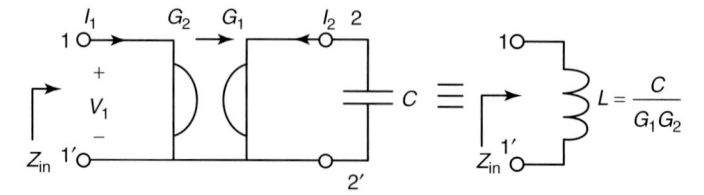

Figure 7.1 Symbol, chain matrix, and [y] of a gyrator.

Figure 7.2 Simulation of a grounded inductor using a gyrator.

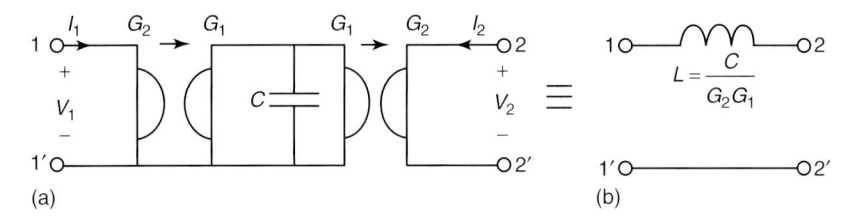

Figure 7.3 Realization of a floating inductor using two matching gyrators.

The chain matrix of the overall two-port is given by

$$[a] = \begin{bmatrix} 0 & \frac{1}{G_1} \\ G_2 & 0 \end{bmatrix} \begin{bmatrix} 1 & 0 \\ sC & 1 \end{bmatrix} \begin{bmatrix} 0 & \frac{1}{G_2} \\ G_1 & 0 \end{bmatrix} = \begin{bmatrix} 1 & (sC)/(G_1 G_2) \\ 0 & 1 \end{bmatrix}$$

which is the chain matrix of a series inductor of value $L = C/(G_1 G_2)$. Thus, a floating inductor of value $L = C/(G_1 G_2)$ is simulated, as shown in Figure 7.3b. It is important to note that this value is realized under the assumption of two matching gyrators, as shown in Figure 7.3. If the gyrators are ideal, $G_1 = G_2 = G$, then the floating inductance realized is $L = C/G^2$.

Even though there are many realizations of gyrators using OAs (Antoniou, 1967), it is much easier to realize a gyrator using OTAs, in view of the fact that the basic equations governing a gyrator are of the form

$$I_1 = G_2 V_2 \tag{7.3a}$$
$$I_2 = -G_1 V_1 \tag{7.3b}$$

which can easily be realized using two OTAs (see Table 5.10, row I); for convenience, it is shown in Figure 7.4, where $G_1 = g_{m1}$ and $G_2 = g_{m2}$. Thus, if a capacitor is connected between terminal 2 and ground, then we realize a grounded inductor at port 1 of value $L = C/(g_{m1} g_{m2})$.

Since the admittance parameters of a practical transconductance device will be functions of frequency, it can be appreciated that an inductance implemented with a transconductance-based gyrator will likewise be frequency dependent. In reality, the inductance appears with a small parasitic resistance in series. This resistance and the inductance exhibit constant values at low frequencies (up to about 10 kHz, depending upon the device characteristics), but start changing with frequency as

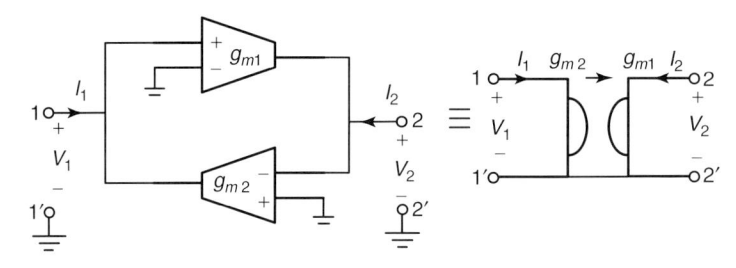

Figure 7.4 A grounded gyrator realization using two OTAs.

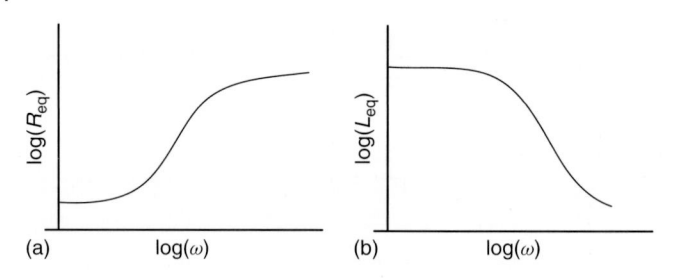

Figure 7.5 Actual inductor characteristics produced by a gyrator implemented with practical semiconductor OTAs: (a) the real part and (b) the imaginary part of the simulated inductor.

the signal frequency becomes higher. Typical characteristics of these components, as a function of frequency, are shown in Figures 7.5a and 7.5b.

A floating inductor can be built using two gyrators as shown in Figure 7.3, wherein each of the gyrators is realized by using the structure of Figure 7.4, thus using four OTAs. The floating inductor could also be implemented by using three OTAs (see Table 5.10, Row J). It should be noted that parasitic components associated with each practical gyrator, however, makes the floating inductances close to ideal only for a limited frequency range. In a later section, we consider some examples of higher-order filter design using the inductance simulation technique that employs gyrators realized using OTAs.

7.1.2 Inductance Simulation Using a Generalized Immittance Converter

We recall that an impedance converter is governed by its chain matrix of the form (see Table 2.2)

$$[a]_{IC} = \begin{bmatrix} K_1 & 0 \\ 0 & K_2 \end{bmatrix} \tag{7.4}$$

If K_1 and K_2 are functions of s, then we call it a *generalized immittance (impedance or admittance) converter* (GIC). It is seen that if a load Z_L is connected at port 2, then the DPI Z_{in} at port 1 is given by

$$Z_{in} = \frac{K_1(s)}{K_2(s)} Z_L \tag{7.5}$$

As an illustration of the application of the GIC, we see that, if $K_1(s) = 1$, $K_2(s) = 1/(Cs)$, and $Z_L = R_L$, then $Z_{in} = s(CR_L)$, thus simulating an inductor of value $L = CR_L$. Similarly, if $K_2(s) = 1$, $K_1(s) = 1/(C_1 s)$, and $Z_L = 1/(C_2 s)$, then $Z_{in} = 1/(C_1 C_2 s^2)$, which corresponds the impedance of an FDNR.

There are many realizations available for a GIC; however, the circuit that is most practical, yet very easy to implement, is due to Antoniou (1969) and uses two OAs, as shown in Figure 7.6. Assuming the OAs to be ideal, the nodes 1, 2, and 4 are all

7.1 Component Simulation Technique

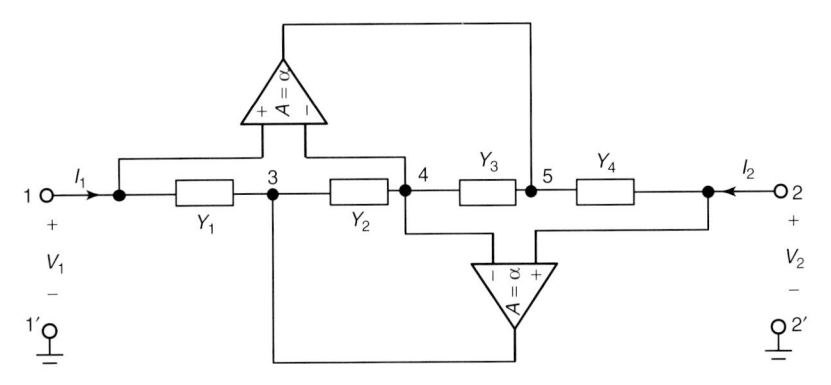

Figure 7.6 Antoniou's generalized immittance converter (GIC) realization using two OAs.

at the same potential, that is,

$$V_1 = V_2 = V_4 \tag{7.6}$$

Also, we have

$$I_1 = (V_1 - V_3)Y_1 \tag{7.7a}$$
$$I_2 = (V_2 - V_5)Y_4 \tag{7.7b}$$
$$(V_3 - V_4)Y_2 + (V_5 - V_4)Y_3 = 0 \tag{7.7c}$$

Substituting Eq. (7.6) in Eq. (7.7a–c) and simplifying, we get

$$I_1 = -\frac{Z_2 Z_4}{Z_1 Z_3} I_2 \tag{7.8}$$

From Eqs. (7.7) and (7.8), we see that the chain matrix of the circuit of Figure 7.6 is given by

$$\begin{bmatrix} 1 & 0 \\ 0 & \frac{Z_2 Z_4}{Z_1 Z_3} \end{bmatrix} \tag{7.9}$$

Thus, the circuit of Figure 7.6 realizes a GIC with $K_1(s) = 1$ and $K_2(s) = (Z_2 Z_4)/(Z_1 Z_3)$. Thus, if the GIC is terminated by a load Z_L at port 2, then from Eq. (7.5), the DPI at port 1 is given by

$$Z_{in} = \frac{Z_1 Z_3}{Z_2 Z_4} Z_L \tag{7.10}$$

If we set $Z_1 = Z_3 = Z_4 = Z_L = R$, and $Z_2 = 1/sC$, we get $Z_{in} = sCR^2$, which simulates an inductance of value $L = CR^2$. This configuration is known as *type-I GIC*. If Z_2 and Z_4 are interchanged, that is, $Z_1 = Z_2 = Z_3 = Z_L = R$, and $Z_4 = 1/sC$, then the GIC is referred to as *type-II GIC*. A detailed analysis assuming practical OAs of finite gain shows that the type-II GIC will simulate an inductance with a high-quality factor (Schaumann, Ghausi, and Laker, 1990) despite small mismatches among the OAs.

212 | 7 Higher-Order Active Filters

It may be mentioned that it is possible to simulate a floating inductor using two GICs (see Problem 7.1). We now consider the application of the GIC-simulated inductances through two examples. One is the realization of an HP filter and the other, the implementation of a coupled-resonator BP filter.

Example 7.1. Implementation of an HP filter: Consider the implementation of an HP CHEB filter with 1 dB ripple in the PB for $f \geq 1400$ Hz. In the attenuation band, the attenuation should be ≥ 35 dB for $f \leq 375$ Hz. Use the component simulation technique for the implementation. The filter is doubly terminated in 1 Ω resistances.

For the given specifications, one can determine the order of the associated normalized LP filter to be 3 and an LC ladder implementation of the LP prototype may be obtained by using Appendix C or the methods given in Chapter 4; the prototype filter so obtained is as shown in Figure 7.7a.

Now applying the LP–HP transformation to covert the LP prototype filter to an HP filter with a PB edge frequency of 1400 Hz (Table 3.2), one gets the HP ladder filter as shown in Figure 7.7b. We now replace the two grounded inductances by GIC-based inductances. The GIC structure for the grounded inductance is shown in

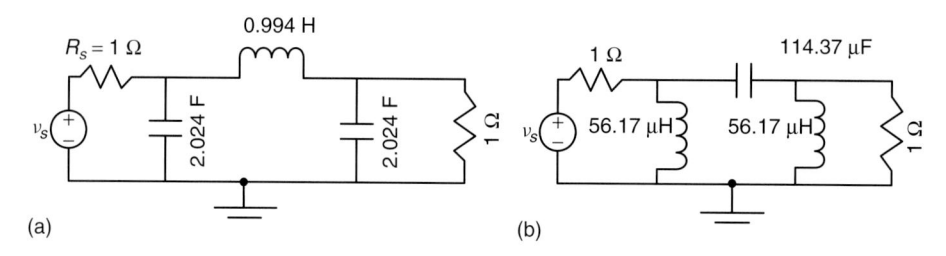

Figure 7.7 (a) An LP LC ladder filter and (b) the associated HP filter after LP to HP transformation.

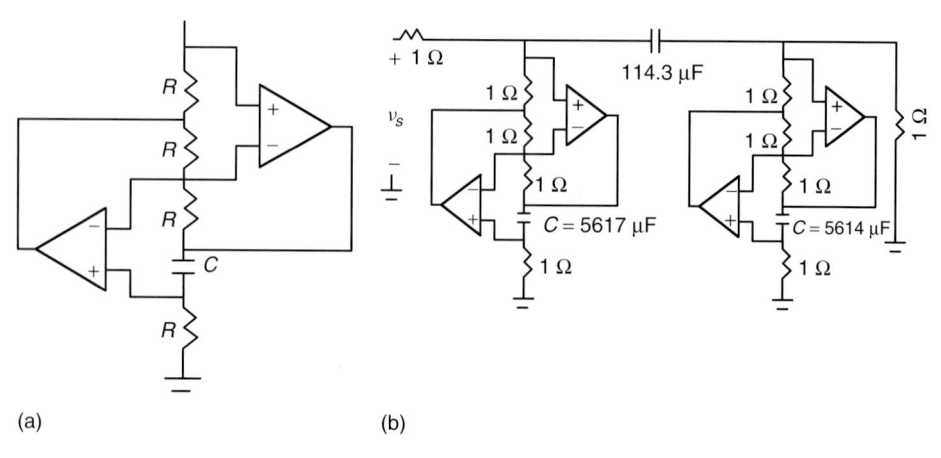

Figure 7.8 (a) GIC-based grounded inductor, and (b) complete realization of the HP filter of Figure 7.7b.

Figure 7.8a. It may be noted that, with the choice of $R = 1\,\Omega$ and $C = 56.17\,\mu F$, the GIC produces an equivalent inductance of $56.17\,\mu H$. The complete implementation of the HP filter of Figure 7.7b, as an active-RC filter, is shown in Figure 7.8b.

7.1.2.1 Sensitivity Considerations

It was mentioned earlier that the excellent sensitivity characteristics of the prototype LC ladder are maintained in the active-RC realization, where the inductances of the HP filter are implemented by GIC networks. This holds true not only for the prototype resistors and capacitors that appear directly in the active realization, but also for the elements of the GIC that realize the inductances. If we consider Eq. (7.4), we can write, in general, $L_{eq} = (sC_4 R_1 R_3 R_L)/R_2$. Then, $S_{R_1,R_3,R_L,C_4}^{L_{eq}} = 1$, $S_{R_2}^{L_{eq}} = -1$. Now, if we consider an arbitrary sensitivity, say, S^y, where y is any usual network function, we can derive that $S_{R_1,R_3,R_5,C_4}^{y} = S_{L_{eq}}^{y} S_{R_1,R_3,R_5,C_4}^{L_{eq}} = S_{L_{eq}}^{y}$. Similarly, $S_{R_2}^{y} = S_{L_{eq}}^{y} S_{R_2}^{L_{eq}} = -S_{L_{eq}}^{y}$. Thus, the absolute value of the sensitivity of any arbitrary function of the ladder network w.r.t. any of the components simulating an inductor is the same as the sensitivity to the inductor that is being simulated.

Example 7.2. Implementation of a coupled BP filter: Although Example 7.1 emphasizes the effectiveness of the GIC to implement HP LC ladder filters, the technique can be used conveniently to implement other filter types that have grounded inductances in their structures. One such case is a coupled-resonator BP filter, as shown in Figure 7.9a (Huelsman, 1993). This represents a narrow-band BP filter with a center frequency of $1\,\text{rad s}^{-1}$, a bandwidth of $0.1\,\text{rad s}^{-1}$, and a maximally flat (Butterworth) magnitude characteristic. The GIC implementation of the normalized filter is shown in Figure 7.9b.

7.1.3
FDNR or Super-Capacitor in Higher-Order Filter Realization

The inductance simulated by GIC technique is very convenient for replacing grounded inductances in the LC ladder filters. The implementation is thus very

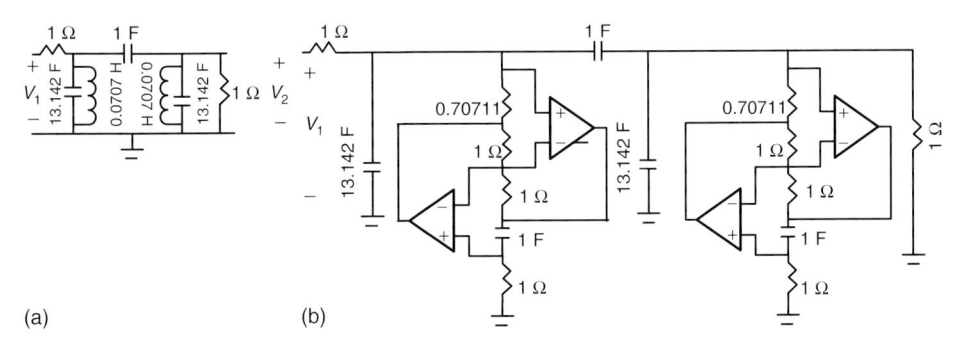

Figure 7.9 (a) RLC-based coupled-resonator BP filter and (b) GIC implementation of the filter in (a).

effective for HP ladder filters. LP ladder filters have inductances in the series branches. Thus, the active-RC implementation would require simulating floating inductances. Floating inductances simulated by gyrator or GIC are not very attractive because of the complexity and the influence of parasitic components. An attractive alternative in this case stems from applying the impedance transformation introduced in Chapter 3 to the elements of the LP LC ladder filter. Thus, if the impedances of the various elements of a doubly terminated LC ladder are multiplied by a factor $1/ks$, that is,

$$Z_i(s) \longrightarrow \frac{1}{ks} Z_i(s) \tag{7.11}$$

then a resistor of value R is transformed to a capacitor of value k/R, an inductor of value L becomes a resistor of value L/k, and a capacitor of value C is transformed into an element whose impedance is $1/s^2 kC = 1/s^2 D$, which corresponds to an FDNR of value $-1/\omega^2 kC$. This transformation is also known as *Bruton's transformation* (Bruton, 1969, 1980). As already mentioned in Chapter 3, such an impedance transformation does not affect either a VTF or a CTF. Hence, for any LP filter subjected to such a transformation, the VTF or CTF will remain unaltered. Thus, if an LP LC ladder filter is transformed by such a transformation, we will have a structure consisting of only resistors, capacitors, and FDNRs. An example of an LP LC ladder and the corresponding structure after the transformation are shown in Figures 7.10a and 7.10b, where D_1 and D_2 correspond to FDNRs of values kC_1 and kC_2. We can realize this as an active-RC filter provided we can implement the FDNR element using active devices and RC elements.

Antoniou's GIC described earlier can readily be used to implement the FDNR. Consider the GIC of Figure 7.6. If we let $Z_1 = Z_L = 1/sC_x$, and $Z_2 = Z_3 = R$ and $Z_4 = R_x$, then, from Eq. (7.10), we see that the DPI at port 1 is given by

$$Z_{in} = \frac{1}{s^2 C_x^2 R_x} = \frac{1}{s^2 D} \tag{7.12}$$

where D, the value of the FDNR (or the $1/s^2$ element), is given by $C_x^2 R_x$. We could also have chosen $Z_1 = Z_3 = 1/sC_x$, $Z_2 = Z_L = R$ and $Z_4 = R_x$. This would also give the input impedance at port 1 to be $Z_{in} = \frac{1}{s^2 C_x^2 R_x} = \frac{1}{s^2 D}$. One may notice that, in view of the impedance transformation Eq. (7.11), the terminal resistances are converted to capacitances, thereby breaking the *DC* path, which, however, must

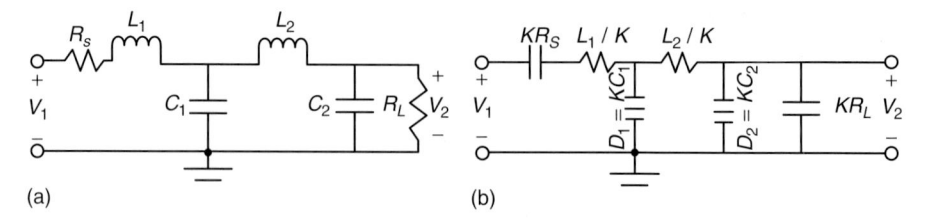

Figure 7.10 (a) A doubly terminated LC LP filter and (b) the same filter after the impedance transformation given by Eq. (7.11).

exist in an LP filter. To overcome this problem, the terminal capacitances are shunted by very high resistances. The values of these resistances are chosen such that the DC gain of the filter using FDNRs is maintained as closely as possible to that in the prototype ladder filter (Figure 7.10a). It should be observed that the FDNR realized above with the GIC is a grounded one and not a floating one. It may be mentioned in passing that it is possible to obtain a floating FDNR using two GICs similar to the realization of a floating inductor using two GICs (see Problem 7.1 for the floating inductor realization). Let us now consider the following example.

Example 7.3. Consider the elliptic LP filter with transmission zeros and terminated in 1 kΩ resistances – see Figure 7.11. Obtain an active-RC implementation using the FDNR technique.

Applying an impedance transformation of $1/(ks)$ with $k = 10^{-6}$, one can find the following correspondence between the given network in Figure 7.11 and the new network elements after applying the FDNR technique.

$$R_s = 1 \text{ k}\Omega \longrightarrow C_s = k/R_s = 1 \text{ nF}, L_1 = 6.9 \text{ mH} \longrightarrow R_1 = L_1/k = 6.9 \text{ k}\Omega,$$
$$L_2 \longrightarrow R_2 = 6.49 \text{ k}\Omega, L_3 \longrightarrow R_3 = 45.55 \text{ k}\Omega, L_4 \longrightarrow R_4 = 0.94 \text{ k}\Omega,$$
$$L_5 \longrightarrow R_5 = 33.9 \text{ k}\Omega, R_L \longrightarrow C_L = 1 \text{ nF}$$

The capacitors C_1 and C_2 get transformed into super-capacitors or FDNRs D_1 and D_2 as follows: $C_1 \rightarrow D_1 = kC_1 = 27.1 \times 10^{-15} \rightarrow C_{41}^2 R_{41}$ in the GIC network. Assuming $C_{41} = 2$ nF, we get $R_{41} = 6.78$ kΩ. Similarly, for $C_2 \rightarrow D_2 = 46.65 \times 10^{-15} \rightarrow C_{42}^2 R_{42}$. With $C_{42} = 1$ nF, $R_{42} = 46.65$ kΩ. For the 12.67-nF capacitor, $R_{43} = 12.67$ kΩ with $C_{43} = 1$ nF.

Figure 7.12a shows one of the FDNR networks, which is a subsystem of the whole filter. The full active-RC implementation is shown in Figure 7.12b.

In Figure 7.12b, the resistances R_a and R_b are inserted to maintain the continuity in the DC path. Because, for equal termination, the DC gain of the ideal LC filter would be $1/2$, one can design R_a and R_b to satisfy this goal approximately. Thus, assuming $R_b = 1$ MΩ, we can design R_a from the equation

$$\frac{R_b}{R_a + R_1 + R_3 + R_5 + R_b} \approx 0.5, \text{ with } R_1 + R_3 + R_5 = 6.9 + 45.55 + 33.9 = 86.35 \text{ k}\Omega$$

Hence, $R_a = 913.65$ kΩ.

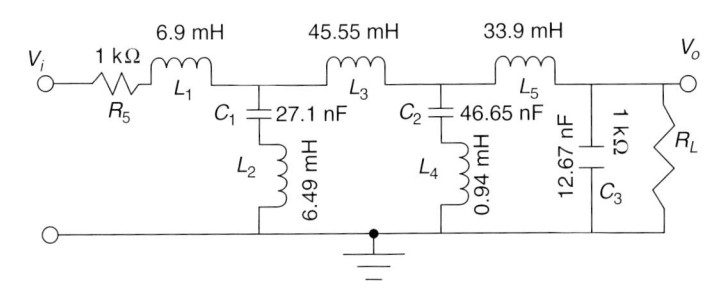

Figure 7.11 A doubly terminated LC LP filter with transmission zeros.

7 Higher-Order Active Filters

(a)

(b)

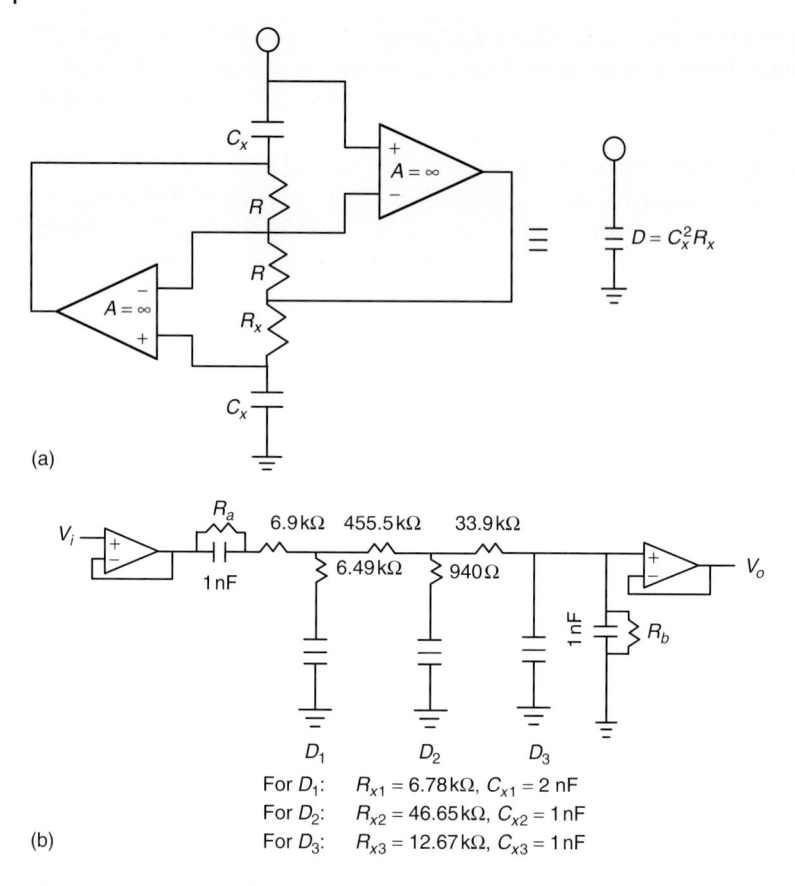

Figure 7.12 (a) A typical FDNR subcircuit to be used in the realization of the LP filter of Figure 7.11 and (b) active-RC implementation of the of filter of Figure 7.11 using the FDNR subcircuits.

For D_1: $R_{x1} = 6.78\,\text{k}\Omega$, $C_{x1} = 2\,\text{nF}$
For D_2: $R_{x2} = 46.65\,\text{k}\Omega$, $C_{x2} = 1\,\text{nF}$
For D_3: $R_{x3} = 12.67\,\text{k}\Omega$, $C_{x3} = 1\,\text{nF}$

Another important aspect of this realization is that the input and output terminals are isolated by unity-gain buffer amplifiers. This is required when the filter serves as a subsystem in a larger overall system. This is because of the fact that the filter elements have been altered by impedance transformation but the other parts of the larger system have not been so altered. Thus, isolation amplifiers are required to prevent unwanted interaction between the filter elements and the rest of the system in which the filter is embedded.

7.1.3.1 Sensitivity Considerations

Like the simulated inductor, the FDNR retains the excellent sensitivity properties of the LP prototype. This is true not only w.r.t. the transformed prototype resistors and capacitors that appear directly in the active realization, but also w.r.t. the elements of the GIC that realize the FDNRs. This can be appreciated from the expression for the super-capacitor $D = \frac{C_1 C_L R_2 R_4}{R_3}$. Obviously, the sensitivity of D with respect to the elements of the FDNR is either $+1$ or -1. Thus, the sensitivity of any network function with respect to the elements of the FDNR will be ± 1 times the sensitivity of the same network function to the prototype capacitor that has been transformed into the super-capacitor.

7.2 Operational Simulation Technique for High-Order Active RC Filters

In the previous sections, we have discussed methods wherein the inductors are avoided by either simulating these using GICs or converting them into resistors using an appropriate impedance transformation. In the latter case, however, the capacitors get converted into super-capacitors (FDNRs), which are then simulated using GICs. In this section, we present an alternative method for the design of high-order active filters. Our starting point is still a prototype LC ladder structure. However, rather than simulating individual inductors or FDNRs in the ladder, we seek a method to simulate the operation of each section of the ladder (Sedra and Brackett, 1978; Huelsman, 1993). In this process, we deal with the I–V relations of each section of the ladder by following the signal flow pattern along the ladder as a sequence of voltage-to-current and current-to-voltage equations. The method is, therefore, sometimes known as the *signal flow graph* or the *operational simulation* method. The basic philosophy of this technique is to find an active circuit that "copies" the operation (of the input current or voltage to produce an output voltage or current, respectively) of an LC ladder prototype filter. Since the I–V equations for an inductor and a capacitor involve derivatives and integrals with respect to time, the functions of these elements are conveniently implemented by using active-RC integrators. Similarly, the KVL equation around a loop and the KCL equation at a node can be implemented by suitable summing amplifiers. Since in practice as the signal flows along the system, the signal mode alternates between voltage and current or vice versa, we can use convenient scaling resistances to convert the current signals into voltage signals so that only voltage-mode building blocks (namely, VCVS-based integrators and summers) are used in the implementation. We illustrate the technique by considering a few representative cases.

7.2.1 Operational Simulation of All-Pole Filters

The design of all-pole active filters, using operational simulation of LC ladder prototypes, is discussed in detail by Girling and Good (1970). The structure has been named the *leapfrog* structure because of the way the schematic of the resulting network appears. Consider Figure 7.13, which depicts an LC ladder structure. Since it is an all-pole network, each of the series and shunt arms contains only one element, either a capacitor or an inductor. The I–V relations for the series and

7 Higher-Order Active Filters

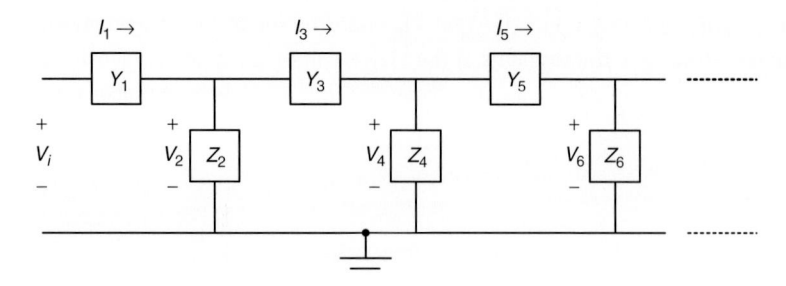

Figure 7.13 Illustration of the formulation of the $I - V$ equations in an LC ladder, where Y_i represents the admittances of the series arms and Z_i the impedances of the shunt arms.

shunt arms can be written in a simple way. Thus,

$$I_1 = (V_1 - V_2)Y_1$$
$$I_3 = (V_2 - V_4)Y_3$$
$$I_5 = (V_4 - V_6)Y_5 \tag{7.13}$$

etc.,

and

$$V_2 = (I_1 - I_3)Z_2$$
$$V_4 = (I_3 - I_5)Z_4 \tag{7.14}$$

etc.

Since we intend to use a voltage-amplifier-based implementation, it will be judicious to convert all the I variables in Eqs. (7.13) and (7.14) to corresponding V variables. This can be easily done by using a suitable scaling resistance (whose value could be 1 Ω). Thus, writing V'_i for $I_i R$, where R is a scaling resistance, we can reorganize Eqs. (7.13) and (7.14) as

$$V'_1 = (V_1 - V_2)Y_1 R$$
$$V_2 = (V'_1 - V'_3)Z_2/R$$
$$V'_3 = (V_2 - V_4)Y_3 R$$
$$V_4 = (V'_3 - V'_5)Z_4/R$$
$$V'_5 = (V_4 - V_6)Y_5 R \tag{7.15}$$

etc.

The expressions in Eq. (7.15) are dimensionless VTFs and hence can be realized with voltage amplifiers. Since we can have both inverting and noninverting voltage amplifiers using OAs, we may rewrite some of the expressions in Eq. (7.15) with a negative sign in front. This way, the implementation can be carried out conveniently with a continuous chain of OA-based inverting and noninverting amplifiers. Two possible ways to rewrite the expressions are as follows:

$$V'_1 = (V_1 - V_2)Y_1 R$$
$$-V'_2 = (V'_1 - V'_3)(-Z_2/R)$$
$$-V'_3 = (-V'_2 + V'_4)Y_3 R$$
$$V'_4 = (-V'_3 + V'_5)(-Z_4/R)$$
$$V'_5 = (V'_4 - V'_6)Y_5 R \tag{7.16a}$$
etc.

or

$$-V'_1 = (V_1 - V_2)(-Y_1 R)$$
$$-V'_2 = (-V'_1 + V'_3)Z_2/R$$
$$V'_3 = (-V'_2 + V'_4)(-Y_3 R)$$
$$V'_4 = (V'_3 - V'_5)Z_4/R$$
$$-V'_5 = (V'_4 - V'_6)(-Y_5 R) \tag{7.16b}$$
etc.

A block diagram for implementing the expressions in Eq. (7.16a) is shown in Figure 7.14. Similarly, Figure 7.15 shows the block diagram for implementing the expressions in Eq. (7.16b). The name "leapfrog" is apparent from the nature of the layout of these block diagrams.

7.2.2
Leapfrog Low-Pass Filters

Figure 7.16 shows the schematic of a fifth-order all-pole LP filter. We intend to implement an active-RC network corresponding to this filter using the leapfrog

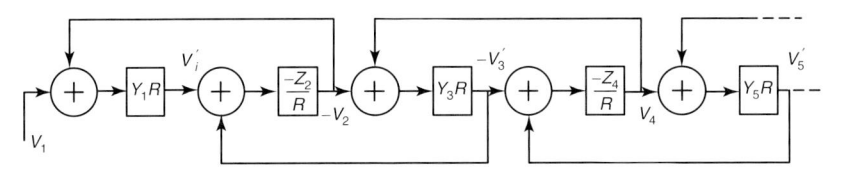

Figure 7.14 Implementation of the $I - V$ relations in Eq. (7.16a) using voltage summing junctions and normalized gain functions $Y_i R$ and $-Z_i/R$.

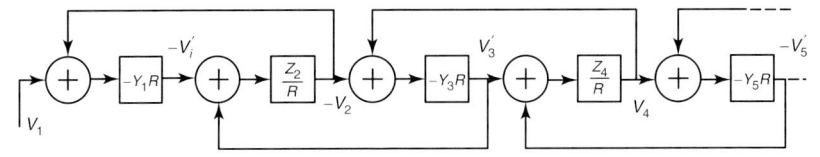

Figure 7.15 An alternative implementation based on the $I - V$ relations in Eq. (7.16b) using voltage summing junctions and normalized gain functions $-Y_i R$ and Z_i/R.

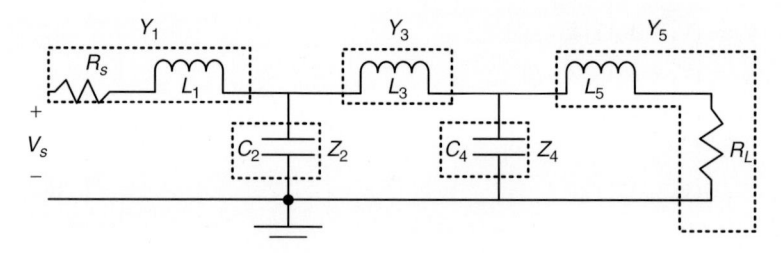

Figure 7.16 Realization of a doubly terminated fifth-order all-pole LC filter using the method of operational simulation.

structure discussed above. For this, we first identify the impedance and admittance elements. The scaled VTFs (i.e., $Y_1 R, Z_2/R, \ldots$) assuming $R = 1$ are as follows:

$$
\begin{aligned}
T_1(s) &\to Y_1 = \frac{1/L_1}{s + R_s/L_1} \\
T_2(s) &\to -Z_2 = -\frac{1}{sC} \\
T_3(s) &\to Y_3 = \frac{1}{sL_3} \\
T_4(s) &\to -Z_4 = -\frac{1}{sC_4} \\
T_5(s) &\to Y_5 = \frac{1/L_5}{s + R_L/L_5}
\end{aligned}
\tag{7.17}
$$

Of the above transfer functions, T_2 and T_4 can be realized by conventional inverting integrators. T_3 can be implemented by a noninverting integrator as shown in Figure 7.17a or a conventional integrator followed by an inverting amplifier of gain unity. The transfer functions T_1 and T_5 can be implemented by noninverting lossy integrators; the noninverting lossy integrator can be realized using an inverting amplifier followed by an inverting lossy integrator, shown in Figure 7.17b.

In order to make the transfer function in Figure 7.17b to comply with T_1, for example, one has to equate the DC gains $1/R_s = R_x/R$, giving $R_x = R/R_s$. Similarly, $1/R_x C_x \to R_s/L_1$, leading to $C_x = L_1/R$. An implementation of the LP filter of Figure 7.16 using the architecture of Figure 7.14 is shown in Figure 7.18. In Figure 7.18, the noninverting integrator has been realized as a cascade of an inverting amplifier and an inverting integrator.

7.2.3
Systematic Steps for Designing Low-Pass Leapfrog Filters

The design procedure for an LP leapfrog filter may be summarized as follows:

1) Design the normalized LP prototype either by using filter tables or by the methods described in Chapter 4.
2) Select one of the general system diagrams from Figure 7.14 or 7.15.

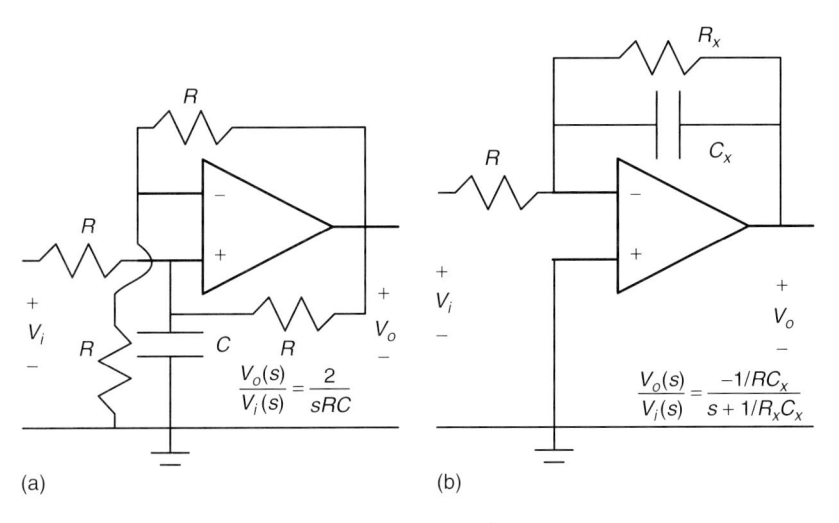

Figure 7.17 (a) Noninverting lossless integrator and (b) inverting lossy integrator for realizing $\pm Y_i R$ and $\pm Z_i / R$ functions with Y_i corresponding to an inductor, and Z_i corresponding to a capacitor.

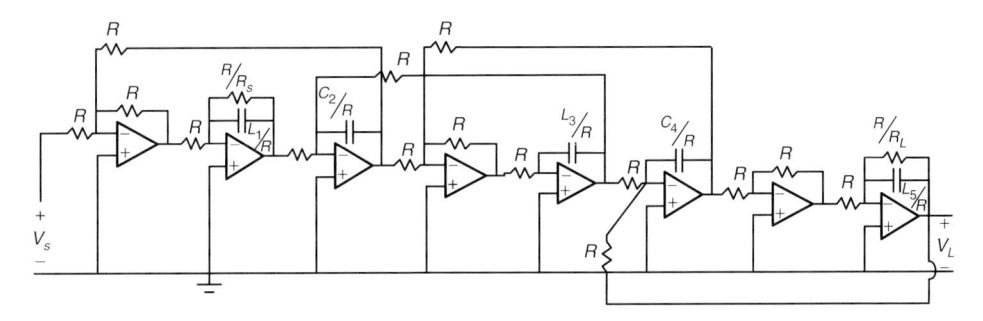

In figure (a): $\dfrac{V_o(s)}{V_i(s)} = \dfrac{2}{sRC}$

In figure (b): $\dfrac{V_o(s)}{V_i(s)} = \dfrac{-1/RC_x}{s + 1/R_x C_x}$

Figure 7.18 The leapfrog structure realization for the filter shown in Figure 7.16.

3) Design the inner elements (i.e., excluding the source and load ends) of the filter using inverting and noninverting integrators and a normalized value of the scaling resistance R equal to unity.
4) Design the terminal elements using the networks in Figure 7.17a or 7.17b.
5) Perform the necessary frequency and impedance denormalization.

Example 7.4. It is required to realize a third-order Butterworth LP filter with a cutoff frequency of 1 k rad s^{-1}. The input source has zero resistance.

This case belongs to the class of singly terminated ladder filters. The prototype ladder is shown in Figure 7.19a. The nominal leapfrog filter structure is shown in Figure 7.19b. A frequency scaling by 10^3 and an impedance scaling by 10^4 will make all the resistances to be equal to 10 kΩ and all the capacitances to get multiplied by 10^{-7}.

222 | 7 Higher-Order Active Filters

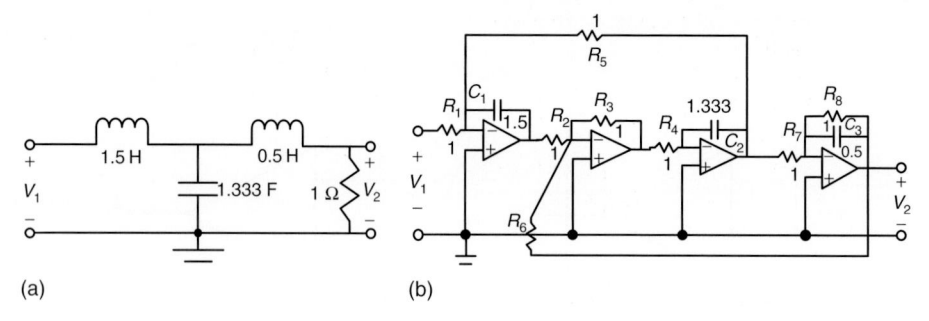

(a) [Circuit: V_1 input, 1.5 H series inductor, 1.333 F shunt capacitor, 0.5 H series inductor, 1 Ω load V_2]

(b) [Circuit: operational simulation with op-amps, resistors R_1–R_8 (values 1.5, 1, 1.333, 1, 0.5), capacitors C_1, C_2, C_3, and R_5, R_6]

Figure 7.19 (a) A third-order low-pass LC filter and (b) operational simulation of the filter in (a).

7.2.4
Leapfrog Band-Pass Filters

The leapfrog technique discussed above is also applicable to BP filters with zeros at the origin and at infinity. This includes, for example, a series resonance network in the series arm and/or a parallel resonance network in the shunt arm of the ladder filter. Figures 7.20a and 7.20b show these two cases with the corresponding Y and Z functions. In the intermediate locations of the ladder, we have $R_i \to 0$ and $R_j \to \infty$.

Thus, instead of first-order RC transfer function networks, as was in the case of an LP filter, we have to use second-order RC-active filter sections in this case to implement the normalized admittance and impedance functions. The second-order RC network must have the capability to produce $Q_p \to \infty$ as will be required for $R_i \to 0$ or $R_j \to \infty$. A Tow–Thomas network with the summing capability at the input, which will also afford this special condition (namely, $Q_p \to \infty$), is shown in Figure 7.21.

The condition $R_i \to 0$ in $Y_i(s)$, and $R_j \to \infty$ in $Z_j(s)$ can be realized from this network by setting $R_1 = \infty$, that is, an open circuit. The leapfrog realization of a prototype BP filter (Figure 7.22a), using the above biquad as a building block, is illustrated in Figure 7.22b.

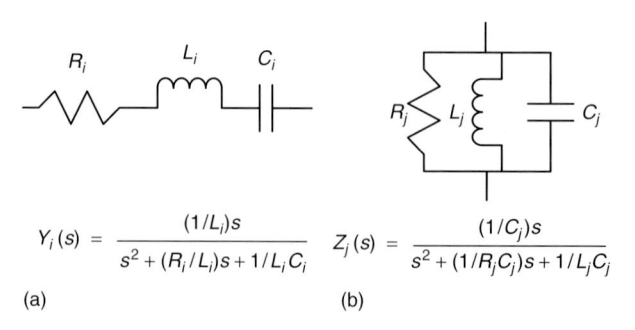

(a) [Series RLC: R_i, L_i, C_i]

$$Y_i(s) = \frac{(1/L_i)s}{s^2 + (R_i/L_i)s + 1/L_iC_i}$$

(b) [Parallel RLC: R_j, L_j, C_j]

$$Z_j(s) = \frac{(1/C_j)s}{s^2 + (1/R_jC_j)s + 1/L_jC_j}$$

Figure 7.20 (a) A third-order low-pass LC filter and (b) operational simulation of the filter in (a).

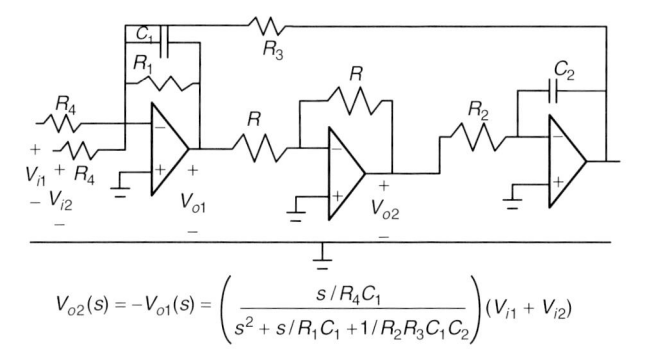

$$V_{o2}(s) = -V_{o1}(s) = \left(\frac{s/R_4C_1}{s^2 + s/R_1C_1 + 1/R_2R_3C_1C_2} \right)(V_{i1} + V_{i2})$$

Figure 7.21 The Tow–Thomas second-order filter that can realize the admittance or the impedance function shown in Figure 7.20.

(a)

(b)

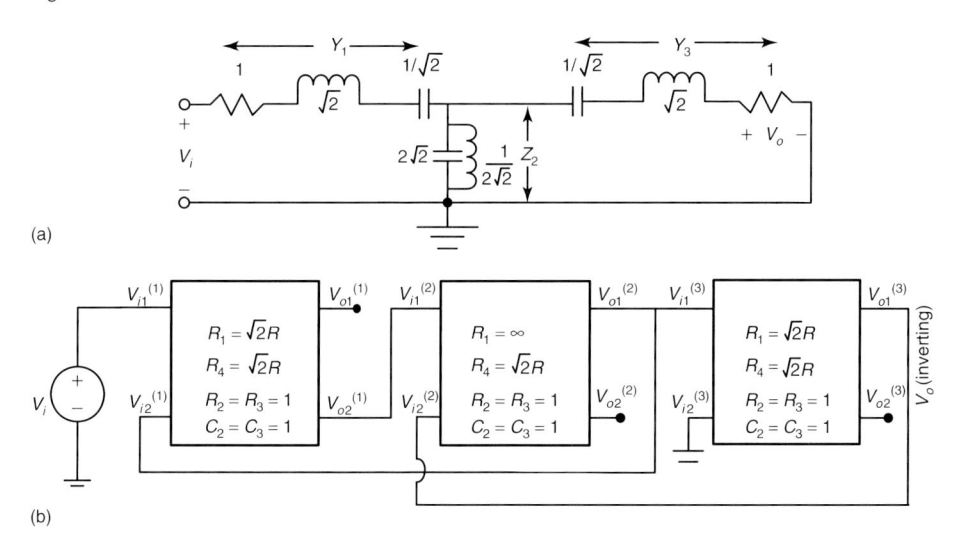

Figure 7.22 (a) A normalized band-pass filter with series LCR and shunt LC elements and (b) leapfrog implementation of the BP filter using the Tow–Thomas biquad, modules (shown as rectangles).

7.2.5
Operational Simulation of a General Ladder Structure

In Section 7.2, we considered the cases of all-pole LP and BP filters. The zeros of the transfer function were either at DC (zero frequency) or at infinity. Finite-transmission zeros occur for elliptic filters, stopband filters, and so on. This implies the presence of a parallel resonant network in the series arm and/or that of a series resonant network in the shunt arm of the ladder network. In general, then, we have to consider a series arm as shown in Figure 7.23a and a shunt arm as in Figure 7.23b.

7 Higher-Order Active Filters

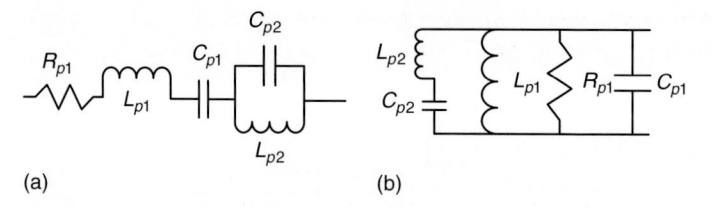

Figure 7.23 (a) The series arm and (b) the shunt arm of a general ladder filter.

The series arm admittance is then

$$Y(s) = \frac{1}{Z(s)} = \frac{1}{R_{p1} + sL_{p1} + \dfrac{1}{sC_{p1}} + \dfrac{1}{sL_{p2} + \dfrac{1}{sC_{p2}}}} \tag{7.18}$$

In the above, we have used the suffix p to indicate the elements to be those of the prototype ladder filter. Similarly, the impedance of the shunt arm is

$$Z(s) = \frac{1}{G_{p1} + sC_{p1} + \dfrac{1}{sL_{p1}} + \dfrac{1}{sL_{p2} + \dfrac{1}{sC_{p2}}}} \tag{7.19}$$

For operational simulation, we have to find active-RC networks, which can simulate the above admittance and impedance functions. Of course, as before, we deal with normalized dimensionless quantities $Y(s)R_p$ and $Z(s)/R_p$, which can be simulated as VTFs. By including the normalized resistance R_p in Eqs. (7.18) and (7.19) and using lower case letters for the normalized quantities, we can write

$$t_y(s) = Y(s)R_p = \frac{1}{r_{p1} + sl_{p1} + \dfrac{1}{sc_{p1}} + \dfrac{1}{sc_{p2} + \dfrac{1}{sl_{p2}}}} \tag{7.20}$$

and

$$-t_z(s) = -\frac{Z(s)}{R_p} = -\frac{1}{G_{p1}R_p + sC_{p1}R_p + \dfrac{1}{sL_{p1}} + \dfrac{1}{(sL_{p2}/R_p) + \dfrac{1}{sC_{p2}R_p}}}$$

$$= -\frac{1}{g_{p1} + sc_{p1} + \dfrac{1}{sl_{p1}} + \dfrac{1}{sl_{p2} + \dfrac{1}{sc_{p2}}}} \tag{7.21}$$

In the above, we have assigned a negative sign in front of the $t_z(s)$ function to imply that this transfer function is going to be realized using an inverting amplifier. The transfer functions given by Eqs. (7.20) and (7.21) appear formidable, but the property that an admittance function in the feedback path of an ideal OA appears inverted in producing the overall closed-loop gain function can be used

7.3 Cascade Technique for High-Order Active Filter Implementation

to implement the above transfer functions. For more details, the reader should consult Schaumann, Ghausi, and Laker (1990) or Schaumann and Van Valkenburg (2001).

7.3 Cascade Technique for High-Order Active Filter Implementation

In the cascade design technique, second-order RC-active networks (and possibly one first-order section) are cascaded to generate a higher-order voltage-mode filter. The method is simple and needs least amount of redesign efforts, because well-behaved and well-characterized first- and second-order systems are available. Each first- and/or second-order filter can be tuned independently and the cascade can be developed in a modular fashion. Of course, this is possible only if each of the sections does not load the output of the preceding one. This is true if the output impedance of each section is much lower than the input impedance of the following section. In this section, we consider the realization of a high-order filter in voltage mode using OAs. Thus, we may write the filter function in the form

$$H(s) = \frac{a_{10}s + a_{00}}{s + b_{00}} \prod_{i=1}^{N} \frac{a_{2i}s^2 + a_{1i}s + a_{0i}}{s^2 + b_{1i}s + b_{0i}} = H_0(s) \prod_{i=1}^{N} H_i(s) \qquad (7.22)$$

The linear term is present only if the filter is of odd order; in the case of an even-order filter, all the sections are biquad sections. The first-order section can be designed using one of the circuits given in Table 5.1 (Chapter 5), while the biquads can be designed using one of the methods discussed in Chapter 5. Let us now consider an example.

Example 7.5. A fourth-order LP Butterworth filter is required to have a gain of 16 and a cutoff frequency of 1 kHz. Obtain a cascade design using OAs. Use 0.01 µF capacitors.

We shall consider the realization in terms of the transfer function $H(s) = H_1(s)H_2(s)$, where $H_1(s) = \frac{1}{s^2 + b_{11}s + b_{01}}$ and $H_2(s) = \frac{1}{s^2 + b_{12}s + b_{02}}$. From the Butterworth filter function table (Appendix A), we find that for the normalized LP filter with a cutoff frequency of 1 rad s^{-1}, $b_{11} = 0.766$, $b_{01} = 1$, $b_{12} = 1.848$, and $b_{02} = 1$. We can use a cascade of two Sallen and Key LP filters, each having the structure as shown in Figure 5.4. For the sake of convenience, the structure is redrawn in Figure 7.24. The transfer function V_2/V_1 is given by

$$\frac{V_2}{V_1} = K \frac{1/(R_1 R_2 C_1 C_2)}{s^2 + s\left[\dfrac{1}{R_1 C_1} + \dfrac{1}{R_2 C_1} + (1-K)\dfrac{1}{R_2 C_2}\right] + \dfrac{1}{R_1 R_2 C_1 C_2}} \qquad (7.23)$$

Letting $C_1 = C_2 = 1$ F, and comparing Eq. (7.23) with $H_1(s) = \frac{1}{s^2 + b_{11}s + b_{01}}$, we get

$$b_{01} = \frac{1}{R_1 R_2}, \quad b_{11} = \frac{1}{R_1} + \frac{2-K}{R_2} \qquad (7.24)$$

7 Higher-Order Active Filters

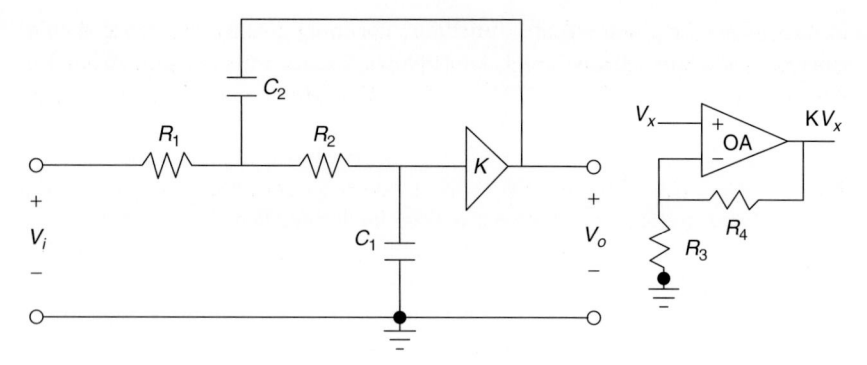

Figure 7.24 Sallen and Key LP structure used in the cascade design of the LP filter of Example 7.5.

Equation (7.24) leads to a quadratic equation in R_2. Solving this, we get

$$R_2 = \left(b_{11} \pm \sqrt{b_{11}^2 - 4b_{01}(2 - K)}\right) \bigg/ 2b_{01}$$

For physically realizable R_2, one must have $[b_{11}^2 - 4b_{01}(2 - K)] \geq 0$. On substituting for b_{01} and b_{11}, one arrives at $K \geq 2.1466$. Let $K = 4$ for each SAB section. Setting $K = 4$, we get $R_2 = 1.85$ and $R_1 = 0.541$. Choose $R_4/R_3 = 3$ to make $K = 1 + R_4/R_3 = 4$.

One can choose R_3 and R_4 with due considerations for DC offset cancellation. Consider Figure 7.25, which is the case for DC (capacitors open circuits!). For ideal OA, $v_x = v_y$ and for zero offset $I_x = I_y$. Then, $\frac{v_x}{R_1+R_2} = \frac{v_y}{R_3||R_4}$, leading to $R_4 = K(R_1 + R_2)$. Then $R_4 = 4(7.1.85 + 0.54) = 9.56$ and $R_3 = 3.18$.

Using a similar procedure for the second stage, we get

$$R_2' = 2.613, \; R_1' = 0.383, \; R_4' = 11.97, \text{ and } R_3' = 3.99.$$

Since $K = 4$ for each of the stages, the overall gain of the filter is 16 as required. The above design is for a normalized cutoff frequency of 1 rad s^{-1} and all capacitors equal to 1 F. To get an $f_c = 1$ kHz and all $C = 0.01$ µF, we have to use both impedance and frequency scaling. An impedance scaling by 10^8 will make all the capacitors to be 0.01 µF, while all the resistors will be multiplied by 10^8. To scale

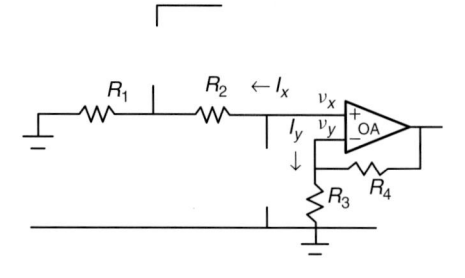

Figure 7.25 DC equivalent of the circuit in Figure 7.24 for offset voltage calculation and cancellation.

to the frequency of $2\pi \times 10^3$ rad s^{-1} without changing the capacitance values, we need to divide the resistors by the same factor. Thus, the overall scale factor for the resistors will be $10^8/2\pi \times 10^3 = 1.59 \times 10^4$. Hence, the final design values are

Stage 1: $C_1 = C_2 = 0.01\ \mu\text{F}$, $R_1 = 8.59\ \text{k}\Omega$, $R_2 = 24.42\ \text{k}\Omega$, $R_3 = 50.56\ \text{k}\Omega$, and $R_4 = 152\ \text{k}\Omega$

Stage 2: $C_1 = C_2 = 0.01\ \mu\text{F}$, $R'_1 = 63.44\ \text{k}\Omega$, $R'_2 = 41.55\ \text{k}\Omega$, $R'_3 = 63.44\ \text{k}\Omega$, and $R'_4 = 190.3\ \text{k}\Omega$.

While cascade design is very straightforward, implementation of high-order filters using a cascade of biquad sections is usually limited to an order of 8 or less. The principal reason is that the cascade designs are often found to be too sensitive to component variations in the PB due to component tolerances in the individual sections. The fact that the individual biquad sections are relatively decoupled from each other makes the system design task easier, and the tuning becomes simpler; but, it makes the control of sensitivity very difficult. This happens because of absence of any feedback between the individual cascaded sections. Usually, several trial designs are required to reduce such sensitivities by careful pairing of the poles and zeros of the given high-order transfer function.

7.3.1
Sensitivity Considerations

It has been found that the sensitivity of a high-order filter to component tolerances depends upon the frequency range of application and the structure chosen to implement the filter. The structure determines the expressions for the pole and zero frequencies and these expressions in turn influence the sensitivity values. In a standard biquadratic transfer function, the overall sensitivity is influenced by the sensitivity due to the numerator minus the sensitivity due to the denominator terms. The numerator terms are produced by products of the form $(s - z_i)$, where z_i is a zero of the transfer function, and the denominator terms are likewise produced by the products of the form $(s - p_j)$, where p_j is a pole of the transfer function. It may be argued that, if the pole-zero pair (p_j, z_i) are close to each other, their influence on the overall sensitivity will tend to cancel each other, thereby leading to low-sensitivity realization. Thus, one would consider implementing the individual biquad sections such that each section contains the pole-zero pair that is close to each other. However, it is not possible to generalize on this argument and classically rather diverse propositions have been advanced (Ghausi and Laker, 1981).

1) Pair the high-Q poles with those zeros farthest away such that $|p_i - z_i|$ is maximum.
2) Pair the high-Q poles with closest zeros such that $|p_i - z_i|$ is minimum.

In practice, one has to evaluate the sensitivities for all possible combinations and ascertain the right kind of pairing, which leads to smallest sensitivity realization. Quite often, computer-aided optimization will be required.

7.3.2
Sequencing of the Biquads

One of the flexibilities of the cascade realization is that the various biquads can be cascaded in a variety of sequences, since the overall transfer function is the product of the individual biquadratic transfer functions. For a moderately complex filter, we can make a judicious choice. It is recommended that the LP filter be placed at the front end, which will cause unwanted high-frequency signals to be attenuated before they reach later sections. Also, it is a good idea to place an HP filter at the end of the cascade. This will help in removing any internally generated noise outside the PB reaching the output. Again, one could make a thorough study of all the possible sequences, and then choose the best sequence.

7.3.3
Dynamic Range Considerations

The *dynamic range* is defined by the ratio of the maximum output signal without causing unacceptable distortion to the minimum detectable signal. The maximum undistorted signal is determined by the saturation of the active devices (depends upon the power supply) and the slew rate of the active devices, whichever leads to a worse-case scenario. The minimum signal level is determined by the noise floor of the system. A signal below the noise floor cannot be detected. For the maximum level, all frequencies are to be considered, while for the minimum signal level only the PB or in-band signals are of primary interest. Active filters typically have a dynamic range between 70 and 100 dB. In a cascaded system, a good dynamic range can be achieved if one can arrange the following:

1) The signal level at all frequencies of interest does not become too large in any section leading to overload of the following sections.
2) The in-band signals do not become too small at any stage and thereby get lost in the noise.

The above two objectives can be well taken care of if the signal level at the output of each section is maintained at the same level. This will need planning regarding the pole-zero pairing, order of cascading, and the choice of gain constants of the constituent sections.

To describe mathematically, if $S_{k\,min}$ is the minimum in-band signal at the output of the kth section within the PB frequencies and $S_{k\,max}$ is the maximum signal level at the output of the kth section for all frequencies (in-band and out-of-band), the maximum value of the ratio $r_j = S_{j\,max}/S_{j\,min}$ over all the $j = 1, 2, \ldots, N$ sections is to be minimized.

For more details concerning pole-zero pairing, sequencing of the biquads, dynamic range optimization, gain adjustment, and sensitivity considerations, the reader is referred to Schaumann, Ghausi, and Laker 1990.

7.4
Multiloop Feedback (and Feed-Forward) System

The low sensitivity of ladder filters is ascribable to the existence of tight feedback between adjacent sections of the filter. This is possible because the adjacent sections are implemented with passive elements and passive elements possess the property of bilateralism. In the cascade of active-RC biquad sections, this bilateralism is lost and the tight feedback is absent. This is the principal reason for the poor sensitivity performance of a cascaded second-order system. Obviously, one may conjecture that the inclusion of negative feedback between the cascaded second-order sections improves the sensitivity performance. This is what is precisely achieved in multiloop feedback and feed-forward systems involving modules of first- or second-order active-RC filters. When the desired transfer function is a high-order symmetrical BP or BR function, multiloop feedback (MLF) realizations result in low sensitivity, OA-efficient designs. It may be mentioned that the leapfrog architecture, which has been already discussed in Section 7.2, is a special case of the MLF implementations.

We now consider some of the MLF realizations using OAs.

7.4.1
Follow the Leader Feedback Structure

The simplest of the MLF structures is the follow the leader feedback (FLF) structure shown in Figure 7.26, where the transfer functions $T_i (i = 1, 2, \ldots, n)$ may be lossless or lossy integrators, bilinear, or biquadratic functions. When all the transfer functions are simple integrators, the structure is also known as the *controller canonic structure* in system theory and has been used to solve differential equations using analog computers.

It is easy to see from Figure 7.26 that

$$KV_1 - \left[f_0 - \left(\frac{f_1}{T_n} \right) - \left(\frac{f_2}{T_{n-1} T_n} \right) + \cdots + \left(\frac{f_{n-1}}{T_2 \ldots T_n} \right) \right] V_2 = \left(\frac{V_2}{T_1 \ldots T_n} \right)$$

Hence,

$$\frac{V_2}{V_1} = \frac{K}{\left[\left(\frac{1}{T_1 \ldots T_n} \right) + \left(\frac{f_{n-1}}{T_2 \ldots T_n} \right) + \cdots + \left(\frac{f_1}{T_n} \right) + f_0 \right]} \quad (7.25a)$$

or

$$\frac{V_2}{V_1} = \frac{K T_1 \ldots T_n}{[1 + f_{n-1} T_1 + f_{n-2} T_1 T_2 + \cdots + f_1 T_1 \ldots T_1 + f_0 T_1 \ldots T_n]} \quad (7.25b)$$

It is clear from Eq. (7.25) that the zeros of the transfer function are located at the zeros of the T_i. Hence, when T_i is a lossless or lossy integrator, then the denominator of Eq. (7.25) will be a polynomial in s, and hence this structure can only realize all-pole functions. It is clear that, in order to realize arbitrary finite-transmission zeros, the various T_i's have to be, in general, biquads. The theory is quite complicated when this is the case and has not been considered here. The interested reader can refer to Laker and Ghausi (1980) and Ghausi and Laker (1981). We only consider the cases when T_i is a lossless or lossy integrator.

7.4.1.1 $T_i = (1/s)$, a Lossless Integrator

In this case, the structure is as shown in Figure 7.27, where $f_0, f_1, \ldots, f_{n-1}$ of Figure 7.26 have been replaced by $b_0, b_1, \ldots, b_{n-1}$. The corresponding transfer function is given by

$$\frac{V_2}{V_1} = \frac{K}{[s^n + b_{n-1}s^{n-1} + b_{n-2}s^{n-2} + \cdots + b_1 s + b_0]} \tag{7.26}$$

which is an LP transfer function. Of course, by taking the outputs at different points, we can also get HP and BP transfer functions.

All the integrators used in Figure 7.27 are noninverting ones. Since inverting integrators are most commonly used in practice, the feedback coefficients have to

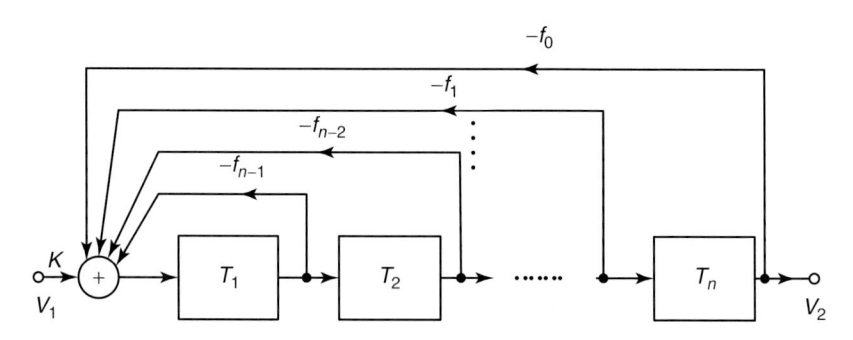

Figure 7.26 Follow the leader feedback structure.

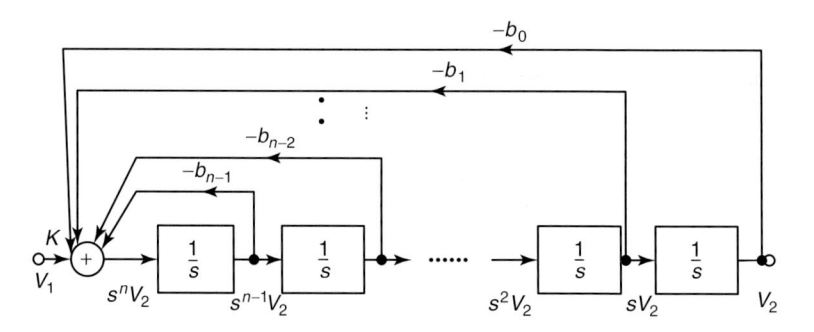

Figure 7.27 FLF structure where each T_i is a lossless integrator.

7.4 Multiloop Feedback (and Feed-Forward) System

be suitably multiplied by ± 1 to realize the structure of Figure 7.27 using OAs. Also, it may be possible to combine the first inverting integrator with an inverted weighted summing integrator. We illustrate such a realization using a fourth-order LP filter.

Example 7.6. Realize the fourth-order LP filter given by

$$H(s) = \frac{V_2}{V_1} = \frac{K}{s^4 + b_3 s^3 + b_2 s^2 + b_1 s + b_0} \tag{7.27}$$

using the FLF structure with lossless inverting integrators.

Equation (7.27) can be rewritten in the form

$$\frac{V_1}{1/K} - \left(\frac{V_2}{1/b_0} + \frac{s^2 V_2}{1/b_2} + s^4 V_2 \right) + \left(\frac{-s V_2}{1/b_1} + \frac{-s^3 V_2}{1/b_3} \right) = 0 \tag{7.28}$$

Equation (7.28) can be realized using four inverting integrators, and two inverting summing amplifier, as shown in Figure 7.28.

However, if we rewrite Eq. (7.28) as

$$\frac{V_1}{1/K} - \left(\frac{V_2}{1/b_0} + \frac{s^2 V_2}{1/b_2} \right) + \left(\frac{-s V_2}{1/b_1} \right) + \left(\frac{-s^3 V_2}{1/b_3} + \frac{-s^3 V_2}{1/s} \right) = 0 \tag{7.29}$$

then Eq. (7.29) can be realized with three inverting integrators, one inverting integrating summer, and one inverting summer, as shown in Figure 7.29.

7.4.1.2 $T_i = 1/(s + \alpha)$, a Lossy Integrator

In this case, we are using lossy integrators in the structure given in Figure 7.26 and Eq. (7.25) reduces to

$$\frac{V_2}{V_1} = \frac{K}{[(s+\alpha)^n + f_{n-1}(s+\alpha)^{n-1} + f_{n-2}(s+\alpha)^{n-2} + \cdots + f_1(s+\alpha) + f_0]} \tag{7.30}$$

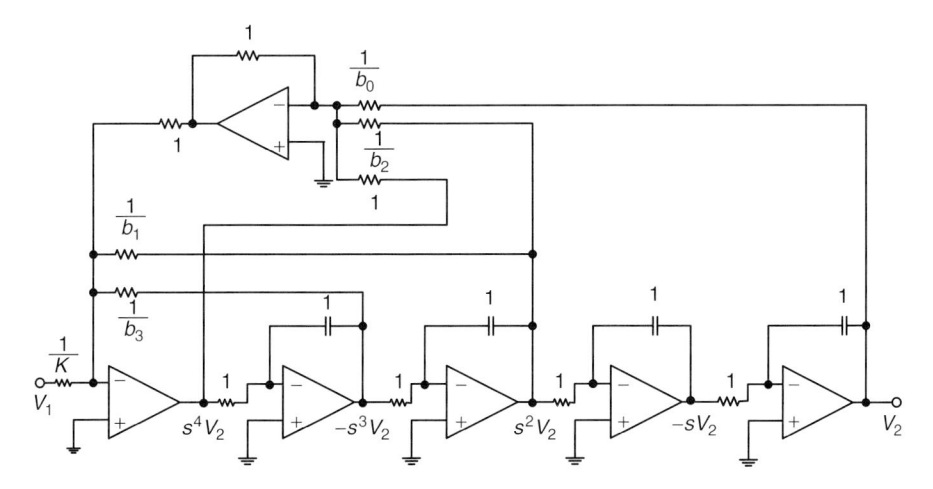

Figure 7.28 FLF realization of the fourth-order LP filter given by Eq. (7.27).

232 | 7 Higher-Order Active Filters

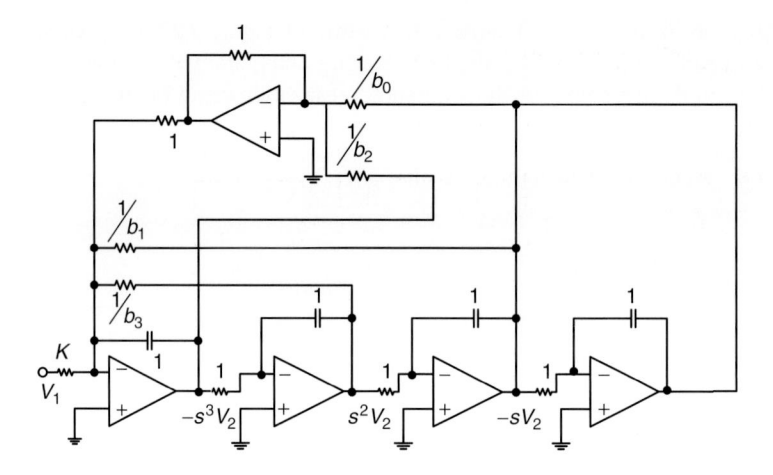

Figure 7.29 Alternative FLF realization of the fourth-order LP filter given by Eq. (7.27).

Comparing Eq. (7.30) with the all-pole LP transfer function given by Eq. (7.26), we have

$$f_{n-1} = b_{n-1} - n\alpha$$
$$f_{n-2} = b_{n-2} - (n-1)\alpha f_{n-1} - \{n(n-1)/2\}\alpha^2$$
$$\vdots$$
$$f_0 = b_0 - \alpha f_1 - \alpha^2 f_2 - \cdots - \alpha^{n-1} f_{n-1} - \alpha^n \qquad (7.31)$$

We may choose a suitable value for α, and find the feedback coefficients $f_{n-1}, f_{n-2}, \ldots, f_1$, and f_0, successively from Eq. (7.31), in terms of b_{n-1}, $b_{n-2}, \ldots, b_1, b_0$, and α. We can now realize Eq. (7.30) using the structure of Figure 7.26, wherein the function $T_i(s) = 1/(s + \alpha)$ is realized using lossy integrators. Since we will be using inverting integrators, we have to associate proper signs to $f_{n-1}, f_{n-2}, \ldots, f_1$, and f_0 in realizing Eq. (7.30). We illustrate this case for a Butterworth filter of order 3. The same procedure could be used for a higher-order filter.

Example 7.7. Realize the third-order LP filter given by

$$H(s) = \frac{V_2}{V_1} = \frac{K}{s^3 + 2s^2 + 2s + 1} \qquad (7.32)$$

using lossy integrators of the type $-1/(s+1)$.

Using Eq. (7.31), we have

$$f_2 = 2 - 3 = 1, \quad f_1 = 2 - 4 + 3 = 1 \text{ and } f_0 = 1 - 2 + 2 - 1 = 0$$

Hence, the given transfer function can be written as

$$\frac{V_2}{V_1} = \frac{K}{[(s+1)^3 - (s+1)^2 + (s+1)]}$$

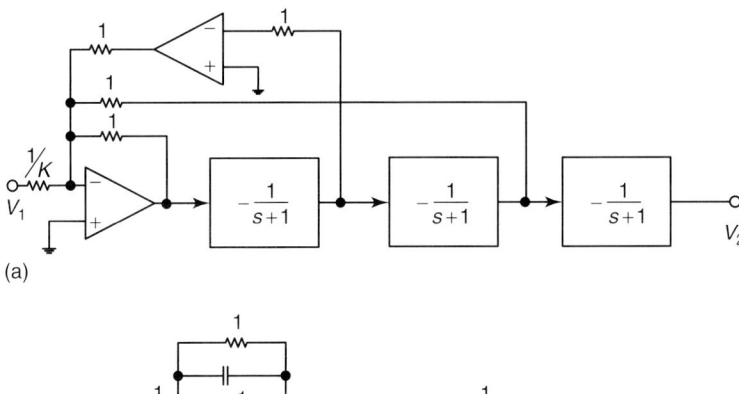

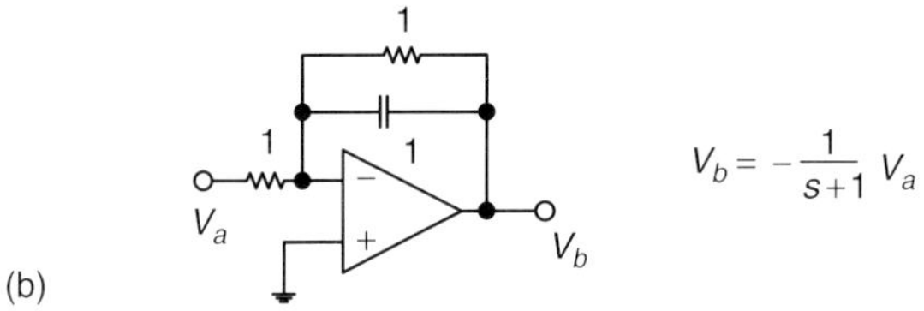

Figure 7.30 FLF realization of the third-order LP filter given by Eq. (7.32).

The above can be easily realized as shown in Figure 7.30a, wherein each of the lossy integrators is realized using the inverting integrator shown in Figure 7.30b.

As mentioned earlier, when T_i is a lossless or lossy integrator, then the denominator of Eq. (7.25) will be a polynomial in s; hence, the structure of Figure 7.26 can only realize all-pole functions. In order to realize finite-transmission zeros using this structure, we will have to use biquads as the building blocks for the various T_i (Laker and Ghausi, 1980; Ghausi and Laker, 1981). However, it is possible to realize finite-transmission zeros using that structure with lossless or lossy integrators, if we provide suitable feed-forward paths. This is considered next.

7.4.2
FLF Structure with Feed-Forward Paths

Consider a general nth-order transfer function of the form

$$H(s) = \frac{V_2}{V_1} = \frac{[a_n s^n + a_{n-1} s^{n-1} + a_{n-2} s^{n-2} + \cdots + a_1 s + a_0]}{[s^n + b_{n-1} s^{n-1} + b_{n-2} s^{n-2} + \cdots + b_1 s + b_0]} \quad (7.33)$$

This can be rewritten as

$$H(s) = H_1(s) H_2(s) \quad (7.34)$$

where

$$H_1(s) = \frac{V_2'}{V_1} = \frac{1}{[s^n + b_{n-1} s^{n-1} + b_{n-2} s^{n-2} + \cdots + b_1 s + b_0]} \quad (7.35)$$

and

$$H_2(s) = \frac{V_2}{V_2'} = [a_n s^n + a_{n-1} s^{n-1} + a_{n-2} s^{n-2} + \cdots + a_1 s + a_0] \quad (7.36)$$

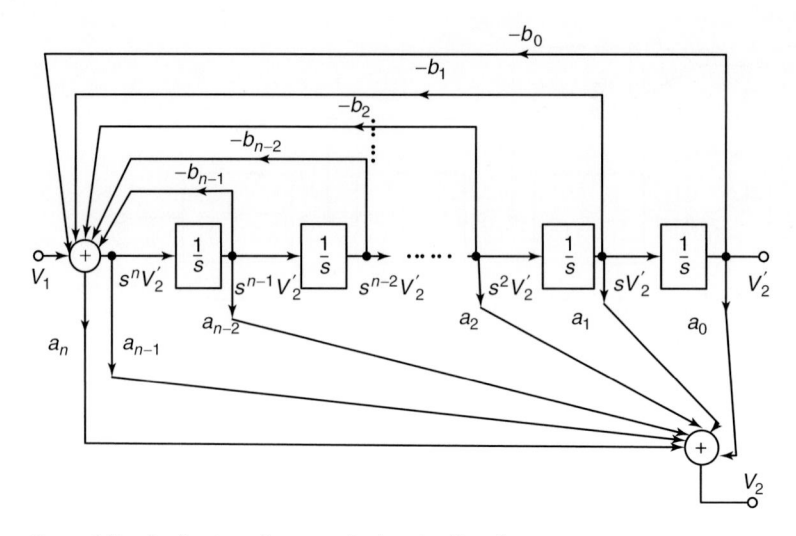

Figure 7.31 Realization of a general nth-order filter function using lossless integrators in an FLF structure with feed-forward paths.

The transfer function $H_1(s)$ can be realized using the structure of Figure 7.27 (with $K = 1$), while $H_2(s)$ can be realized by adding the outputs at various points in the structure of Figure 7.27 after multiplying these outputs by appropriate values. The overall realization of $H(s)$ given by Eq. (7.35) is shown in Figure 7.31. Again, it should be remembered that, since we are going to be using inverting integrators, the signs for the various coefficients a_i and b_i have to be properly chosen. A similar procedure can be used when lossy integrators are used instead of lossless ones shown in Figure 7.31.

As an illustration, the realization of a general fourth-order transfer function given by

$$H(s) = \frac{V_2}{V_1} = \frac{a_4 s^4 + a_3 s^3 + a_2 s^2 + a_1 s + a_0}{s^4 + b_3 s^3 + b_2 s^2 + b_1 s + b_0} \quad (7.37)$$

is shown in Figure 7.32a using lossless integrators, wherein each of the integrators is realized using the inverting integrator shown in Figure 7.32b.

7.4.3
Shifted Companion Feedback Structure

We saw in Section 7.1 that by choosing all the T_i's to be lossy integrators of the form $T_i = 1/(s + \alpha)$, we could realize an all-pole LP filter given by

$$\frac{V_2}{V_1} = \frac{K}{[s^n + b_{n-1} s^{n-1} + b_{n-2} s^{n-2} + \cdots + b_1 s + b_0]} \quad (7.38)$$

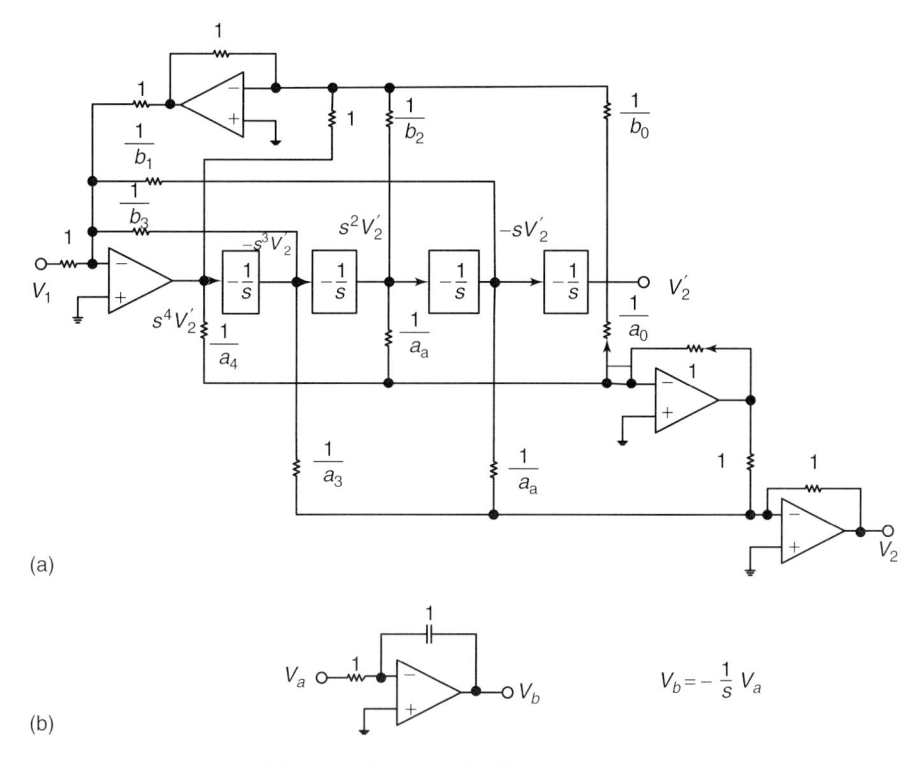

Figure 7.32 Realization of the general fourth-order filter function given by Eq. (7.37) using the FLF structure with feed-forward paths shown in Figure 7.31.

by rewriting it as

$$\frac{V_2}{V_1} = \frac{K}{[(s+\alpha)^n + f_{n-1}(s+\alpha)^{n-1} + f_{n-2}(s+\alpha)^{n-2} + \cdots + f_1(s+\alpha) + f_0]} \quad (7.39)$$

where the various f_is are give by Eq. (7.31). If we choose

$$\alpha = b_{n-1}/n \quad (7.40)$$

then

$$f_{n-1} = 0 \quad (7.41)$$

that is, there is no feedback path from the output of the first integrator. Such a structure is called the shifted companion feedback (SCF) structure, and is shown in Figure 7.33. It may be mentioned that, by providing suitable feed-forward paths from the various points in this structure and adding them at the output as we did in the case of lossless integrators (see Figure 7.31), it is possible to realize a general VTF of the form Eq. (7.33) using lossy integrators.

7 Higher-Order Active Filters

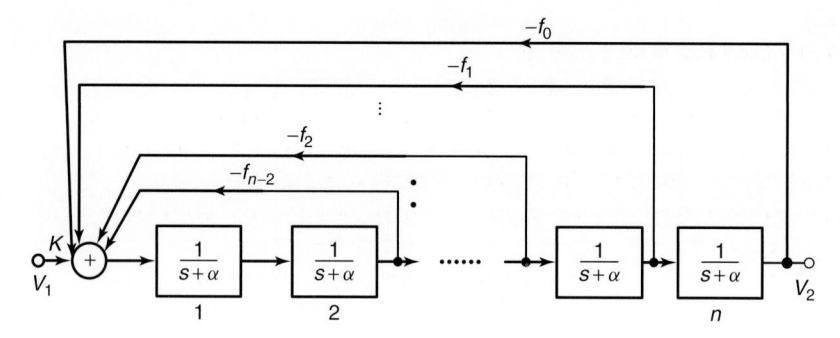

Figure 7.33 Shifted companion feedback structure.

As an illustration, consider the third-order Butterworth filter given by Eq. (7.32) and let $\alpha = 2/3$, then we find from Eq. (7.31) that $f_2 = 0$, $f_1 = 2/3$, and $f_0 = 7/27$. Hence, we can rewrite Eq. (7.32) as

$$\frac{V_2}{V_1} = \frac{K}{\left[\left(s+\frac{2}{3}\right)^3 + \left(\frac{2}{3}\right)\left(s+\frac{2}{3}\right) + \frac{7}{27}\right]} \tag{7.42}$$

This can be realized as shown in Figure 7.34a, wherein each of the lossy integrators is obtained in the inverting form as shown in Figure 7.34b.

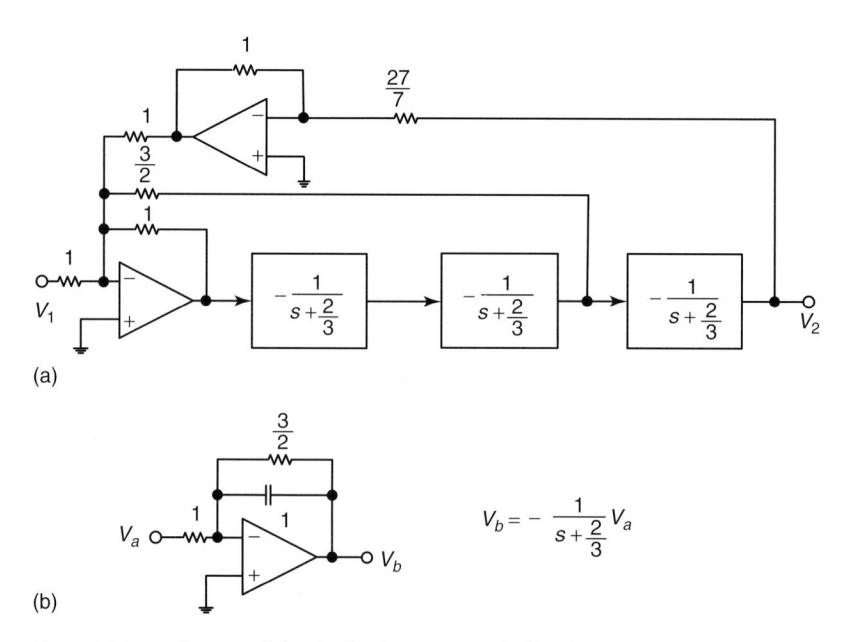

Figure 7.34 Realization of the third-order Butterworth filter by SCF structure.

7.4.4
Primary Resonator Block Structure

It was mentioned earlier in Section 7.4.1 that the VTF V_2/V_1 of the structure Figure 7.26 can have transmission zeros only at the zeros of the various T_i. Hence, biquads are needed to realize arbitrary finite-transmission zeros for V_2/V_1. We also saw in the previous subsections as to how one could realize an all-pole VTF using lossy integrators for the various T_i. We now show how to realize a high-order BP filter using the LP to BP transformation in conjunction with the structure of Figure 7.33.

Let the all-pole LP filter given by

$$\frac{V_2}{V_1} = \frac{K}{[s^n + b_{n-1}s^{n-1} + b_{n-2}s^{n-2} + \cdots + b_1 s + b_0]} \quad (7.43)$$

rewritten as

$$\frac{V_2}{V_1} = \frac{K}{[(s+\alpha)^n + f_{n-1}(s+\alpha)^{n-1} + f_{n-2}(s+\alpha)^{n-2} + \cdots + f_1(s+\alpha) + f_0]} \quad (7.44)$$

and realized by the SCF structure of Figure 7.33. Then,

$$\alpha = b_{n-1}/n \quad \text{and} \quad f_{n-1} = 0. \quad (7.45)$$

The remaining feedback coefficients $f_{n-2}, \ldots, f_1, f_0$ can be obtained in terms of α and the various b_i using Eq. (7.31). Now introduce the normalized LP to BP transformation

$$s = Q\frac{s^2 + 1}{s} \quad (7.46)$$

Then

$$(s + \alpha) \rightarrow \left(Q\frac{s^2+1}{s} + \alpha\right) = \frac{Q}{s}\left[s^2 + \frac{\alpha}{Q}s + 1\right] \quad (7.47)$$

Hence, each of the lossy integrators in Figure 7.33 is replaced by the second-order BP filter

$$\frac{s/Q}{s^2 + (\alpha/Q)s + 1} = \frac{(H_0/Q_p)s}{s^2 + (1/Q_p)s + 1} \quad (7.48)$$

where $Q_p = Q/\alpha$ is the pole Q and $H_0 = 1/\alpha$ is the gain at the center frequency of the second-order BP filter. Each of the BP biquad may be realized by any of the methods discussed in Chapter 5. Thus, a high-order BP filter may be realized by coupling a number of identical BP biquads. These identical biquads are called *primary resonant blocks* (Hurtig, 1973) and the structure itself the *primary resonator block* (PRB) structure. We illustrate the method by the following example.

Example 7.8. Design an eighth-order Butterworth BP filter with a center frequency of 3000 Hz, a bandwidth of 600 Hz, and gain of unity at its center frequency using the PRB structure.

7 Higher-Order Active Filters

Since an eighth-order Butterworth BP filter is required, we start off with the fourth-order Butterworth LP filter

$$H_{LP}(s) = \frac{V_2(s)}{V_1(s)} = \frac{1}{s^4 + 2.613126s^3 + 3.414214s^2 + 2.613126s + 1} \quad (7.49)$$

Then, from Eq. (7.40),

$$\alpha = \frac{b_3}{4} = 0.65329$$

We can get the values of f_2, f_1, and f_0 using Eq. (7.31) as

$$f_2 = 0.8536, \quad f_1 = 0.3838, \quad \text{and} \quad f_0 = 0.2036.$$

From the given specifications, the Q of the required eighth-order BP filter is $(3000/600)$, that is, $Q = 5$. Therefore, Q_p of the PRB is given by $(Q/\alpha) = 7.654$. Hence, from Eq. (7.48), the transfer function of the PRB is given by

$$\frac{(H_0/Q_p)s}{s^2 + (1/Q_p)s + 1} \quad \text{with} \quad H_0 = 1.5307 \quad \text{and} \quad Q_p = 7.654$$

Since the required center frequency is 3000 Hz, the pole frequency for the PRB is 6000π. Hence, the denormalized transfer function of the PRB is given by

$$T_{PRB} = \frac{(H_0/Q_p)s}{s^2 + (1/Q_p)s + (6000\pi)^2} \quad \text{with} \quad H_0 = 1.5307 \quad \text{and} \quad Q_p = 7.654 \quad (7.50)$$

We may use any of the methods discussed in Chapter 5 to design the T_{PRB} given by Eq. (7.50). Then, the overall realization of the required BP filter is as shown in Figure 7.35, assuming that each of the PRBs is realized using a noninverting amplifier.

It should be mentioned that, in addition to structures that we have considered, there are many other multiple-loop structures such as the modified leapfrog structure, inverse follow the leader feedback (IFLF) structure, and minimum sensitivity feedback (MSF) structure. For details, one may refer to Laker and Ghausi (1980) and Ghausi and Laker (1981).

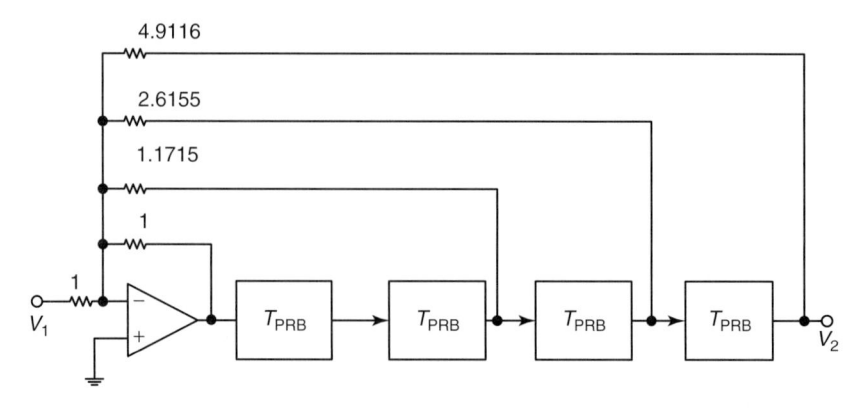

Figure 7.35 Realization of the transfer function of Example 7.8 by the PRB structure.

7.5
High-Order Filters Using Operational Transconductance Amplifiers

The design of high-order transconductance–C filters proceeds essentially along the same lines as the design of OA-based high-order filters. Basically, two approaches can be identified, namely, (i) cascade design and (ii) ladder simulation. In the case of ladder simulation, one can simulate the inductance as an element or one can simulate the $I - V$ relations of the ladder arms (i.e., operational simulation).

7.5.1
Cascade of OTA-Based Filters

Cascade of OTA-based second-order filters can be used to realize high-order filters in the same way as the cascade of OA-based filters. The only important concern is the high output resistance of an OTA compared with that of an OA. This high impedance node can combine with the input parasitic capacitance of the following stage to produce a low-frequency parasitic pole, which would vitiate the desired transfer function. However, the design capacitance can be adjusted to take care of this parasitic pole. Since the output of an OTA does not have a low resistance like an OA, the loading effect of two consecutive OTA stages may be an important concern. However, as long as the input impedance of an OTA is at least 10 times or more than the output impedance of the preceding OTA, the effect will be negligible.

7.5.2
Inductance Simulation

OTAs are very well poised for implementing an inductance. It has been already discussed under Section 7.1.1 that the OTA-based gyrators can be used to implement both grounded and floating inductances. In Chapter 5 (see Table 5.10), we have also shown how OTAs can be used to implement various important network elements like a resistance, an integrator, and a gyrator. This knowledge can be easily used to implement a high-order LC ladder filter. Figure 7.36a shows a passive third-order elliptic LP filter with source and load-terminating resistances R_1 and R_2. Figure 7.36b shows the filter with the source transformation applied at the input terminal. Figure 7.36c shows an OTA-based implementation, where the floating inductance has been simulated by the OTAs numbered 3, 4, 5, and 6 and the capacitance C_L.

7.5.3
Operational Simulation Technique

The procedure follows on the same lines as in the case of the operation simulation using OAs. We illustrate it by considering the same ladder of Figure 7.36a, which is shown in Figure 7.37a with the $I-V$ variables labeled on the schematic. The capacitor C_2 is taken into consideration later. The remainder of the ladder can be

7 Higher-Order Active Filters

(a)

(b)

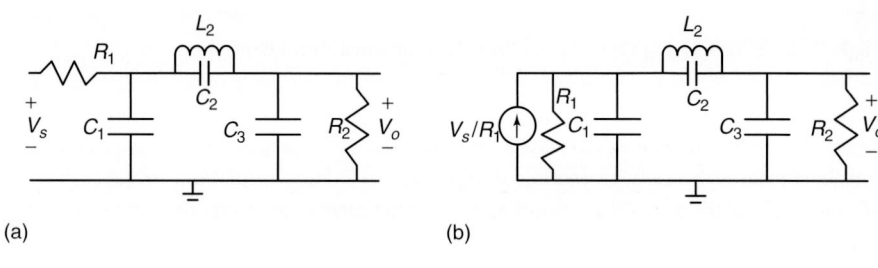

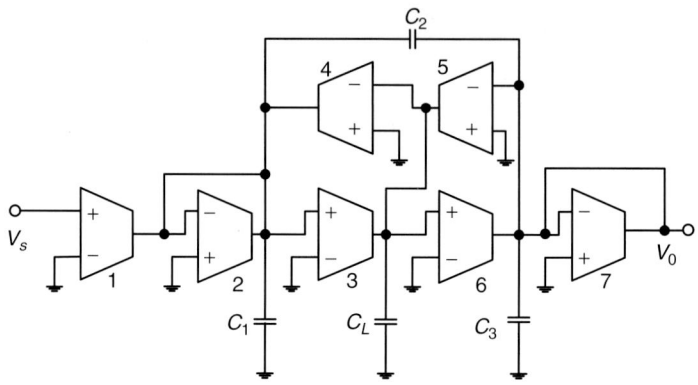

(c) $g_{m1} = g_{m2} = G_1$, $g_{m3} = g_{m4} = g_{m5} = g_{m6} = g$, $C_L = g^2 L$, $g_{m7} = G_2$; g_{mx} is the transconductance of the OTA numbered x.

Figure 7.36 (a) A third-order elliptic low-pass filter, (b) the filter after source transformation, and (c) realization of the filter in (b) using a gyrator-based floating inductor.

described by the following equations:

$$V_1 = \frac{I_s - I_2}{G_1 + sC_1}, \quad I_2 = \frac{V_1 - V_3}{sL_2}, \quad \text{and} \quad V_3 = \frac{I_2}{G_2 + sC_3} \qquad (7.51)$$

We may write first of the equations in Eq. (7.51) as

$$V_1 = \frac{R}{R}\frac{I_s - I_2}{G_1 + sC_1} = \frac{(R/R_1)(V_s) - RI_2}{R(G_1 + sC_1)}$$

By letting $R = R_1$ and $RI_2 = \hat{V}_2$, the above equation reduces to

$$V_1 = \frac{(V_s - \hat{V}_2)G}{G_1 + sC_1} \qquad (7.52a)$$

As a consequence, the other equations in Eq. (7.51) become

$$\hat{V}_2 = \frac{V_1 - V_3}{sL_2/R} = \frac{(V_1 - V_3)}{sL_2 G} = \frac{(V_1 - V_3)G}{sL_2 G^2} = \frac{(V_1 - V_3)G}{sC_L} \qquad (7.52b)$$

7.5 High-Order Filters Using Operational Transconductance Amplifiers

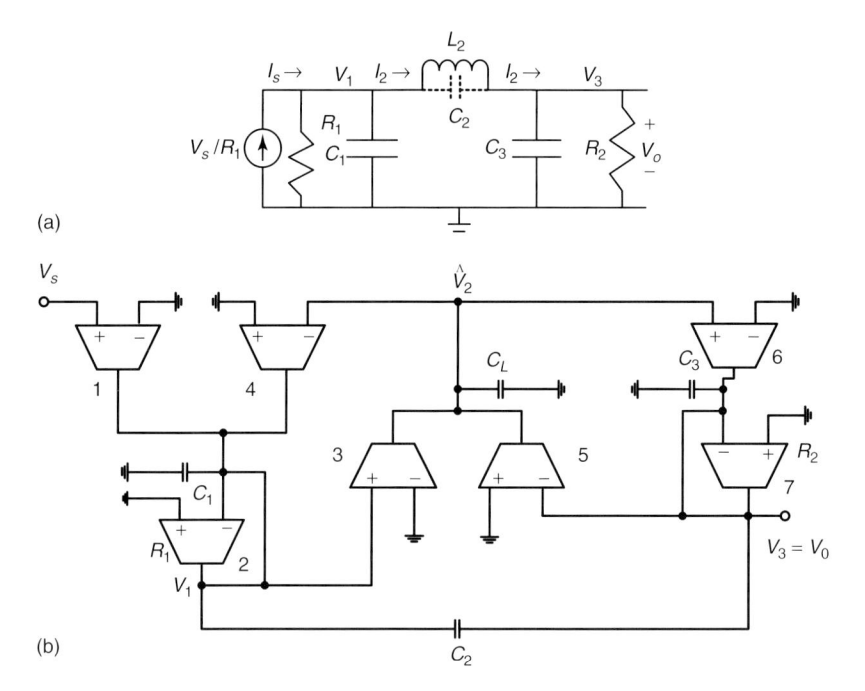

Figure 7.37 (a) Doubly terminated LC third-order elliptic filter. (b) OTA-based implementation of the filter using operational simulation technique.

and

$$V_3 = \frac{\hat{V}_2 G}{G_2 + sC_3} \tag{7.52c}$$

where $C_L = L_2 G^2$.

Equations (7.52a) and (7.52c) represent operations of lossy integrators with differential inputs, while Eq. (7.52b) is an ideal integrator with a differential input. These operations can be very easily accomplished by differential OTAs. However, we give here the realization with single-ended OTAs. Finally, the capacitance C_2 is inserted across the V_1 and V_3 nodes, since the current through it is equal to $(V_1 - V_3)sC_2$. The final OTA-based implementation is shown in Figure 7.37b, where we have used $g_{m1} = g_{m2} = G_1 = 1/R_1 = g_{m3} = g_{m4} = g_{m5} = g_{m6} = G$, and $g_{m7} = G_2 = 1/R_2$, g_{mx} being the transconductance of the OTA numbered x. A careful scrutiny will reveal that the structures in Figures 7.36c and 7.37b are indeed the same, but have been arranged in slightly different ways. As mentioned earlier, the latter structure can be realized using differential-output OTAs and thus economize on the number of OTAs.

7.5.4
Leapfrog Structure for a General Ladder

We now consider a general passive ladder structure and show how it can be realized in a systematic fashion as a leapfrog structure employing OTAs. For illustrative purposes, we consider a ladder with six elements as shown in Figure 7.38, but the same procedure can be adopted for any general ladder.

For the ladder network, we may write the I–V relations as

$$\begin{aligned} I_1 &= Y_1(V_{in} - V_2) & V_2 &= Z_2(I_1 - I_3) \\ I_3 &= Y_3(V_2 - V_4) & V_4 &= Z_4(I_3 - I_5) \\ I_5 &= Y_5(V_4 - V_6) & V_{out} &= V_6 = Z_6(I_5) \end{aligned} \qquad (7.53)$$

We may rewrite the above equations in the following form so that all the current variables are converted into equivalent voltage variables:

$$\begin{aligned} V'_1 &= \frac{Y_1}{g}(V_{in} - V_2) & V_2 &= gZ_2(V'_1 - V'_3) \\ V'_3 &= \frac{Y_3}{g}(V_2 - V_4) & V_4 &= gZ_4(V'_3 - V'_5) \\ V'_5 &= \frac{Y_5}{g}(V_4 - V_6) & V_{out} &= V_6 = gZ_6(V'_5) \end{aligned} \qquad (7.54)$$

where

$$V'_j = (I_j/g), j = 1, 3, 5 \qquad (7.55)$$

and g is an arbitrary scaling factor. Hence,

$$V'_j = \frac{Y_j}{g}(V_{j-1} - V_{j+1}) \quad \text{for odd } j \qquad (7.56)$$

$$V_j = gZ_j(V'_{j-1} - V'_{j+1}) \quad \text{for even } j \qquad (7.57)$$

Consider the OTA circuit of Figure 7.39a. It is very easy to see that we have

$$V'_j = g_{mj}Z'_j(V_{j-1} - V_{j+1}) \qquad (7.58)$$

Comparing Eqs. (7.58) and (7.56), we see that (7.56) can be realized by the OTA circuit of Figure 7.39a provided $(Y_j/g) = g_{mj}Z'_j$, that is,

$$Z'_j = \frac{1}{gg_{mj}}Y_j \quad \text{for odd } j \qquad (7.59)$$

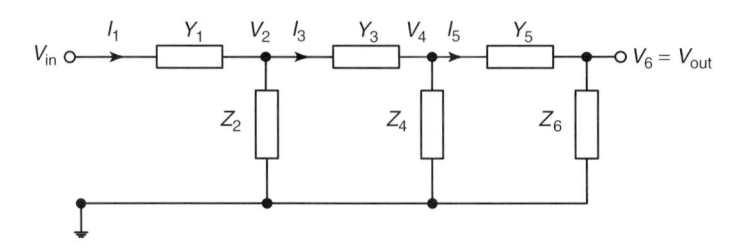

Figure 7.38 A general passive ladder structure.

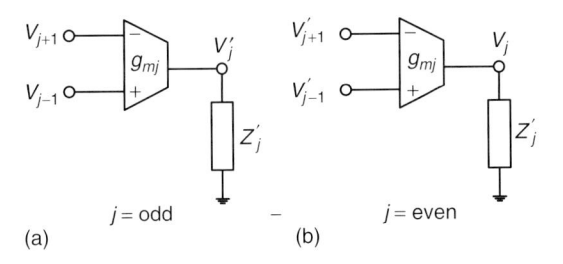

7.5 High-Order Filters Using Operational Transconductance Amplifiers

Figure 7.39 OTA structures realizing Eqs. (7.52) and (7.53).

Similarly, it can be shown that the OTA circuit of Figure 7.39b will realize Eq. (7.57) provided $gZ_j = g_{mj}Z'_j$, that is

$$Z'_j = \frac{g}{g_{mj}} Z_j \quad \text{for even } j \tag{7.60}$$

Thus, the set of I–V relations given by Eq. (7.53), and hence the ladder of Figure 7.38, can be realized by the leapfrog structure utilizing OTAs, as shown in Figure 7.40. It is noted that the various Z'_j (j odd or even) are due to R, L, and C elements, which can be easily realized using OTA-C structures. We now illustrate the procedure by an example.

Example 7.9. Realize the third-order elliptic LP filter shown in Figure 7.41a by the leapfrog structure discussed above.

Using the method discussed above, we see that the filter can be realized by the structure of Figure 7.40, where the various impedances $Z'_1, Z'_2, Z'_3,$ and Z'_4 are given by Eqs. (7.59) and (7.60). Hence,

$$Z'_1 = \frac{1}{gg_{m1}} G_1, \quad Z'_3 = \frac{1}{gg_{m3}}\left(sC_3 + \frac{1}{sL_3}\right)$$

$$Z'_2 = \frac{g}{g_{m2}} \frac{1}{sC_2}, \quad Z'_4 = \frac{g}{g_{m2}} \frac{1}{(G_4 + sC_4)}$$

Figure 7.40 Realization of the ladder network of Figure 7.38 by a leapfrog structure.

(a)

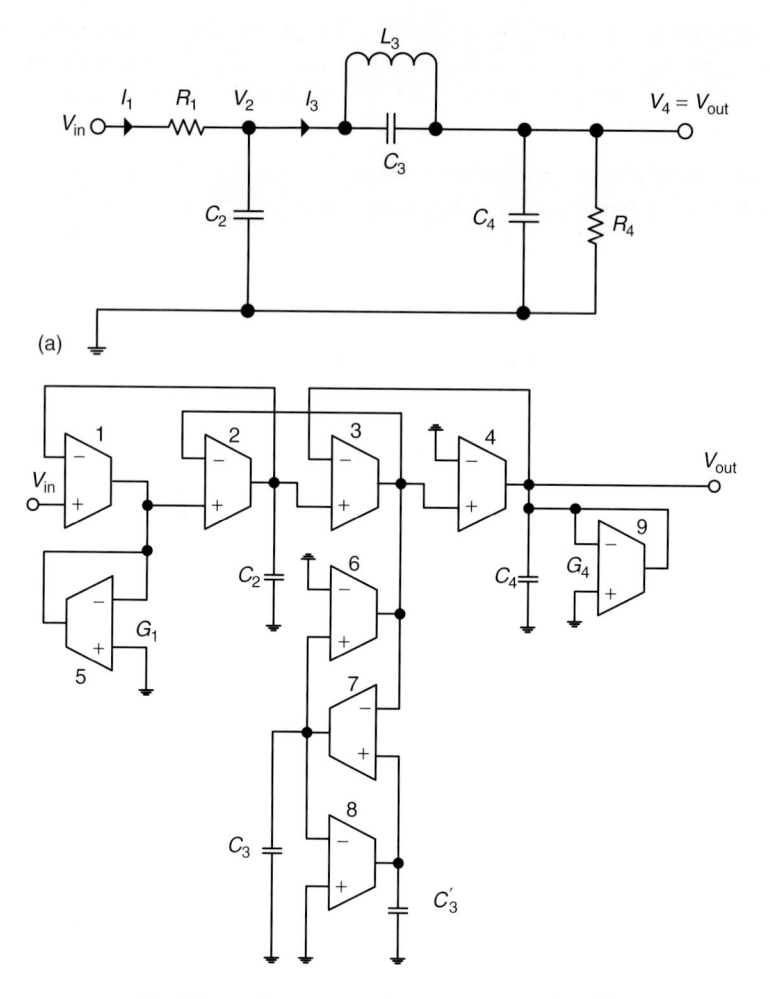

(b) g_{mj} ($j = 1,2, \ldots, 8$) $= g = g_m = G_1$, $g_{m9} = G_4$, $C'_3 = g^2_m L_3$

Figure 7.41 Leapfrog realization of the elliptic filter of Example 7.9.

Let us choose $g_{m1} = g_{m2} = g_{m3} = g_{m4} = g = g_m = G_1$; then the above expressions become

$$Z'_1 = R_1, \quad Z'_2 = \frac{1}{sC_2}, \quad Z'_3 = s\frac{C_3}{g^2_m} + \frac{1}{s}\left(\frac{1}{g^2_m L_3}\right), \quad Y'_4 = (G_4 + sC_4)$$

We see that Z'_1 is a grounded resistor of value R_1, while $Z'_4 = 1/Y'_4$ is a parallel combination of a grounded capacitor C_4 and a grounded resistor R_4. The two grounded resistors R_1 and R_4 can be easily realized using OTAs of transconductances G_1 and G_4, respectively. The impedance Z'_3 is a series combination of an inductor of value C_3/g^2_m and a capacitor of value $g^2_m L_3$. We have the option of having the inductor or the capacitor grounded. Even though the former arrangement uses only two OTAs, the latter arrangement is preferable since in an IC implementation, it is better to have grounded capacitors. In such a case, the ungrounded inductor can be realized using three OTAs and a grounded capacitor (see Table 5.10, Row J). The complete structure realizing the ladder network is shown in Figure 7.41b.

There are a number of other multiloop structures utilizing OTAs that can be used to realize a general high-order filter, and the reader is referred to Deliyanis, Sun, and Fidler (1999) for more details.

7.6 High-Order Filters Using Switched-Capacitor (SC) Networks

In SC filters, we can follow, for high-order filter realization, a scenario similar to that in the case of continuous filters. Thus, we can build a given high-order filter using a cascade of second- and first-order SC networks. We can also use second-order sections in a feedback structure to generate a given z-domain transfer function. We could use SC integrators and inverters to simulate the operation of a given LC ladder filter. For efficient realization of a high-order SC filter, which is the counterpart of a given active-RC filter under bilinear $s \leftrightarrow z$ transformation, it is necessary that the SC integrators that are counterparts of active-RC integrators under bilinear $s \leftrightarrow z$ transformation be available. Bilinear SC integrator configurations that preserve the PI features, such as the PI inverting and noninverting SC integrators introduced in Chapter 6, are discussed next.

7.6.1 Parasitic-Insensitive Toggle-Switched-Capacitor (TSC) Integrator

The TSC is one of the earliest configurations of SC networks used to produce an equivalent resistance. The TSC lost its popularity because of the presence of the parasitic capacitance at the top plate. In a biphase (i.e., two-phase) clocking scheme, the effect of this parasitic capacitance on the charge conservation equation cannot be discarded and hence a PI integrator cannot be built from a TSC. However, if one takes recourse to a three-phase clocking scheme, it is possible to render the TSC to a PI network (Bermudez and Bhattacharyya, 1982). The basic network and the timing diagram of the three-phase clock are shown in Figures 7.42a and 7.42b. The symbol $\phi_1 \oplus \phi_3$ in Figure 7.42 represents the logical-or operation between the clock signals ϕ_1 and ϕ_3.

During the phase when ϕ_3 is ON, the capacitor C_1 is charged from the voltage source V_1. During the ON phase of ϕ_2, the top-plate parasitic capacitance is discharged, while the signal voltage on C_1 is reversed, since the bottom plate is lifted off from the ground (ϕ_1 and ϕ_3 being both off). At clock phase ϕ_3, the bottom plate of C_1 is grounded again, while the top plate of C_1 is connected to the virtual ground input of the OA. The charge transfer on to the integrating capacitor C_2 thus becomes PI. If the output is sampled at each period of ϕ_1, the above operation will lead to a z-domain transfer function $(V_o^{(1)})/(V_1^{(1)}) = -(C_1/C_2)z^{-1}/(1 - z^{-1})$.

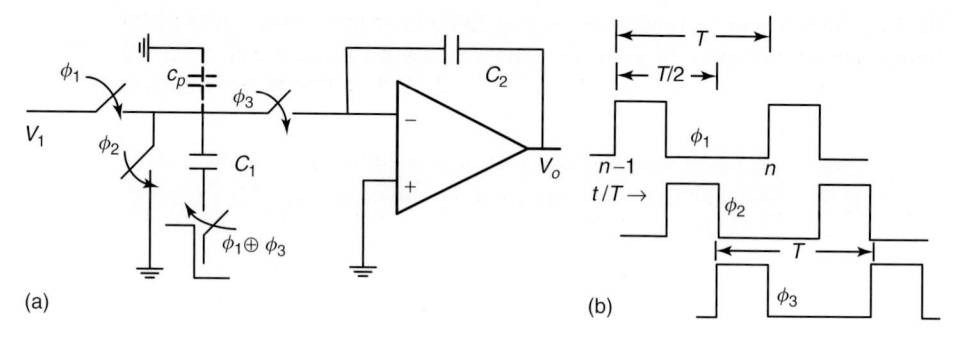

Figure 7.42 (a) Parasitic-insensitive toggle-switched-capacitor (TSC) inverting integrator using a three-phase clock signal and (b) the clock signals.

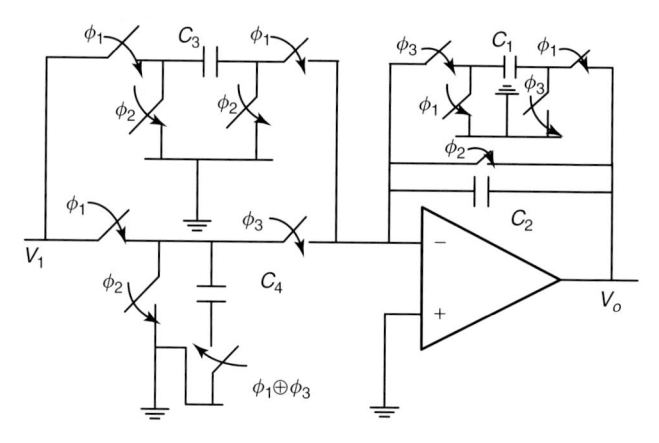

Figure 7.43 Use of TSC inverting integrator to realize a general bilinear first-order sampled-data transfer function.

By adding a few extra switches and capacitors, as shown in Figure 7.43, a first-order z-domain transfer function can be obtained.

The expression for the transfer function is

$$\frac{V_o^{(1)}}{V_1^{(1)}} = -\frac{(C_3/C_2) + (C_4/C_2)z^{-1}}{1 - (C_1/C_2)z^{-1}} \qquad (7.61)$$

with the output taken during ϕ_1. Clearly, if we set $C_1 = C_2$ and $C_3 = C_4$, the transfer function given by Eq. (5.57) assumes the form

$$-\frac{C_3}{C_2}\frac{1+z^{-1}}{1-z^{-1}}$$

This is in compliance with the bilinear $s \leftrightarrow z$ transformation, where the substitution

$$s \to \frac{2}{T}\frac{1-z^{-1}}{1+z^{-1}}$$

is used. Thus, the network of Figure 7.43 can be used for an SC inverting integrator implementing bilinear $s \leftrightarrow z$ transformation. It may be noted that if C_1 and C_2 are unequal, while $C_3 = C_4$, Eq. (7.61) represents an inverting lossy bilinear SC integrator.

7.6.2
A Stray-Insensitive Bilinear SC Integrator Using Biphase Clock Signals

A bilinear SC integrator transfer function can be realized with biphase clock signal using the network shown in Figure 7.44. The transfer function in the z-domain is given by (Mohan, Ramachandran, and Swamy 1982, 1995) as

$$\frac{V_o^{(2)}}{V_1^{(1)}} = -\frac{C_1(1+z^{-1})}{(C_6 + C_{10})(1 - z^{-1}\frac{C_{10}}{C_6 + C_{10}})} \qquad (7.62)$$

In arriving at Eq. (7.62), the condition $C_3 = C_1 + C_4$ is necessary. The above represents an inverting lossy bilinear integrator. A true bilinear integrator is obtained when $C_6 = 0$ (i.e., when C_6 is absent). By maintaining a sample-and-hold condition for the input signal, that is, $V_1^{(2)} = z^{-1/2} V_1^{(1)}$, a bilinear integrator can be obtained with fewer capacitors and switches. This is discussed next.

7.6.3
A Stray-Insensitive Bilinear Integrator with Sample-and-Hold Input Signal

The SC network pertaining to the present case is shown in Figure 7.45 (Knob, 1980). Note that the SC network following the sample-and-hold (S/H) network is the same as the one discussed in Chapter 6 (Section 6.5.3, Figure 6.15). If we use

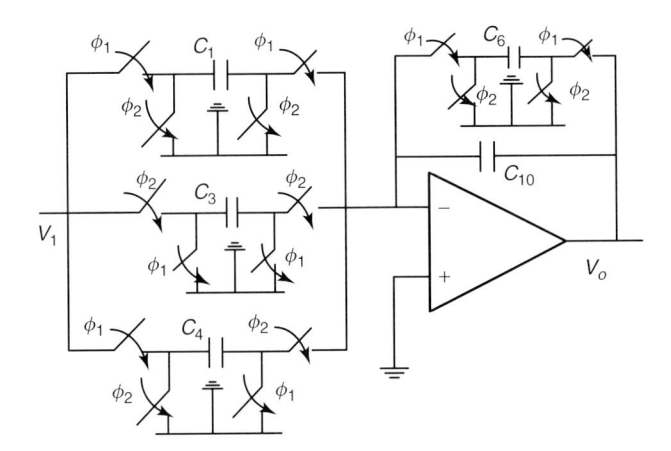

Figure 7.44 A parasitic-insensitive switched-capacitor using two-phase clock and realizing an inverting bilinear first-order sampled-data transfer function.

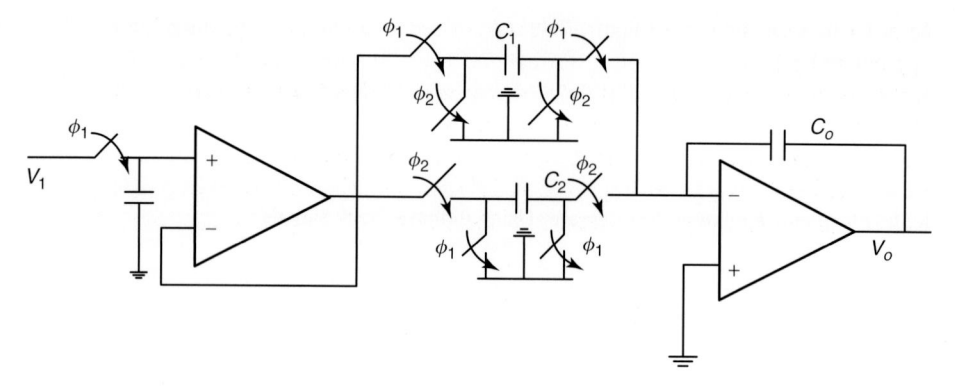

Figure 7.45 A parasitic-insensitive switched-capacitor bilinear integrator with sample-and-hold input signal.

Eq. (6.36) and solve for $V_o^{(1)}$, we get

$$V_o^{(1)} = -\frac{C_1 V_1^{(1)} + C_2 z^{-1/2} V_2^{(2)}}{C_o(1 - z^{-1})} \quad (7.63)$$

In the context of Figure 7.45, $V_2 = V_1$. Then, by invoking the S/H condition that is, $V_1^{(2)} = z^{-1/2} V_1^{(1)}$, the above expression will lead to an inverting bilinear integrating function. A true bilinear function is obtained if $C_1 = C_2$. It may be noted that, if $C_1 = C_2$ is assumed, then the bilinear transfer function can be obtained by using only one capacitor (i.e., C_1), which is switched at a rate twice the clock rate of the S/H network. This will result in the saving of one capacitor and two switches (Knob, 1980).

An example of a high-order SCF implementation using cascading approach is considered next.

7.6.4
Cascade of SC Filter Sections for High-Order Filter Realization

In the cascade approach, the first step involves the application of the prewarping to get the prewarped s-domain transfer function. Then, we apply the $s \leftrightarrow z$ transformation to determine the $H(z)$ to be synthesized. The next step is to decompose the $H(z)$ so determined into a product of first- and second-order sections. Suitable SC networks are then obtained to realize these constituent sections. The last step will be to put these sections in a cascade.

Example 7.10. We intend to realize an SC LP filter satisfying the following specifications (Schaumann and Van Valkenburg, 2001):
PB attenuation $A_p \leq 0.9$ dB, for $0 \leq f = f_p = 3.3$ kHz
Stopband attenuation $A_a \geq 22$ dB for $f \geq 4.5$ kHz
The clock frequency is $f_c = 32$ kHz
The first step will be to apply the prewarping formula $\hat{\omega} = 2 f_c \tan \frac{\omega}{2 f_c}$, to get the new critical frequencies $\hat{\omega}_p = 2\pi \times 3420.5$ rad s^{-1} and $\hat{\omega}_a = 2\pi \times 4817.6$ rad s^{-1}.

7.6 High-Order Filters Using Switched-Capacitor (SC) Networks

In order to keep the order of the filter as low as possible, we choose the elliptic approximation.

We then calculate $\alpha_1 = 10^{-(0.9/20)} = 0.9$, $\alpha_2 = 10^{-(22/20)} = 0.08$, and $\omega_s = 4817.6/3420.5 = 1.41$. On examining the tables of elliptic filter transfer functions in Appendix A, we can select a third-order transfer function with $\alpha_1 = 0.9$, $\alpha_2 = 0.077$, and $\omega_s = \frac{\hat{\omega}_a}{\hat{\omega}_p} = 1.40$. This will meet the PB specification, and will exceed the selectivity required in the stopband. The frequency normalized prewarped transfer function is

$$T(s) = \frac{0.2745(s^2 + 2.41363)}{(s + 0.63584)(s^2 + 0.36139s + 1.04183)} \tag{7.64}$$

After denormalizing with respect to $\hat{\omega}_p = 2\pi \times 3420.5$ rad s^{-1} and applying the bilinear $s \leftrightarrow z$ transformation formula, we arrive at the z-domain transfer function

$$T(z) = \frac{0.0779(1+z^{-1})(1-1.1442z^{-1}+z^{-2})}{(1-0.6481z^{-1})(1-1.4247z^{-1}+0.8041z^{-2})} \tag{7.65}$$

The cascadable sections then can be chosen as

$$T_1(z) = 0.176 \frac{1+z^{-1}}{1-0.6481z^{-1}} \tag{7.66}$$

and

$$T_2(z) = 0.443 \frac{1-1.1442z^{-1}+z^{-2}}{1-1.4247z^{-1}+0.8041z^{-2}} \tag{7.67}$$

In the above, we have chosen 0.0779 as 0.176×0.443 so that each of T_1 and T_2 has a gain of nearly 1 at DC (i.e., $z^{-1} = 1$). The first-order transfer function calls for a bilinear function in z^{-1}. This can be obtained, for example, by employing the bilinear integrator described in Section 7.6.2 (Figure 7.44). In order to realize $T_1(z)$ as in Eq. (7.66), one can set $C_1 = 1$ pF, $C_6 = 2$ pF, and $C_{10} = 3.86$ pF. Further, $C_3 = C_1 = 1$ pF, and $C_4 = 0$ (i.e., C_4 absent).

For the second-order section, we can use, for example, the network discussed in Chapter 6 (Figure 6.20), which is redrawn in Figure 7.46. The capacitors L and K have been set to zero and capacitors D, A, and B have been set equal to C' in Figure 7.46. The transfer function $T_2(z)$ is, in this case, given by (see Eq. (6.39) in Chapter 6)

$$\frac{V_2}{V_i} = -\frac{I + z^{-1}(G - I - J) + z^{-2}(J - H)}{C' + F + z^{-1}(E + C - F - 2C') + z^{-2}(C' - E)} \tag{7.68}$$

Equation (7.68) can be rewritten in the form

$$\frac{V_2}{V_i} = \frac{I}{C'+F} \frac{1+z^{-1}\frac{G-I-J}{I}+z^{-2}\frac{J-H}{I}}{1+z^{-1}\frac{E+C-F-2C'}{C'+F}+z^{-2}\frac{C'-E}{C'+F}} \tag{7.69}$$

By comparing with the coefficients of $T_2(z)$, we get

$$\frac{I}{C'+F} = 0.443, \quad \frac{G-I-J}{I} = -1.1442, \quad \frac{J-H}{I} = 1 \tag{7.70}$$

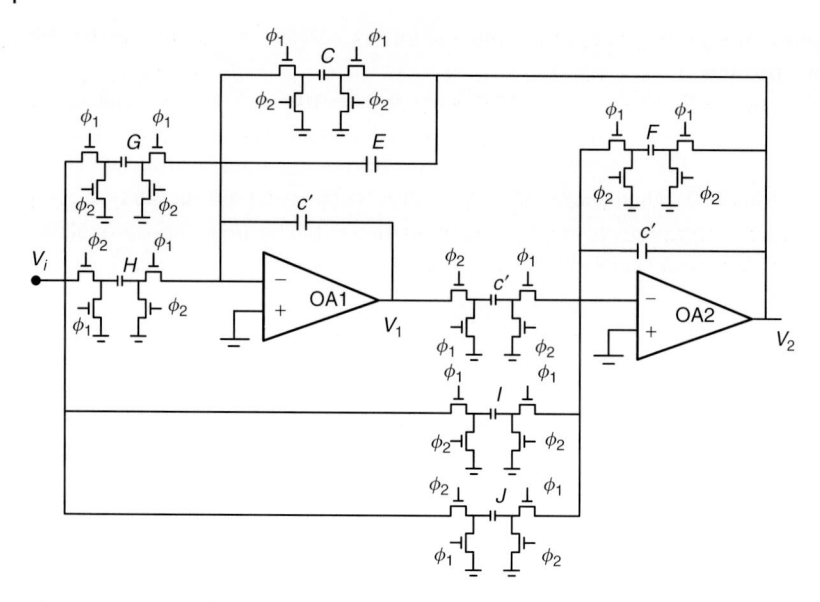

Figure 7.46 A switched-capacitor network for realization of a general second-order sampled-data transfer function.

$$\frac{E+C-F-2C'}{C'+F} = -1.4247, \quad \frac{C'-E}{C'+F} = 0.8041$$

In the above, there are five equations but eight unknowns. We shall simplify the situation by assuming $E = 0$ (i.e., E capacitor absent) in Figure 7.46, and $H = 0$. Then, using the expressions in Eq. (7.70), we get

$$J = I, G = 0.8558I, F = 0.2436C', C = 0.4718C', I = 0.5509C'.$$

The minimum valued capacitor is F. If we want to make $F = 1$ pF, then $C' = 4.1$ pF. Subsequently, $I = J = 2.26$ pF, $C = 1.93$ pF, $G = 1.93$ pF. Thus, all the capacitors are designed. Putting Figures 7.43 and 7.46 in cascade completes the design. The output of T_2 (i.e., V_2 in Figure 7.46) is to be sampled at phase ϕ_1. It may be noted that, in the realization of $T_2(z)$, the capacitors (J,H), (C,F), (G,I) can share the switches, thereby saving few MOS transistor switches.

7.6.5
Ladder Filter Realization Using the SC Technique

A given ladder filter network can be designed by a corresponding SC network in the same manner as is done for the active-RC implementation. To begin with, one needs the transfer function of the ladder filter together with the LC ladder structure. Since in the sampled-data domain, prewarping plays an important role, the necessary prewarping formula has to be applied to the s-domain transfer function and the consequent modifications in the coefficients of the transfer function need be reflected in the values of the LC ladder structure. This second ladder structure, which can be named as the prewarped ladder filter, is the one which has to be adopted for implementation as an SC filter network. An alternative method to obtain the prewarped ladder will be to work from the prewarped transfer function and then develop the ladder network elements by the methods discussed in Chapter 4.

An important step that will be encountered in simulating the series and shunt arms of the ladder is the implementation of the associated integrators in the sampled-data domain. We have seen that, in the continuous-time domain, the integrators are either ideal or lossy RC-active integrators. We have to use similar integrators implemented with active SC network elements. The sampled-data domain integrators built with SC elements should conform to the same $s \leftrightarrow z$ transformation as pertinent to the prewarping formula used before. From our knowledge of several popular $s \leftrightarrow z$ transformations that are being used in relating the continuous-time domain and sampled-data domain filters, we can infer that the resulting SC integrators will have the form $N(z)/D(z)$ where $N(z)$ and $D(z)$ will contain, in the most general case, terms involving z^{-1} and $z^{-1/2}$ in the sample data domain variable $z = \exp(j\Omega T)$ as introduced in Chapter 6. A particular case will be an SC integrator with bilinear transfer function pertinent to the bilinear $s \leftrightarrow z$ transformation. The PI bilinear SC integrators discussed in Sections 7.6.1 and 7.6.2 will be useful for the implementation of the active SC equivalent to the LC ladder filter.

Practice Problems

7.1 Consider a GIC network N, whose chain matrix is given by

$$\begin{bmatrix} 1 & 0 \\ 0 & \frac{z_2 z_4}{z_1 z_3} \end{bmatrix} = \begin{bmatrix} 1 & 0 \\ 0 & \frac{1}{F(s)} \end{bmatrix}$$

(a) What is the chain matrix of the network N_R obtained by reversing the terminals of the input and output?
(b) Consider the network shown in Figure P7.1. Show that if $F(s) = s$, then we realize a floating inductor L of value $L = R$.
(c) Choose a set of values for z_1, z_2, z_3, and z_4 in Antoniou's GIC so that $F(s) = s$.
(d) If $F(s) = 1/s^2$, show that we realize a floating FDNR D. Find its value in terms of R.
(e) Choose a set of values for z_1, z_2, z_3, and z_4 in Antoniou's GIC so that $F(s) = 1/s^2$

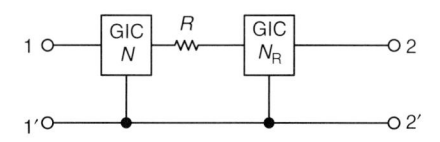

Figure P7.1

7.2 Design a GIC-based third-order HP filter with a Butterworth magnitude characteristic and a 3-dB frequency of 10 kHz. Use a doubly terminated network with terminations of 1 Ω each. In the final filter, all the capacitors should have a value of 1 nF.

7.3 Design a fourth-order singly terminated HP filter with a Butterworth magnitude characteristic. The terminating resistance should have a value of 1000 Ω. The 3-dB frequency is 500 Hz. The resistors in the GIC network should be equal to 10 kΩ each.

7.4 Design an FDNR realization for a third-order Butterworth LP voltage transfer function. The prototype normalized filter should have equal 1 Ω terminations. In the final realization, the bandwidth should be 1000 rad s^{-1} and all the capacitors should have the value of 0.1 μF.

7.5 Using the FDNR technique, design a fourth-order LP Butterworth voltage transfer function. The cutoff frequency should be 5 kHz and the prototype source and load resistors should have a value of 500 Ω. All the resistors in the FDNR sections should be 7500 Ω.

7.6 Use the leapfrog method to design a filter having a fifth-order LP Butterworth characteristic. The cutoff frequency should be 1 k rad s^{-1}. All resistors should have a value of 10 kΩ.

7.7 A leapfrog filter is shown in Figure P7.7. Find the voltage transfer function for the filter. Assume $C = 10^{-8}$ F and $R = 10^4$ Ω.

7.8 Find a leapfrog realization of a normalized fourth-order BP function with a maximally flat magnitude characteristic, a center frequency of 1 rad s^{-1}, and a bandwidth of 1 rad s^{-1}. Use a doubly terminated lossless ladder filter as a prototype filter. Each of the terminating resistors is 1 Ω.

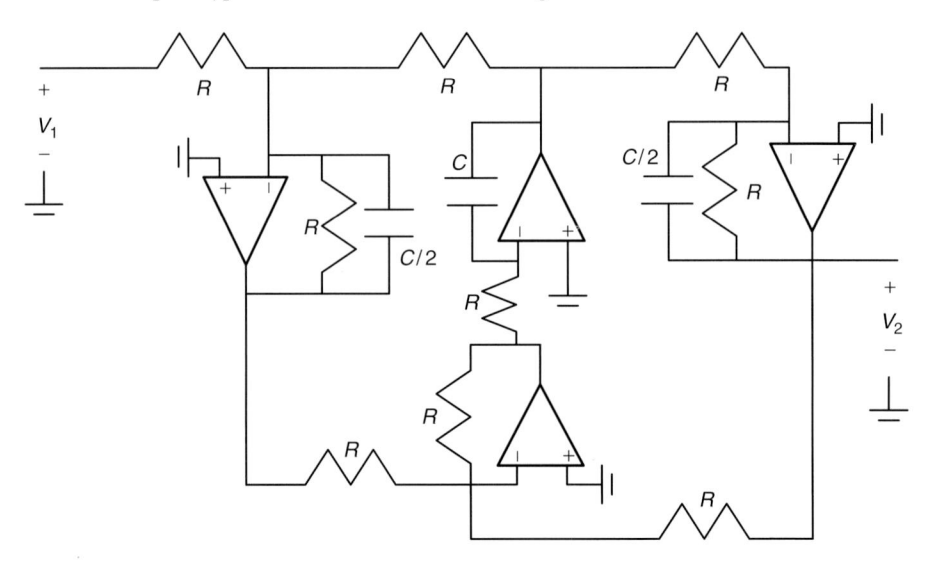

Figure P7.7

7.9 Use the PRB technique to design a normalized fourth-order BP filter with a center frequency of 1 rad s^{-1}, a bandwidth of 0.1 rad s^{-1}, and a Butterworth magnitude characteristic. The gain at resonance should be unity. Use the infinite-gain SAB second-order filters as the building blocks.

7.10 Design a sixth-order BP filter using the PRB technique. Start with a third-order Butterworth LP network function. The resulting BP structure should have a gain of 2 at the normalized center frequency of 1 rad/s. The bandwidth should be 0.1 rad/s. Make the coefficient $a_1 = 0$. Do not realize the individual second-order sections; just indicate their parameters, namely, Q_p and H_o. Let $R_f = 1$ and draw the overall configuration showing the values of the resistors.

7.11 Consider Problems 4.4–4.8 in Chapter 4. For each one, show an active-RC realization using any of the techniques you have learnt in this chapter. Verify your work using network simulation. You may use the subcircuit of any standard commercial OA (such as µA 741 or LM741) available in the simulation database.

7.12 Use operational simulation to design an eighth-order CHEB low-pass filter with 0.1 dB maximum ripple in the passband $0 \leq f \leq 128$ kHz. The prescribed load resistor is 1450 Ω. Verify your design using a simulation tool. You may use LM741 or similar OA available in the simulation tool.

7.13 The LC ladder implementation of a Bessel–Thomson filter with $D = 30$ µs delay and $\leq 1\%$ delay error of 13 kHz is shown in Figure P7.13. The filter has less than 2-dB attenuation in $f \leq 110$ kHz. Realize the network using the operational simulation method. Verify the design using simulation.

7.14 The circuit of Figure P7.14 is the LC ladder implementation of a BP filter with a 1.7 dB equiripple passband in 10 kHz $\leq f \leq$ 18 kHz. The minimum stopband attenuation is 42 dB for $f > 26$ kHz. Produce an active-RC implementation using operational simulation method.

7.15 Figure P7.15 shows a band-reject filter, which has equal-ripple passbands in $f \leq 80$ kHz and $f \geq 180$ kHz with $A_p|_{max} = 1$ dB. The stopband attenuation is $A_a|_{min} = 20$ dB over 100 kHz $\leq f \leq$ 150 kHz. Implement the network using operational simulation technique.

7.16 Consider a fourth-order Chebyshev LP filter with a passband from 0 to 1 rad s^{-1}, with $\alpha_p = 1$ dB and a dc gain of unity. Realize this filter by an FLF structure. Show how you can get a HP output.

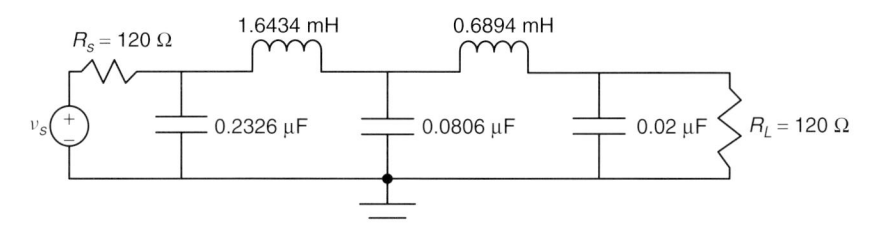

Figure P7.13

7 Higher-Order Active Filters

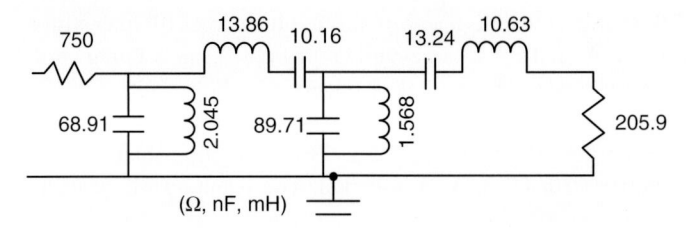

Figure P7.14

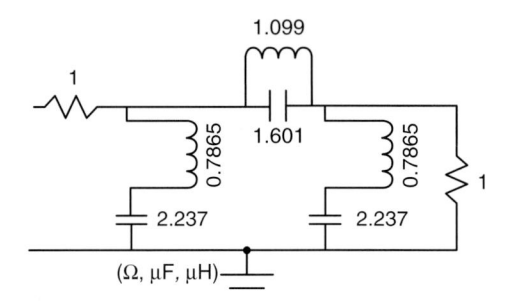

(Ω, nF, mH)

Figure P7.15

(Ω, μF, μH)

7.17 Realize the transfer function

$$H(s) = \frac{V_2}{V_1} = \frac{a_4 s^4 + a_3 s^3 + a_2 s^2 + a_1 s + a_0}{s^4 + b_3 s^3 + b_2 s^2 + b_1 s + b_0}$$

by an FLF structure with feed-forward paths, but using lossy integrators of the form $-1/(s + \alpha)$.

7.18 Figure P7.18 shows an LC ladder network. Design the network using a suitable OTA-C structure. You may consider floating inductor simulation using OTA.

7.19 Produce OTA-C-based implementations for Problems 7.2, 7.3, 7.6, 7.8–7.15. Verify the designs using a suitable network simulation tool. You may use LM 13 700 for a practical OTA.

7.20 Design a fourth-order switched-capacitor LP filter with a Butterworth magnitude characteristic, a cutoff frequency of 1 kHz, and a clock frequency of 100 kHz. Choose the minimum capacitance of 1 pF.

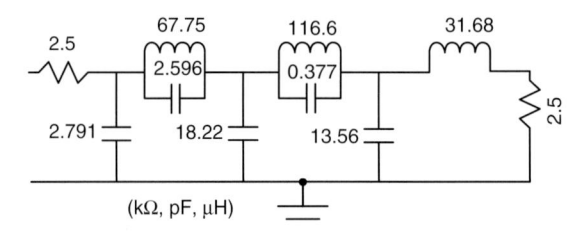

(kΩ, pF, μH)

Figure P7.18

8
Current-Mode Filters

The advancements in IC technology together with the demand for smaller and low-power devices have ushered in the era of IC filters. While the initial goal of microminiaturization was to produce filters for audio frequency applications, the focus has progressively changed toward higher-than- audio frequency filters. The advent of submicron IC technological processes (0.5 μm and smaller) has facilitated realization of filters in the VHF frequency band (30–300 MHz) (Raut and Guo, 1997). Together with higher frequency of operation, reduction in the power consumption by the various electronic devices has also been of concern to present-day researchers. In view of this, attention is being paid toward signal processing in terms of currents rather than voltages. This new type of signal processing is known as *current-mode signal processing*.

It can be easily appreciated that CM signal processing will lead to higher frequency of operation, since the signal current is delivered into a small (ideally short circuit) load resistance. The parasitic pole due to such a small resistance (pole frequency being inversely proportional to the resistance) will be very high and hence, a high-frequency signal can be processed without substantial impairment due to the presence of such a high-frequency parasitic pole. In this chapter, we first introduce the basic principles of CM signal processing. Realization techniques for second-order CM filters using several CCs (Sedra and Smith, 1970) will be presented next. This will be followed by presenting techniques that can be used to convert an existing voltage-mode (VM) filter to a CM filter using generalized duals (GDs) and transposes; thus, the wealth of knowledge that already exists in the classical area of VM filters can be readily and efficiently used. The discussion will then continue with SI filters, the analog of SC filters for CM signal processing.

8.1
Basic Operations in Current-Mode

8.1.1
Multiplication of a Current Signal

The magnification or attenuation (i.e., multiplication by a constant) of a current signal is one of the fundamental operations required in CM filtering. Several alternatives exist to achieve this operation and a few of these are now briefly discussed.

8.1.1.1 Use of a Current Mirror

A current mirror is the simplest device that can be used to multiply a current signal by a constant. The ideal mirror will function like a CCCS with zero input resistance and an infinite output resistance. This is depicted in Figure 8.1. In a CMOS IC technology, the multiplication value is proportional to the ratio of the area of the MOS transistor at the output to the area of the MOS transistor at the input. If we consider a simple CMOS current mirror as shown in Figure 8.2a, with its AC equivalent circuit as shown in Figure 8.2b, the current signal transfer ratio at low frequency will be given by

$$\frac{I_o}{I_i} = \frac{g_{mn2} + g_{mp2}}{g_{mn1} + g_{mp1} + g_{dp1} + g_{dn1}} \tag{8.1}$$

The condition of zero input resistance (inversely proportional to the sum $g_{mn1} + g_{mn2} + g_{dp1} + g_{dp2}$) and infinite output resistance (inversely proportional to $g_{dp2} + g_{dn2}$) are seldom achieved in practice. Special circuit configurations are used to achieve conditions close to these ideal ones.

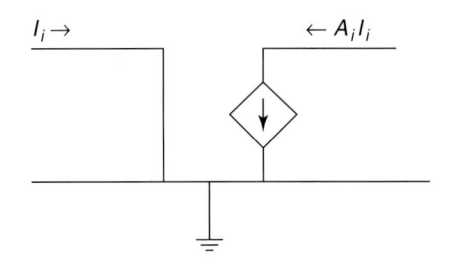

Figure 8.1 A current-controlled current source (CCCS).

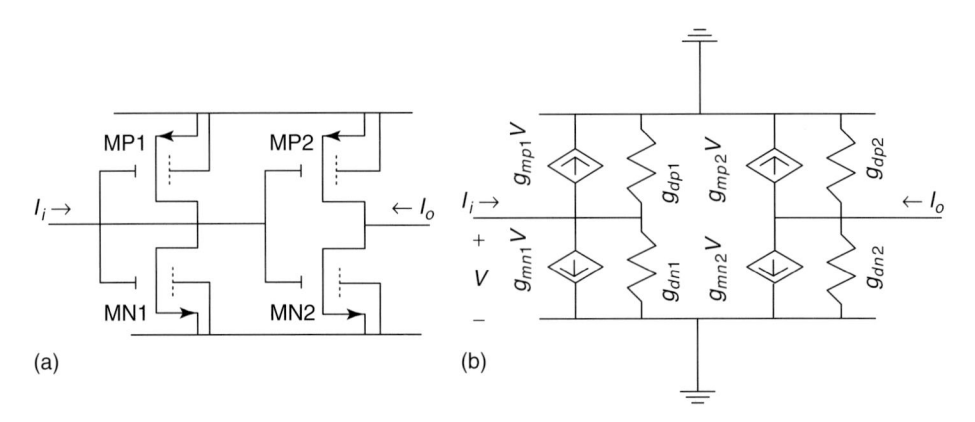

Figure 8.2 (a) Current mirror using CMOS transistors and (b) its AC equivalent circuit, where g_{dni}, g_{dpi}, $i = 1, 2$ are conductances.

8.1.1.2 Use of a Current Conveyor

The functional characteristic of a CC has been introduced in Chapter 5. CCII (Sedra and Smith, 1970) has become popular for practical applications since the 1990s (Toumazou, Lidgey, and Haigh, 1990). The schematic for a CCII and the matrix showing its terminal $i-v$ relationships are repeated here for convenience (Figure 8.3).

According to the relation $i_z = \pm i_x$, the CCII is further categorized as positive (+) and negative (−) CCII, that is, CCII+ and CCII−. Figures 8.4a and 8.4b show the schematics of typical positive and negative gain CCs implemented in CMOS IC technology (Toumazou, Lidgey, and Haigh, 1990). Figure 8.4c shows how a CCII+ and a CCII− can be used to multiply a current signal by a constant value.

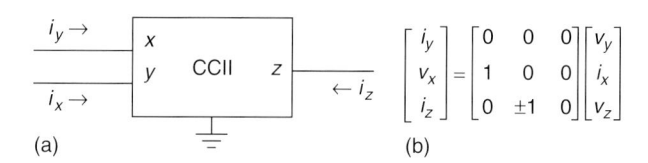

Figure 8.3 (a) Symbol of a CCII and (b) $i-v$ relations of the CCII.

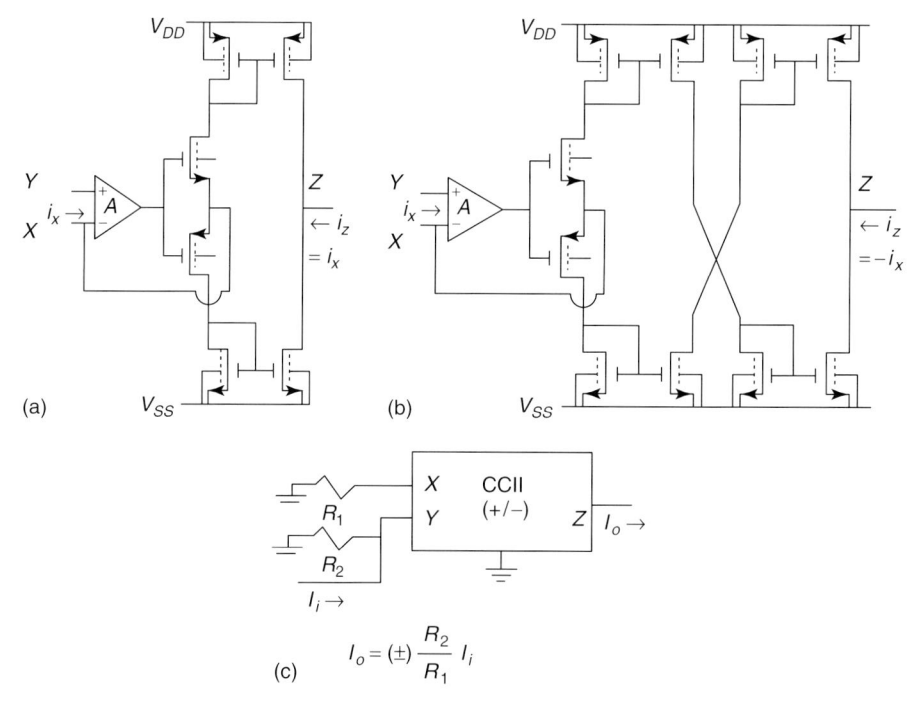

$$\begin{bmatrix} i_y \\ v_x \\ i_z \end{bmatrix} = \begin{bmatrix} 0 & 0 & 0 \\ 1 & 0 & 0 \\ 0 & \pm 1 & 0 \end{bmatrix} \begin{bmatrix} v_y \\ i_x \\ v_z \end{bmatrix}$$

$$I_o = (\pm)\frac{R_2}{R_1} I_i$$

Figure 8.4 (a) Implementation of a CCII−, (b) implementation of a CCII+, and (c) a current multiplier using CCII−. (a) and (b) taken from Toumazou, Lidgey, and Haigh 1990.

8.1.1.3 Use of Current Operational Amplifier

A COA acts in the same way on the current signals as an OA acts on a voltage signal. It ideally approaches the characteristics of a CCCS. Thus, the input resistance needs to be small while the output resistance is required to be very high. Similarly, the short circuit current gain is required to be high. A COA typically uses a cascade of a transimpedance amplifier (for low input resistance) followed by a transconductance amplifier (for high output impedance). Figure 8.5 shows the schematic of a tunable COA implemented in CMOS technology. The input is applied to the common-gate MOS amplifier stages of a transimpedance amplifier (Bruun, 1994) followed by a tunable transconductance stage (Assi, Sawan, and Raut, 1996). The circuit has been implemented using differential-in differential-out (Bruun, 1991) topology in a 0.8-μm BiCMOS IC technology. A summary of the simulated response characteristics is presented in Table 8.1. More details can be found in Assi, Sawan, and Raut (1997).

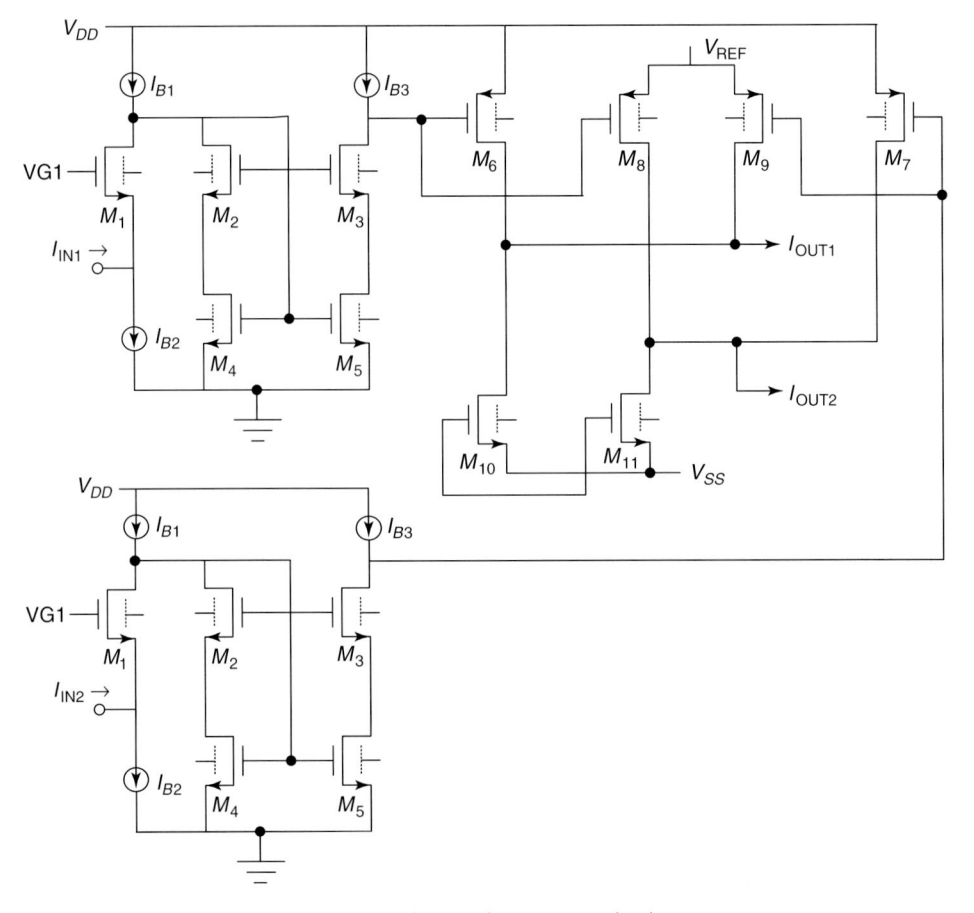

Figure 8.5 A tunable COA implemented in CMOS technology.

Table 8.1 Simulated response of the tunable COA of Figure 8.5.

Gain (voltage controllable)	70–96 dB
Output current (peak-to-peak)	280 µA
GBW	145 MHz
Supply voltages	$V_{DD} = -V_{SS} = 1.5$ V
Power consumption	<0.5 mW

8.1.2
Current Addition (or Subtraction)

Current addition is obtained by simply connecting the paths on to a single node. For subtraction, the pertinent current signal has to be reversed before subjecting to addition. Crossing the wire is the conventional technique for reversing the current. Figure 8.6 presents a typical circuit for achieving this operation in CMOS IC technology (Toumazou, Lidgey, and Haigh, 1990).

8.1.3
Integration and Differentiation of a Current Signal

A simple current integrator is depicted in Figure 8.7a. The current I_i from an ideal source forms a voltage across the capacitor C. The OTA with a transconductance g_m converts this voltage to a current at the output.

The transfer characteristic, in the ideal case, is given by

$$I_o = \frac{g_m}{sC} I_i \tag{8.2}$$

If the input current source has a finite resistance $R_i = 1/G_i$, and the output is not a short circuit (for small signal), that is, $R_L \neq 0$, then the response characteristic

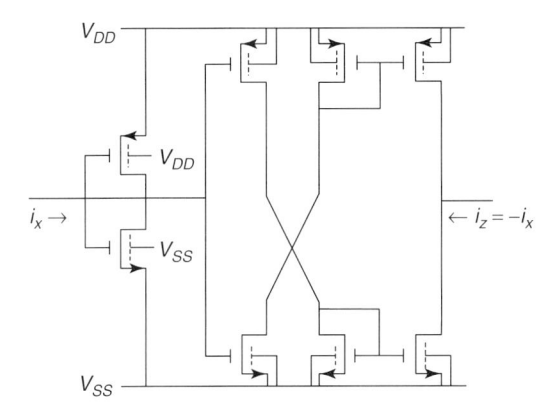

Figure 8.6 Implementation of the reversal of a current signal.

8 Current-Mode Filters

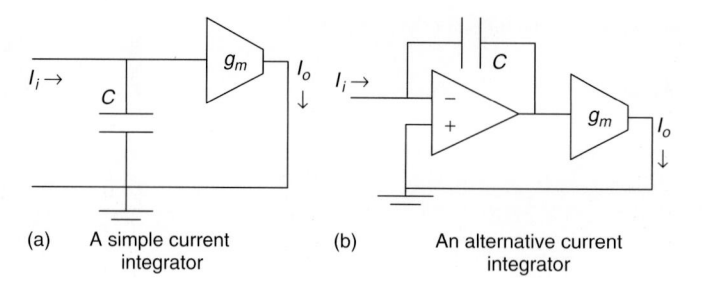

Figure 8.7 Integration of a current signal.

(a) A simple current integrator
(b) An alternative current integrator

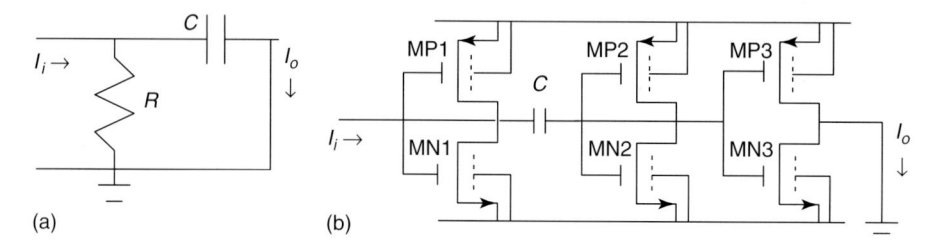

Figure 8.8 Current differentiator using (a) RC elements and (b) CMOS transistors.

will be that of a lossy integrator:

$$\frac{I_o}{I_i} = \frac{g_m R_o}{R_o + R_L} \cdot \frac{1}{G_i + sC} \tag{8.3}$$

where R_o is the output resistance of the transconductor.

A better alternative for the CM integrator is shown in Figure 8.7b. Since the current source I_i feeds to the virtual ground terminal of the OA, the effect of the finite value of G_i will be insignificant. For finite R_o and R_L, the transfer relation is given by

$$\frac{I_o}{I_i} = \frac{g_m R_o}{R_o + R_L} \cdot \frac{1}{sC} \tag{8.4}$$

A CM integrator can also be obtained using a CCII+ or a CCII−; these are shown in Table 8.2.

A basic differentiator network for current signal is shown in Figure 8.8a. The product RC should be ideally zero for perfect differentiation. An IC version in CMOS technology is shown in Figure 8.8b.

Considering finite transconductances of the MOS transistors, the transfer characteristic equation is given by

$$\frac{I_o}{I_i} = k \frac{sC/G_{m1}}{1 + sC(G_{m1} + G_{m2})/G_{m1}G_{m2}} \tag{8.5}$$

In the above, G_{m1} is the equivalent input conductance of the (MP1, MN1) transistor pair, G_{m2} that of the (MP2, MN2) transistor pair, and k is the current mirroring ratio between the pairs (MP3, MN3) and (MP2, MN2). We have ignored the parasitic capacitances of the MOS transistors in deriving Eq. (8.5).

Table 8.2 Some basic building blocks using CCII.

	CCII circuits	Transfer function	Remarks
A		$I_o = (\pm)\dfrac{R_2}{R_1} I_i$	Noninverting and inverting current amplifier
B		$I_o = (\mp)\displaystyle\sum_{j=1}^{n} I_j$	Inverting and noninverting current summer
C		$I_o = (\pm)\dfrac{1}{sCR} I_i$	Noninverting and inverting current integrator
D		$I_o = (\pm)sCR\, I_i$	Noninverting and inverting current differentiator

(*continued overleaf*)

Table 8.2 (continued).

CCII circuits	Transfer function	Remarks
E	$[a] = \begin{bmatrix} 1 & 0 \\ 0 & 0 \end{bmatrix}$	Unity-gain VCVS
F	$[a] = \begin{bmatrix} 0 & 0 \\ 0 & 1 \end{bmatrix}$	Unity-gain CCCS
G	$[y] = \begin{bmatrix} 0 & 0 \\ (\mp)g & 0 \end{bmatrix}$	VCCS
H	$[z] = \begin{bmatrix} 0 & 0 \\ (\pm)r & 0 \end{bmatrix}$	CCVS

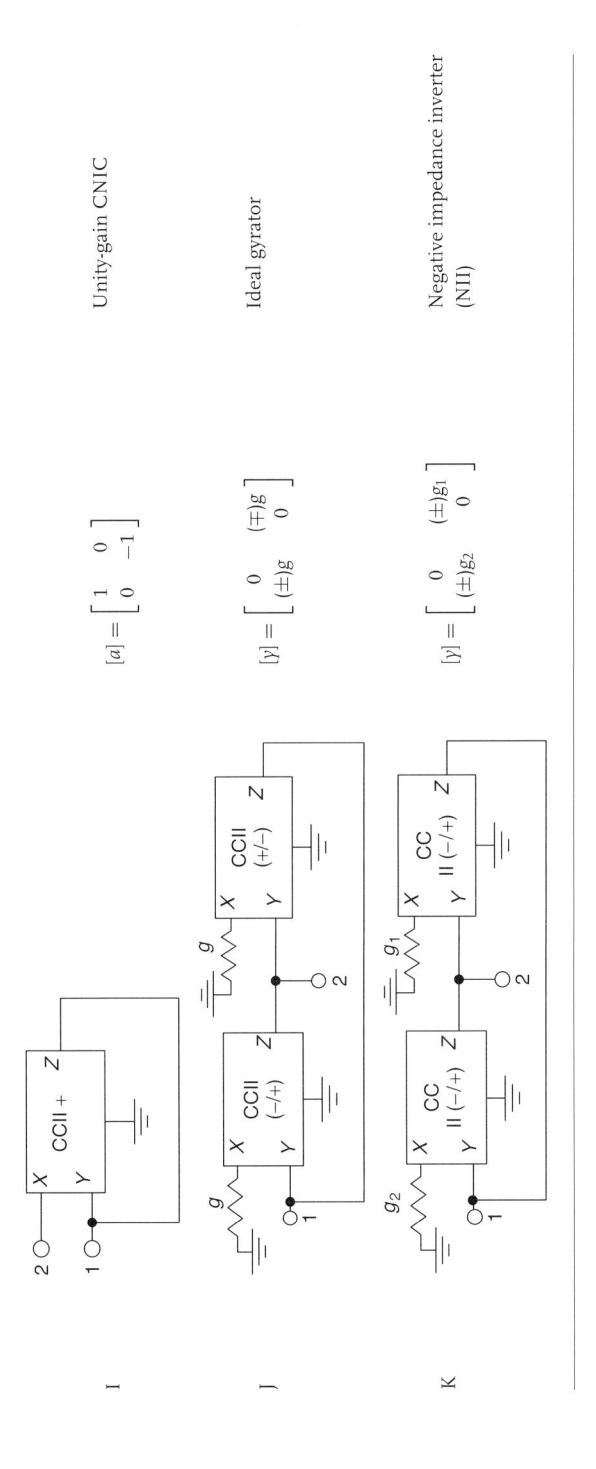

8.1 Basic Operations in Current-Mode

I	(CCII+ circuit diagram)	Unity-gain CNIC	$[a] = \begin{bmatrix} 1 & 0 \\ 0 & -1 \end{bmatrix}$
J	(CCII (-/+) and CCII (+/-) with g)	Ideal gyrator	$[y] = \begin{bmatrix} 0 & (\mp)g \\ (\pm)g & 0 \end{bmatrix}$
K	(CCII (-/+) and CCII (-/+) with g_1, g_2)	Negative impedance inverter (NII)	$[y] = \begin{bmatrix} 0 & (\pm)g_1 \\ (\pm)g_2 & 0 \end{bmatrix}$

8.2
Current Conveyors in Current-Mode Signal Processing

8.2.1
Some Basic Building Blocks Using CCII

In Section 8.1.1, it has been shown how CCII+ and CCII− can be used as current signal multiplying devices. Some examples of signal processing operations using CCII are presented in Table 8.2.

8.2.2
Realization of Second-Order Current-Mode Filters

In the following sub-sections, we discuss several cases of implementation of second-order CM filters using CCs.

8.2.2.1 Universal Filter Implementation
Since a CC can transport current signals, it can be used together with passive-RC components to realize CM transfer functions in the same way as an OA is used to produce VM transfer functions. Figure 8.9 presents a universal CM filter with a single input and three outputs. The configuration uses both CCII+ and CCII−

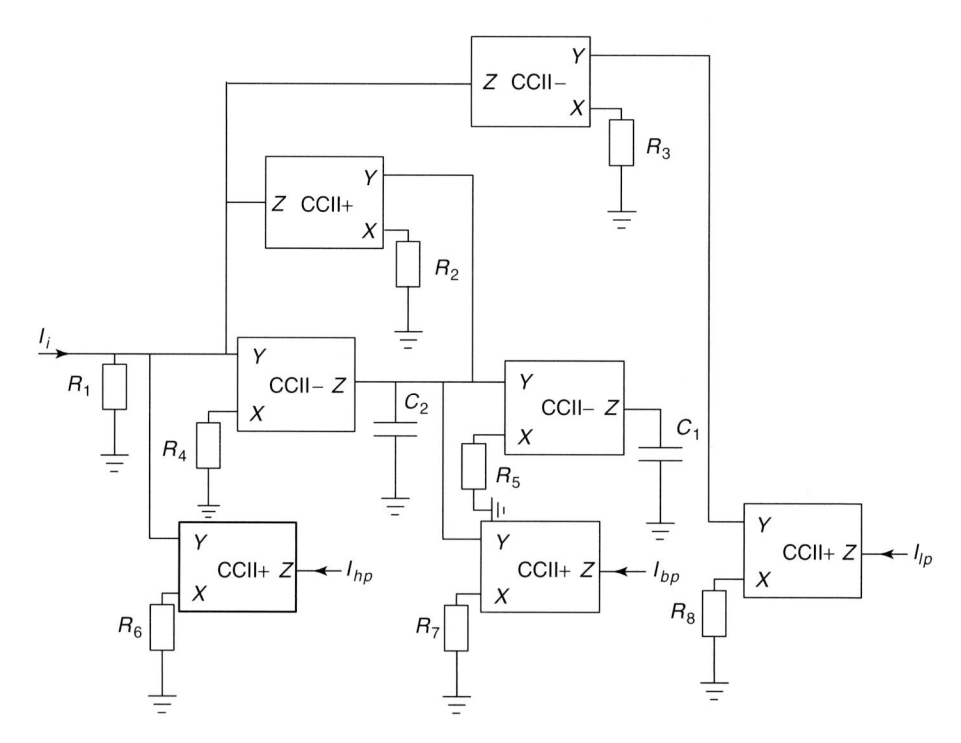

Figure 8.9 A universal second-order CM filter implemented with CCII+ and CCII−.

as the active building blocks. The three primary transfer functions are given by (Chang, 1993), with $G_i = 1/R_i (i = 1, 2, 3, \ldots)$:

$$\frac{I_{hp}}{I_i} = \frac{s^2 C_1 C_2 G_6}{s^2 C_1 C_2 G_1 + s C_1 G_2 G_4 + G_3 G_4 G_5} \tag{8.6}$$

$$\frac{I_{bp}}{I_i} = \frac{-s C_1 G_4 G_7}{s^2 C_1 C_2 G_1 + s C_1 G_2 G_4 + G_3 G_4 G_5} \tag{8.7}$$

$$\frac{I_{lp}}{I_i} = \frac{G_4 G_5 G_8}{s^2 C_1 C_2 G_1 + s C_1 G_2 G_4 + G_3 G_4 G_5} \tag{8.8}$$

By connecting the I_{hp} and I_{lp} outputs, we get the notch filter response. Similarly, by adding the three outputs and assuming $G_6 = G_1$, $G_7 = G_2$, and $G_8 = G_3$, we get an AP characteristic (Senani, 1992). The pertinent transfer functions are

$$\frac{I_{notch}}{I_i} = \frac{s^2 C_1 C_2 G_6 + G_4 G_5 G_8}{s^2 C_1 C_2 G_1 + s C_1 G_2 G_4 + G_3 G_4 G_5} \tag{8.9}$$

$$\frac{I_{ap}}{I_i} = \frac{s^2 C_1 C_2 G_1 - s C_1 G_2 G_4 + G_3 G_4 G_5}{s^2 C_1 C_2 G_1 + s C_1 G_2 G_4 + G_3 G_4 G_5} \tag{8.10}$$

The pole frequency ω_p and Q_p the pole Q are given by

$$\omega_p = (G_3 G_4 G_5 / C_1 C_2 G_1)^{1/2}$$
$$Q_p = (1/G_2)(G_1 G_3 G_5 C_2 / C_1 C_4)^{1/2} \tag{8.11}$$

It may be noted that the ω_p and Q_p are tunable by adjusting R_3 and R_2, in that order. The CCs can be implemented using commercial OAs such as, LF356N, followed by current mirrors composed of transistor arrays (CA3096AE).

8.2.2.2 All-Pass/Notch and Band-Pass Filters Using a Single CCII

Figure 8.10 shows a CM AP/notch and BP filter using a single negative gain CCII. The network needs two grounded capacitors, and (at most) four resistors. The CTF

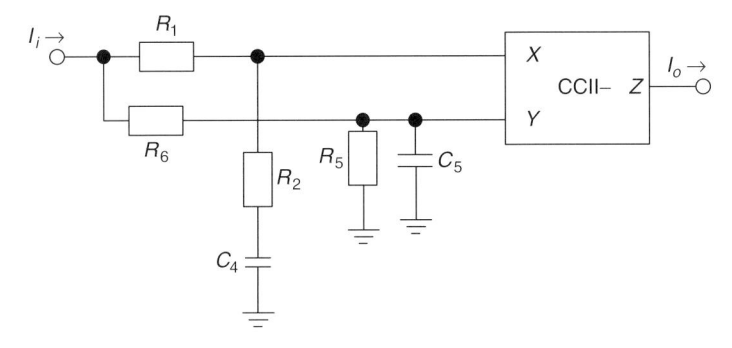

Figure 8.10 All-pass, band-pass, and notch filters using a single CCII−.

is given by (Chang, 1991)

$$\frac{I_o}{I_i} = \frac{s^2 G_1 C_4 C_5 + s(G_1 C_4 C_5 + G_1 G_2 C_5 - G_2 G_6 C_4) + G_1 G_2 G_5}{(G_1 + G_2)[s^2 C_4 C_5 + s(G_5 C_4 + G_2 C_5) + G_2 G_5]} \quad (8.12)$$

The BP case arises on selecting $G_1 = 0$ (i.e., R_1 open circuit). The notch filter realization is possible with $G_6 = (C_4 G_5 G_1 + C_5 G_2 G_1)/G_2 C_4$, and the AP filter is realized by the choice $G_6 = 2(C_4 G_5 G_1 + C_5 G_2 G_1)/G_2 C_4$. The pole frequency ω_p and the pole Q are given by

$$\omega_p = \sqrt{\frac{G_2 G_5}{C_4 C_5}}, \quad \frac{1}{Q_p} = \sqrt{\frac{C_4 G_5}{C_5 G_2}} + \sqrt{\frac{C_5 G_2}{C_4 G_5}} \quad (8.13)$$

8.2.2.3 Universal Biquadratic Filter Using Dual-Output CCII

A dual-output current conveyor type 2 (DOCCII) is a more versatile building block than a single-output CCII and an implementation using CMOS technology has already been given in Figure 5.33. The operation of a DOCCII is described by the matrix equation

$$\begin{bmatrix} V_x \\ I_y \\ I_{z1} \\ I_{z2} \end{bmatrix} = \begin{bmatrix} 0 & 1 & 0 & 0 \\ 0 & 0 & 0 & 0 \\ 1 & 0 & 0 & 0 \\ k & 0 & 0 & 0 \end{bmatrix} \begin{bmatrix} I_x \\ V_y \\ V_{z1} \\ V_{z2} \end{bmatrix} \quad (8.14)$$

If $k = 1$, then DOCCII+ is defined and if $k = -1$, then DOCCII− is defined.

Four DOCCII+ configured as in Figure 8.11 can function as a universal CM biquadratic filter (Minaei, Kuntman, and Cicekoglu, 2000). Analysis of the circuit leads to

$$\frac{I_{o1}}{I_{in}} = -y_2 y_4/\Delta, \quad \frac{I_{o2}}{I_{in}} = -y_4 y_5/\Delta, \quad \frac{I_{o3}}{I_{in}} = -y_3 y_5/\Delta \quad (8.15a)$$

where

$$\Delta = y_4 y_6 + y_3 y_5 - y_1 y_4 \quad (8.15b)$$

Suitable selection of the admittance elements $y_1, y_2, \ldots, y_6$ will lead to realization of several second-order filters. Choosing $y_1 = 0, y_2 = G_2, y_3 = sC_3, y_4 = G_4, y_5 = sC_5$, and $y_6 = G_6 + sC_6$ will lead to

$$\frac{I_{o1}}{I_{in}} = -(G_2 G_4/C_3 C_5)/\Delta, \quad \frac{I_{o2}}{I_{in}} = -s(G_4 C_5/C_3 C_5)/\Delta, \quad \frac{I_{o3}}{I_{in}} = -s^2/\Delta \quad (8.16a)$$

where

$$\Delta = s^2 + (G_4 C_6/C_3 C_5)s + (G_4 G_6/C_3 C_5) \quad (8.16b)$$

The above equations represent, respectively, an LP, a BP, and an HP filter. By adding the outputs I_{o1} and I_{o3} and making $G_2 = G_6$, one can get a regular notch

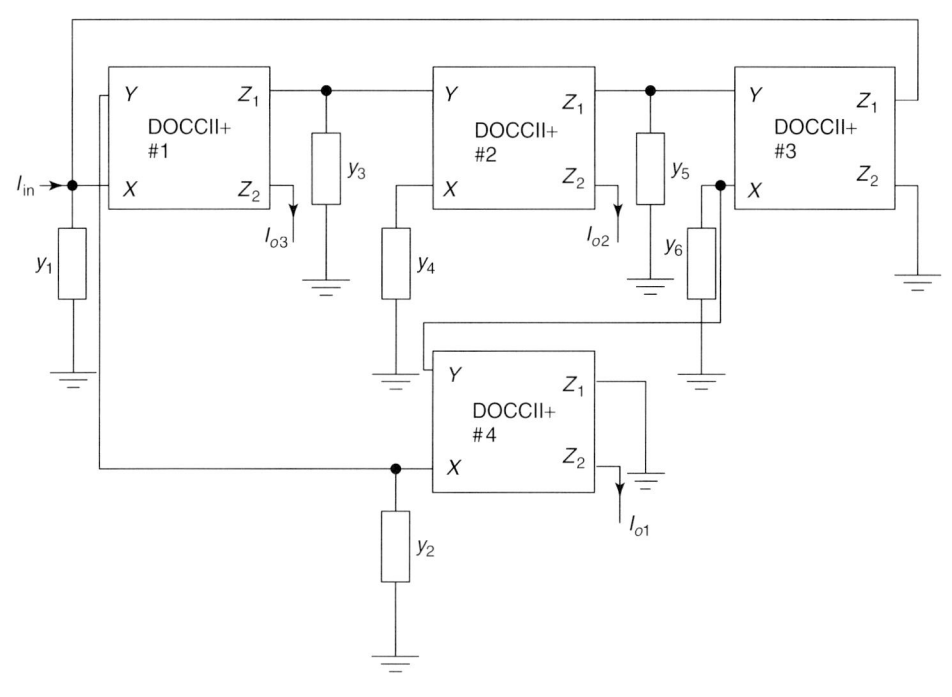

Figure 8.11 A universal second-order CM filter using dual-output current conveyors.

filter. The pole frequency and pole Q are given by

$$\omega_p = \sqrt{\frac{G_4 G_6}{C_3 C_5}}, \quad Q_p = \frac{1}{C_6}\sqrt{\frac{C_3 C_5 G_6}{G_4}} \tag{8.17}$$

It is seen that ω_p and Q_p can be tuned orthogonally by adjusting grounded passive elements. It is also readily seen that the sensitivity of Q_p w.r.t. G_2 as well as that of ω_p w.r.t. C_6 and G_2 are zero, that of Q_p w.r.t. C_6 is -1, while the magnitudes of the sensitivities of ω_p and Q_p w.r.t. the remaining passive elements are 0.5. Thus, the sensitivities w.r.t. the passive elements are not greater than unity.

8.3
Current-Mode Filters Derived from Voltage-Mode Structures

As mentioned earlier, the IC technology has made CM signal processing an area of immense interest. However, it should be pointed out that the concept of CM filters itself is not new and goes back to the 1950s when Thomas (1959) proposed structures for realizing CTFs using voltage- and current-inverting NICs.

The concept of deriving a CM structure from a VM one goes back historically to 1971, when Bhattacharyya and Swamy (1971) introduced the concept of network transposition, whereby given a network N with an admittance matrix [y], one could easily convert it to another network N^T whose admittance matrix is the transpose of [y]. They also derived the transposes of standard two-port elements such as the controlled sources, impedance converters, and impedance inverters. Their basic intention was to obtain a structure realizing a CTF, which is identical to the VTF of a given VM structure, as well as to obtain alternate structures for DPIs. They also pointed out that the transpose was essentially the same as the adjoint network of Director and Rohrer (1969), who had defined the latter two years earlier in connection with the calculation of sensitivities. Swamy, Bhusan, Bhattacharyya (1974, 1976) also defined generalized duals (GDs) and generalized dual transposes (GDTs), and gave a graphical method of obtaining the GD for a two-port network consisting of one-ports and three-terminal two-ports as subnetworks. Some of this material has been treated in Chapters 2 and 3. It is only after the advent of IC technology, have these ideas become useful. In the late 1980s and early 1990s, Roberts and Sedra, unaware of these ideas, resuggested the use of adjoints in obtaining a network whose CTF is the same as the VTF of a network realized using OAs (Roberts and Sedra, 1989, 1992). Later, Carlosena and Moschytz presented the concept of nullor representation of the active devices to arrive at a CTF from a given VTF (Carlosena and Moschytz, 1993). They specifically addressed the case of VM to CM conversion by considering VCVS (or OA)-based VM filters (Moschytz and Carlosena, 1994).

It was shown in Chapter 3 that if the chain matrix of a given network N is

$$[a]_N = \begin{bmatrix} A & B \\ C & D \end{bmatrix} \quad (8.18)$$

then the chain matrix of the reversed transposed network N_R^T, obtained by transposing N and reversing the input and output ports of the transposed network, is given by

$$[a]_{N_R^T} = \begin{bmatrix} D & B \\ C & A \end{bmatrix} \quad (8.19)$$

Also, the chain matrix of the GD network N_D has been defined in Chapter 3 as

$$[a]_{N_D} = \begin{bmatrix} D & Cf(s) \\ \frac{B}{f(s)} & A \end{bmatrix} \quad (8.20)$$

where $f(s)$ is an arbitrary function of s. It is seen from Eqs. (8.18), (8.19), and (8.20), and as already pointed out in Chapter 3, that the CTF of N_R^T as well as that of N_D is the same as the VTF of N. Thus, we can derive two structures for the CTF from that of a VTF through the GD and transposition operations.

In the next two sections, we will deal with the transformation of a VM filter N to a CM filter using the GD and transposition operations.

8.4
Transformation of a VM Circuit to a CM Circuit Using the Generalized Dual

To illustrate the procedure, let us consider the simple circuit of Figure 8.12a, which realizes the first-order VTF given by

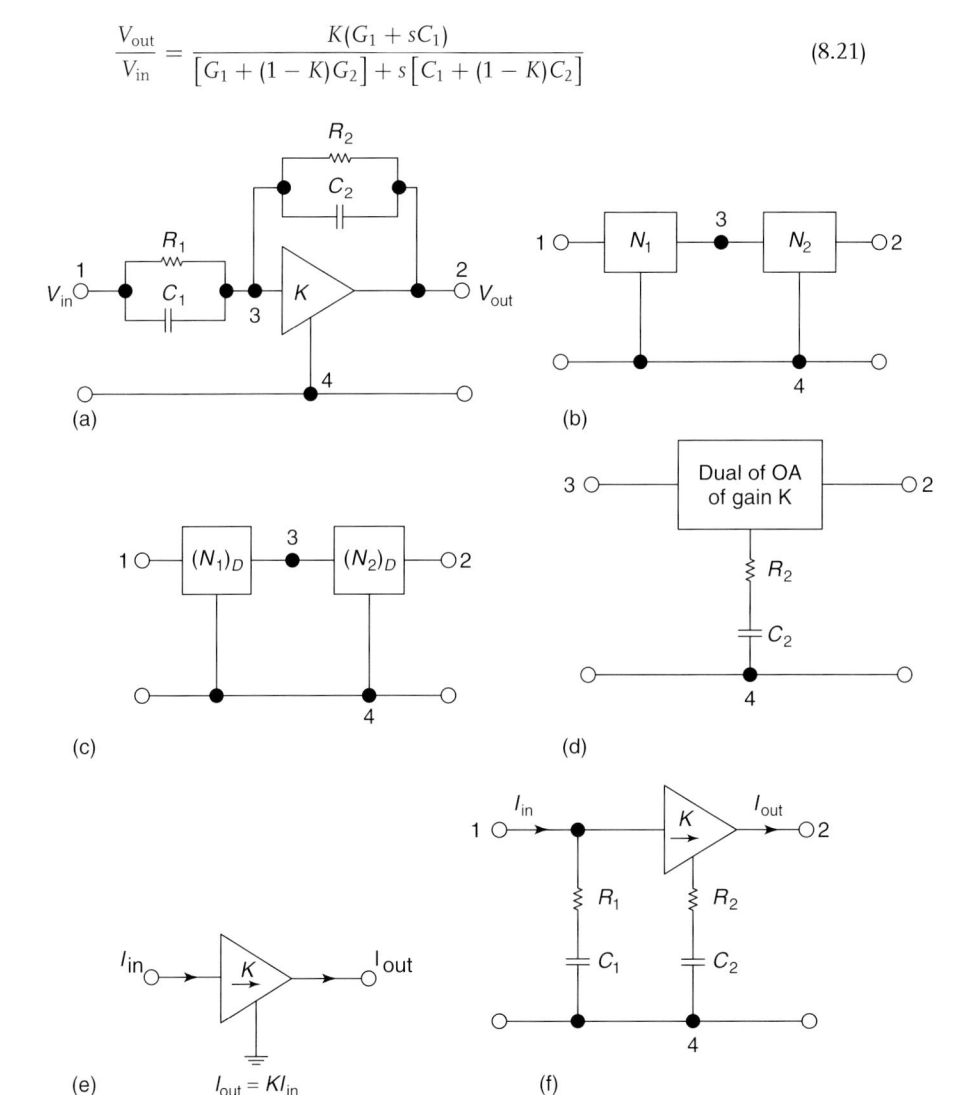

$$\frac{V_{out}}{V_{in}} = \frac{K(G_1 + sC_1)}{\left[G_1 + (1-K)G_2\right] + s\left[C_1 + (1-K)C_2\right]} \qquad (8.21)$$

Figure 8.12 (a) A first-order VM circuit using a voltage amplifier (VA) of gain K; (b) the circuit in (a) being considered as a cascade of two three-terminal two-ports; (c) dual of the network in (b); (d) $(N_2)_D$, the capacitive dual of the network N_2; (e) current amplifier (CA) of gain K, the capacitive dual of a VA of gain K; and (f) the dual of the VM circuit shown in (a).

The circuit of Figure 8.12a can be thought of as a cascade of two networks N_1 and N_2, as shown in Figure 8.12b. We know from Theorem 3.1 that the GD of a cascade of two-ports is the cascade of the GDs of the individual two-ports; hence, we may find the GD of the circuit of Figure 8.12b to be as shown in Figure 8.12c. From Theorem 3.2, we have the result that the GD of a series element of impedance $z(s)$ is a shunt element of impedance $f(s)/z(s)$. If we now choose $f(s) = 1/s$, (i.e., if we take the capacitance duals), we see that $(N_1)_D$ is nothing but a shunt element of impedance $z_{1D} = (1/s)(G_1 + sC_1) = (1/sR_1) + C_1$; thus, z_{1D} is the impedance of a shunt element consisting of a series combination of resistor of C_1 Ω and a capacitor of R_1 farads. Further, from the result of Problem 3.2, we see that $(N_2)_D$ is the capacitive dual of the VA of gain K in series with the dual of R_2 and C_2 in parallel. This is shown in Figure 8.12d. Now, the chain matrix of a VA of gain K is given by

$$\begin{bmatrix} \frac{1}{K} & 0 \\ 0 & 0 \end{bmatrix}$$

Hence, using Eq. (8.20), the capacitive dual of the VA has the chain matrix

$$\begin{bmatrix} 0 & 0 \\ 0 & \frac{1}{K} \end{bmatrix}$$

which is nothing but a CA of gain K, which is symbolized as shown in Figure 8.12e. Thus, the dual of the given network is as shown in Figure 8.12f.

From this very simple example, we can observe the following:

1) The dual of an RC-VA circuit can be obtained as an RC-CA circuit by employing the capacitive dual.
2) The active element, namely, the VA, which was grounded in the original network, has become a floating active element in the dual network. This is due to the fact that the dual of the feedback arm in the original circuit appears in series with the active element in the dual network. This is always the case in that whenever we have a feedback path for the active element in the original network, the grounded active element is converted to a floating dual active element in the dual network. This makes the approach of obtaining CM circuits from VM circuits using GDs not a very attractive one for implementation in IC technology, in view of the attendant influence of parasitic capacitances and enhancement of common-mode noise.
3) If the active element in the original circuit is an OTA of transconductance g_m or a operational transresistance amplifier (OTRA) of transresistance r, then their respective GD elements would be an OTRA of transresistance $f(s)g_m$ and an OTA of transconductance $r/f(s)$, respectively. If $f(s)$ is chosen as $1/s$ to make sure that the passive-RC network part is converted into another RC network, then the transadmittances of the OTAs and the transimpedances of the OTRAs in the dual network are no longer resistive in nature, and become capacitive. On the other hand, if $f(s)$ is chosen as unity so as to make the transadmittances of the OTAs and the transimpedances of the OTRAs in the dual network to be resistive in nature, then the passive resistor and

capacitor elements would be transformed into resistors and inductors in the dual network. Thus, the GD is not beneficial for use in obtaining CM filters from VM filters if we have active elements such as OTAs, OTRAs, or impedance inverters.

Hence, the GD does not play an important role in converting VM circuits to CM circuits. We will next consider the application of transposition in deriving CM circuits from VM circuits.

8.5 Transformation of VM Circuits to CM Circuits Using Transposition

We saw in Chapter 3 that the transpose of any given network N can be very simply obtained by replacing the nonreciprocal subnetwork of N by its respective transpose. It was also seen that the transpose of a VCVS is nothing but a CCCS with its input and output ports reversed and whose current gain is the same as the voltage gain of the VCVS. Hence, for any circuit N that has been designed using finite gain VAs, we may obtain its transpose N^T by simply replacing each of the VAs in N by a CA of the same gain, but with its input and output ports interchanged. Then, the reverse CTF of the transposed network is identical to the forward VTF of the original network and vice versa. It should also be noted that the corresponding sensitivities of the CTF and VTF w.r.t. the various elements would be the same.

The importance of the transpose is particularly apparent in the case of circuits that employ OTAs. Since the transpose of a VCCS is itself with its input–output ports reversed, when a network N containing OTAs is transposed, the resulting network will also consist of only OTAs (i.e., no new active element is necessary). It is obvious that similar results hold good for any VM network designed using OTRAs.

The single-input single-output relationship between the VTF of a VM circuit and the reverse CTF of its transpose can be extended easily to multiport networks. Consider a network N consisting of $(n + m + 1)$ terminals with n input ports and m output ports realizing a VTF matrix $[T_v]$. We can easily find its transpose N^T using the procedure outlined above. Then the roles of the input and output ports are reversed in the resulting network (i.e., the m output ports and the n input ports in N now become the m input and n output ports respectively). Then the CTF matrix $[T_i]$ of this m-input n-output port network is the same as $[T_v]^T$. Interested readers can refer to Bhattacharyya and Swamy (1971) for the proof. We will later take an example to illustrate the usefulness of this result.

Finally, it should be emphasized that the transpose (or adjoint) operation is useful for not only deriving structures for CTFs, but also for obtaining alternate structures for driving point functions as well as for transfer impedance and transfer admittance functions.

8.5.1
CM Circuits from VM Circuits Employing Single-Ended OAs

As mentioned earlier, the transpose of a VCVS of gain A is a CCCS of gain A with its input and output ports interchanged. For an infinite-gain OA ($A \to \infty$), its transpose is a COA of infinite gain. This is symbolically shown in Figure 8.13.

We now illustrate the method of obtaining CM circuits from OA-based VM circuits. We first take the case of a filter that employs a finite gain VA (which can be realized using an OA), and then two filters that employ infinite-gain OAs.

8.5.1.1 CM Biquads Derived from VM Biquads Employing Finite Gain Amplifiers

Consider the LP Sallen and Key VM structure of Figure 5.4, shown as Figure 8.14a for convenience. We know from Eq. (5.13) that its VTF is given by

$$\frac{V_o}{V_i} = \frac{\dfrac{K(G_1 G_3)}{C_2 C_4}}{s^2 + s\left\{\left(\dfrac{G_1}{C_2}\right) + \left(\dfrac{G_3}{C_2}\right) + (1-K)\left(\dfrac{G_3}{C_4}\right)\right\} + \dfrac{G_1 G_3}{C_2 C_4}} \tag{8.22}$$

One can easily obtain the transpose by replacing the OA of gain K by a COA of gain K with its ports reversed and leaving the other elements intact. The resulting CM circuit is shown in Figure 8.14b. It can easily be verified that the reverse short circuit CTF of Figure 8.14b is the same as the forward VTF given by Eq. (8.22).

Similarly, it can be shown that the type 1 and type 2 sensitivity-compensated active networks proposed by Daggett and Vlach (1969) for CTF and VTF are, in fact, transposes of each other.

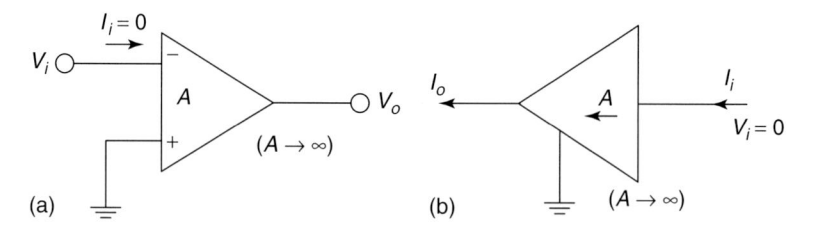

Figure 8.13 (a) An infinite-gain OA and (b) its transpose, a COA of infinite gain.

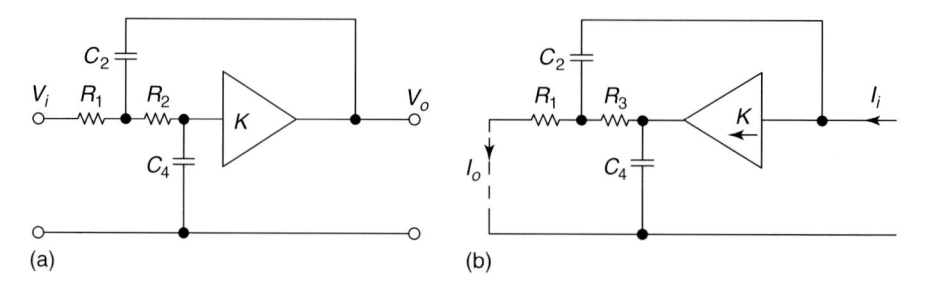

Figure 8.14 (a) Sallen and Key LP VM filter and (b) its corresponding CM filter obtained using transposition.

8.5.1.2 CM Biquads Derived from VM Biquads Employing Infinite-Gain Amplifiers

Consider the general VM single-OA biquad of Figure 5.6, which is redrawn as shown in Figure 8.15a for the sake of convenience. It can be shown that the VTF of Figure 8.15a is given by

$$T_v(s) = \frac{V_o}{V_i} = \frac{-Y_1 Y_3}{[Y_5(Y_1 + Y_2 + Y_3 + Y_4) + Y_2 Y_3] + \frac{1}{A} \times [(Y_1 + Y_2 + Y_4)(Y_3 + Y_5) + Y_3 Y_5]}$$

(8.23)

As $A \to \infty$, the above reduces to

$$T_v(s) = \frac{V_o}{V_i} = \frac{-Y_1 Y_3}{[Y_5(Y_1 + Y_2 + Y_3 + Y_4) + Y_2 Y_3]}$$

(8.24)

We can easily get the transpose of the VM circuit by replacing the OA in the circuit by a reversed COA of infinite gain as shown in Figure 8.15b.

We can verify that for the transposed network, the reverse CTF, $T_i(s)$, is the same as the VM VTF, $T_v(s)$ given by Eq. (8.24). It should be pointed out that $A \to \infty$ only in theory. In practical situations, A is large but finite, and is a function of the complex frequency s. Hence, in order to study the behavior of $T_v(s)$ or $T_i(s)$ under practical conditions, one has to substitute the appropriate expressions for the gain $A(s)$ for the OA or the COA, as the case may be.

The same procedure can be used to transpose any VM circuit that employs multiple OAs with infinite gain. For example, if one considers the VM Fleischer–Tow

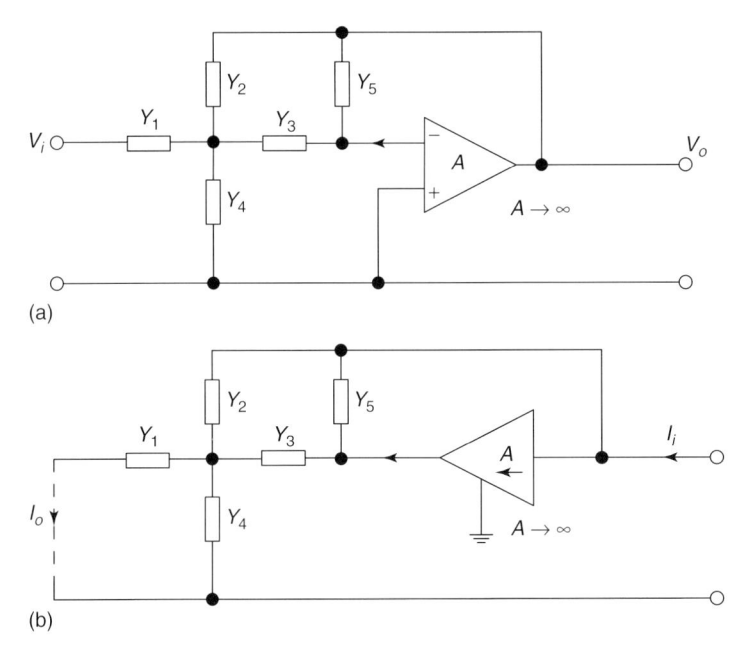

Figure 8.15 (a) A single-amplifier VM general biquad and (b) its transpose.

8 Current-Mode Filters

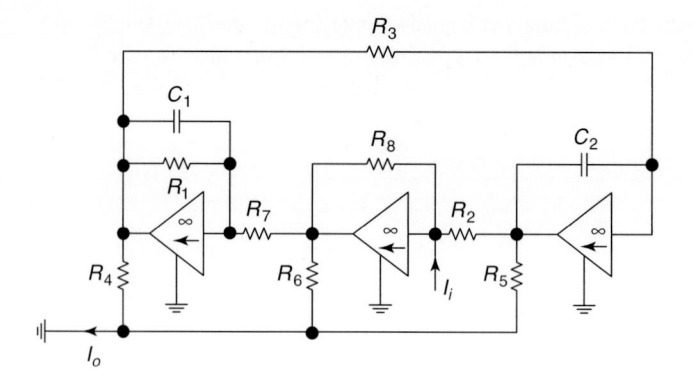

Figure 8.16 Transpose of the Fleischer–Tow VM biquad.

biquad of Figure 5.10, one can easily obtain the corresponding CM structure by simply replacing the OAs by reversed COAs of infinite gain, as shown in Figure 8.16. It can be verified that the CTF of the CM circuit is the same as the VTF of the original, as given by Eq. (5.47).

8.5.2
CM Circuits from VM Circuits Employing OTAs

8.5.2.1 VM Circuits Using Single-Ended OTAs

If one of the input terminals of the OTA (see Figure 5.17) is grounded, say the negative terminal, then the OTA and its equivalent circuit are as shown in Figure 8.17. Its admittance matrix is

$$[y]_N = \begin{bmatrix} 0 & 0 \\ -g_m & 0 \end{bmatrix}$$

Hence, the admittance matrix of its transpose is

$$[y]_N^T = \begin{bmatrix} 0 & -g_m \\ 0 & 0 \end{bmatrix}$$

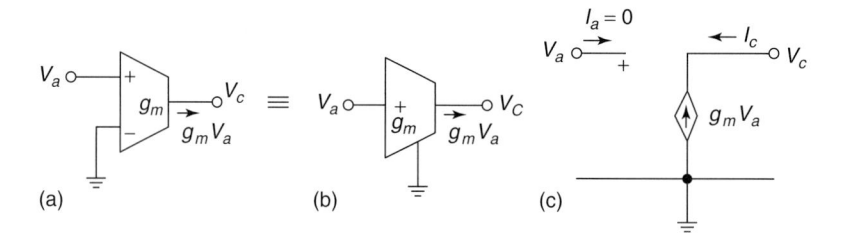

Figure 8.17 (a) and (b) Symbols for an OTA whose positive terminal is grounded, and (c) its AC equivalent circuit.

8.5 Transformation of VM Circuits to CM Circuits Using Transposition

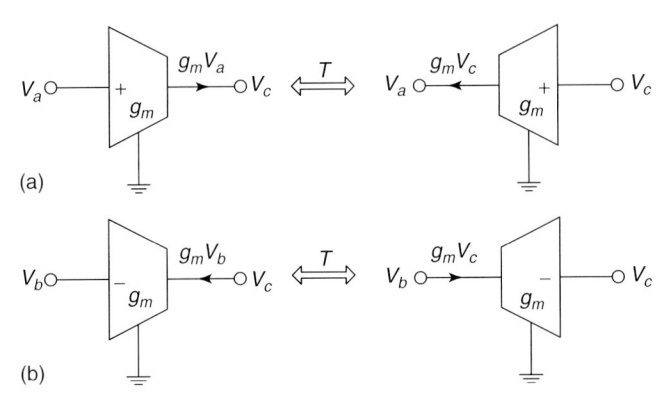

Figure 8.18 (a) OTA with its negative input terminal grounded and its transpose (b) OTA with its positive input terminal grounded and its transpose.

which corresponds to nothing but the original OTA with its input and output ports interchanged. Similarly, if the positive input terminal of the OTA is grounded, then the corresponding transpose would be itself with its input and output ports interchanged. These are symbolically shown in Figure 8.18. Thus, the transpose of a single-ended OTA is itself and is not a new element. This is particularly useful since the same element can be used for both the VM and CM circuits.

As an example (Swamy and Raut, 2002), consider the VM g_m-C BP filter employing single-ended OTAs (Ghausi and Laker, 1981), as shown in Figure 8.19a. The corresponding CM filter can readily be obtained using transposition and is shown in Figure 8.19b. It is easily verified that the VTF of the former and the reverse CTF of the latter are indeed the same and are given by (Senani, 1992)

$$\frac{V_o}{V_i} = \frac{I_o}{I_i} = \frac{sC_1 g_{m2}}{s^2 C_1 C_2 + sC_1 g_{m3} + g_{m1} g_{m2}} \qquad (8.25)$$

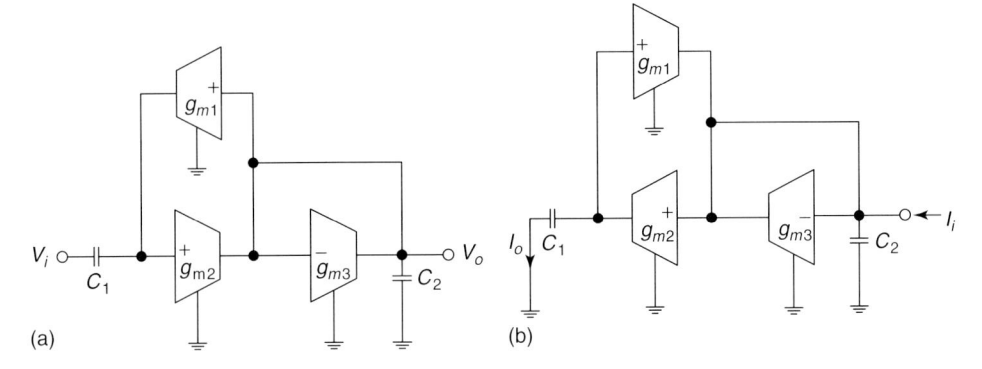

Figure 8.19 (a) VM g_m-C BP filter employing single-ended OTAs and (b) the corresponding CM filter (transpose of the circuit in (a)).

8 Current-Mode Filters

Obviously the sensitivities of the pole frequency and pole Q of both the VTF and CTF are the same w.r.t. the various parameters.

We now consider an example of obtaining a CTF matrix from a network realizing a VTF matrix (Swamy and Raut, 2002). Figure 8.20 shows a general VM biquad structure that realizes LP, BP, and HP functions depending on whether the input is V_A, V_B, or V_C (Geiger and Sanchez-Sinencio, 1985). This structure can be considered as a three-input single-output network realizing the VTF matrix $[T_v]$ given by

$$V_o = [T_v] \begin{bmatrix} V_A \\ V_B \\ V_C \end{bmatrix} = \frac{1}{D(s)} \begin{bmatrix} g_{m2}g_{m5} & sC_1 g_{m4} & s^2 C_1 C_2 \end{bmatrix} \begin{bmatrix} V_A \\ V_B \\ V_C \end{bmatrix} \quad (8.26a)$$

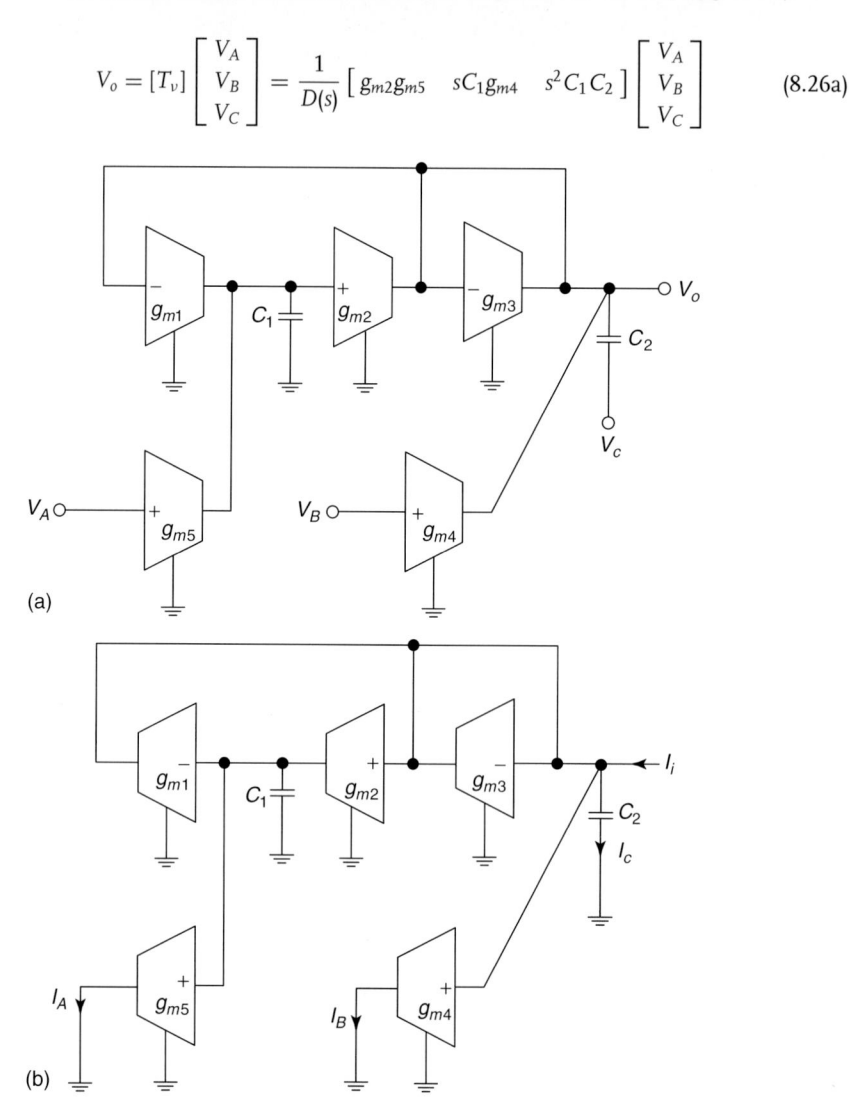

Figure 8.20 (a) A three-input single-output VM biquad structure realizing LP, BP, and HP functions and (b) its transpose, a single-input three-output CM biquad structure.

where

$$D(s) = s^2 C_1 C_2 + sC_1 g_{m3} + g_{m1} g_{m2} \tag{8.26b}$$

The transpose of the given network is as shown in Figure 8.20b. This is a single-input three-output network realizing the CTF matrix $[T_i] = [T_v]^T$. Thus, the CM filter realizes the CTF matrix

$$\begin{bmatrix} I_A \\ I_B \\ I_C \end{bmatrix} = \frac{1}{s^2 C_1 C_2 + sC_1 g_{m3} + g_{m1} g_{m2}} \begin{bmatrix} g_{m2} g_{m5} \\ sC_1 g_{m4} \\ s^2 C_1 C_2 \end{bmatrix} I_i \tag{8.27}$$

This realization is the same as the one given by Senani (1992).

8.5.2.2 VM Circuits Using Differential-Input OTAs

In the above section, we derived CM circuits from VM circuits that employ single-ended OTAs. We now consider the transformation of VM circuits employing differential-input single-output (DISO) OTAs to CM circuits. For this, we first derive the transpose of a DISO-OTA, as shown in Figure 8.21a.

The $[y]$ of the differential-input OTA is given by

$$\begin{bmatrix} I_A \\ I_B \\ I_C \end{bmatrix} = \begin{bmatrix} 0 & 0 & 0 \\ 0 & 0 & 0 \\ -g_m & g_m & 0 \end{bmatrix} \begin{bmatrix} V_A \\ V_B \\ V_C \end{bmatrix} \tag{8.28}$$

Hence,

$$[y]^T = \begin{bmatrix} 0 & 0 & -g_m \\ 0 & 0 & g_m \\ 0 & 0 & 0 \end{bmatrix} \tag{8.29}$$

Thus, in the transposed element, the current–voltage relations are

$$I_A = -g_m V_C, \quad I_B = g_m V_C, \quad I_C = 0 \tag{8.30}$$

Thus the transpose of a DISO-OTA is a single-input dual-output (SIDO) OTA; it is symbolically shown as in Figure 8.21b. A SIDO-OTA can be obtained by simply

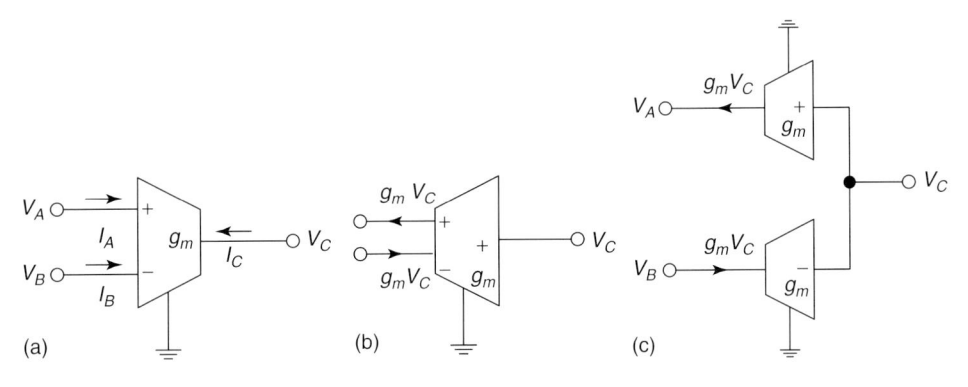

Figure 8.21 (a) A DISO-OTA, (b) its transpose, a SIDO-OTA, and (c) a realization of a SIDO-OTA.

applying a current replica in a standard OTA. A simple realization using two single-ended OTAs is shown in Figure 8.21c. Using the transpose of a DISO-OTA, we can easily derive CM circuits from VM circuits that employ such OTAs. We illustrate this with two examples.

Example 8.1. SIDO-OTA universal CM filter: Consider the VM DISO-OTA universal filter shown in Figure 8.22a (Deliyanis, Sun, and Fidler, 1999). It can be shown that the VTF of this circuit is given by

$$T_v(s) = \frac{V_o}{V_i} = \frac{g_m Y_2 Y_4}{D(s)} \tag{8.31a}$$

where

$$D(s) = (Y_1 Y_2 + Y_2 Y_3 + Y_3 Y_1)(Y_4 + Y_5) + (Y_1 Y_5 + Y_2 Y_5 + g_m Y_2) Y_4 \tag{8.31b}$$

By replacing the DISO-OTA by its transpose, we get the CM filter shown in Figure 8.22b, proposed by Al-Hashimi and Fidler (1988). It can easily be shown that the CTF $T_i(s) = \frac{I_o}{I_i}$ is given by

$$T_i(s) = \frac{I_o}{I_i} = \frac{g_m Y_2 Y_4}{D(s)}$$

which is the same as the $T_v(s)$ of the original network. Various filters such as LP, BP, and HP second-order filters can be obtained by appropriately choosing the values of the various admittances.

Example 8.2. Leapfrog structure: As a second example, we consider the derivation of the CM leapfrog structure from that of a VM leapfrog structure. Consider the ladder network of Figure 7.38. The realization of this ladder for its VTF using DISO-OTAs is shown in Figure 7.40 and is redrawn as Figure 8.23 for convenience. The various impedances $Z_1', Z_2', \ldots, Z_6'$ in Figure 8.23 are related to the series and shunt passive impedances $Z_1, Z_2, \ldots, Z_6$ of the ladder of Figure 7.40 by the relations

$$Z_j' = \frac{1}{g g_{mj}} \frac{1}{Z_j}, \quad \text{for } j = 1, 3, 5 \tag{8.32a}$$

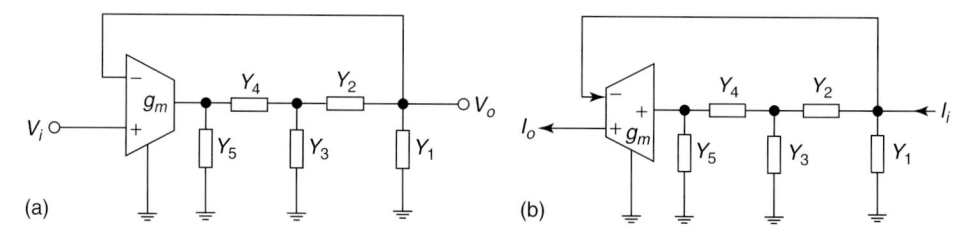

Figure 8.22 (a) A universal VM filter using a DISO-OTA, (b) its transpose, the CM counterpart of the circuit given in (a).

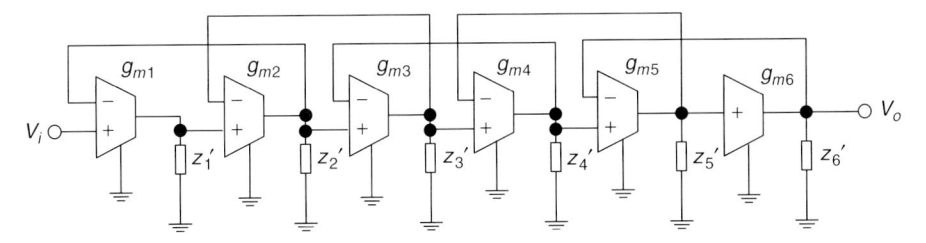

Figure 8.23 VM leapfrog structure for the ladder network of Figure 7.48 using OTAs.

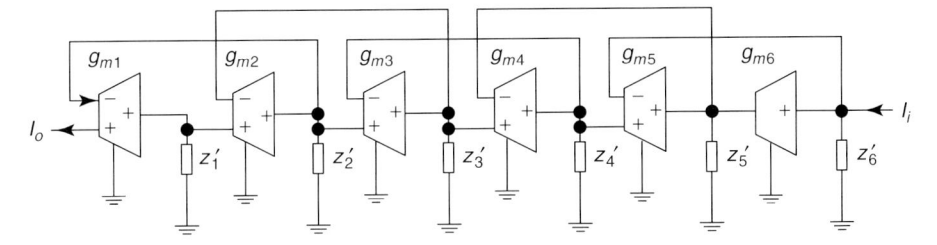

Figure 8.24 Transpose of Figure 8.23, the corresponding CM leapfrog structure.

and

$$Z'_j = \frac{1}{g_{mj}} Z_j, \quad \text{for } j = 2, 4, 6 \tag{8.32b}$$

Since these elements are unchanged in the transpose and only the DISO-OTAs are replaced by their transposed elements, which are SIDO-OTAs, the CM leapfrog structure can readily be drawn, as shown in Figure 8.24. The CTF $\frac{I_o}{I_i}$ of the CM structure is the same as the VTF of the VM structure of Figure 8.23.

Thus, any VM structure that is realized using OTAs can easily be transformed to derive corresponding CM structures that also use OTAs. In this connection, it is to be noted that the various canonical structures derived for CM structures (Sun and Fidler, 1995) and for VM structures (Sun and Fidler, 1997) are simply transposes of one another (Tang, 2004). Even though the subject matter of this book is on filter design, we would like to point out that transposition can also be used to obtain alternate structures for oscillators from known oscillator structures (Bhattacharyya, Sundaramurthy, and Swamy, 1981; Swamy, Raut, and Tang, 2004; Tang, 2004). We next consider the application of transposition in conjunction with the concept of nullors to obtain structures for CTFs that employ single-ended OAs from VM circuits realized using single-ended OAs.

8.6 Derivation of CTF Structures Employing Infinite-Gain Single-Ended OAs

In Section 8.5.1, we gave a method of transforming VM circuits using single-ended OAs into CM circuits; however, the CM circuits need to use COAs of infinite gain. We will now show, using the principle of nullors along with the concept of transposes, how to derive structures for CTFs that utilize infinite-gain OAs from VTF structures that also employ infinite-gain OAs (Raut, Swamy, and Tian, 2007).

Consider a structure N that uses ideal single-ended OAs for the realization of a given VTF. Since an ideal OA can be considered as an ideal VCVS of infinite gain, we may replace the VCVSs by their nullor equivalents. We now use the fact that the nullor equivalent of an ideal CCCS is exactly the same as that of the ideal VCVS (Bruton, 1980). Hence, we can replace each of the OAs in the VTF structure by an ideal CCCS, and the VTF realized by the resulting structure N' is the same as that of N. If we now transpose the VTF structure N' to obtain $[N']^T$, then each of the CCCSs in N' would be replaced by a VCVS (infinite-gain OA) with its input and output ports reversed. Utilizing the property of a transposed network, we see that the reverse CTF of $[N']^T$ is the same as the VTF of N. Since all the passive elements in N are unchanged during these operations, it is clear that the CTF structure is obtained simply by reversing the input–output ports of the OAs. We will now illustrate the usefulness of this result by the following examples.

8.6.1
Illustrative Examples

8.6.1.1 Single-Amplifier Second-Order Filter Network

Consider the SAB VM filter using an ideal OA (an ideal VCVS), as shown in Figure 8.25. The VTF of this network is given by

$$\frac{V_o}{V_i} = -\frac{Y_3 Y_4}{Y_1 Y_2 + Y_2 Y_3 + Y_1 Y_3 + Y_1 Y_4} \tag{8.33}$$

If we now replace the ideal VCVS by its nullor equivalent, and use the fact that the nullor equivalents of all the four ideal controlled sources are the same, we can replace the ideal VCVS in Figure 8.25a by an ideal CCCS to obtain the VM circuit using a CCCS, as shown in Figure 8.25b (where A tends to infinity), whose VTF should be the same as that of the circuit of Figure 8.25a. In fact, it can easily be verified that this is true.

If we now take the transpose of the circuit of Figure 8.25b, we get the circuit shown in Figure 8.26, whose reverse CTF would be the same as the VTF of that of

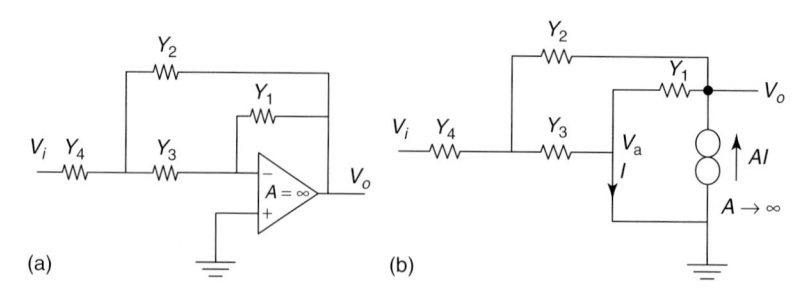

Figure 8.25 (a) A single-amplifier VM filter using an ideal OA and (b) its nullor equivalent using a CCCS.

8.6 Derivation of CTF Structures Employing Infinite-Gain Single-Ended OAs

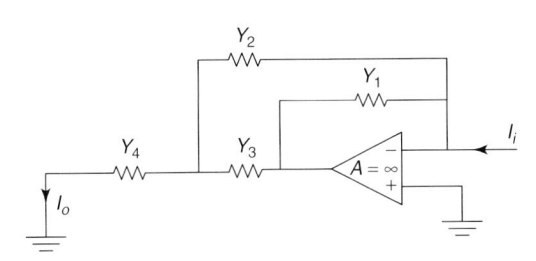

Figure 8.26 Circuit obtained from Figure 8.25a with its CTF same as that of the VTF of the latter.

Figure 8.25b, that is, the same as that of Figure 8.25a. It can indeed be verified that the reverse CTF I_o/I_i of the CM filter of Figure 8.26 is given by

$$\frac{I_o}{I_i} = -\frac{Y_3 Y_4}{Y_1 Y_2 + Y_2 Y_3 + Y_1 Y_3 + Y_1 Y_4} \tag{8.34}$$

As mentioned earlier, the circuit of Figure 8.26 could have been obtained directly from the VTF circuit of Figure 8.25a by simply turning around the input and output ports of the VCVS.

8.6.1.2 Tow-Thomas Biquad

The Tow–Thomas VM filter N_V, employing three OAs is shown in Figure 8.27. Its VTF is given by

$$\frac{V_o}{V_i} = -\frac{\dfrac{C_1}{C}s^2 + \dfrac{1}{RC}\left(\dfrac{R}{R_1} - \dfrac{r}{R_3}\right)s + \dfrac{1}{C^2 R R_2}}{s^2 + \dfrac{1}{CR_4}s + \dfrac{1}{C^2 R^2}} \tag{8.35}$$

The corresponding CTF network N_C can be obtained directly by reversing the input/output ports of each of the VCVS elements in Figure 8.27, and is shown in Figure 8.28. It can be verified that the reverse CTF of this circuit is given by

$$\frac{I_o}{I_i} = -\frac{\left[\dfrac{C_1}{C}s^2 + \dfrac{1}{CR}\left(\dfrac{R}{R_1} - \dfrac{r}{R_3}\right)s + \dfrac{1}{C^2 R R_2}\right]}{s^2 + \dfrac{1}{CR_4}s + \dfrac{1}{C^2 R^2}} \tag{8.36}$$

which is the same as the VTF of the VM filter of Figure 8.27

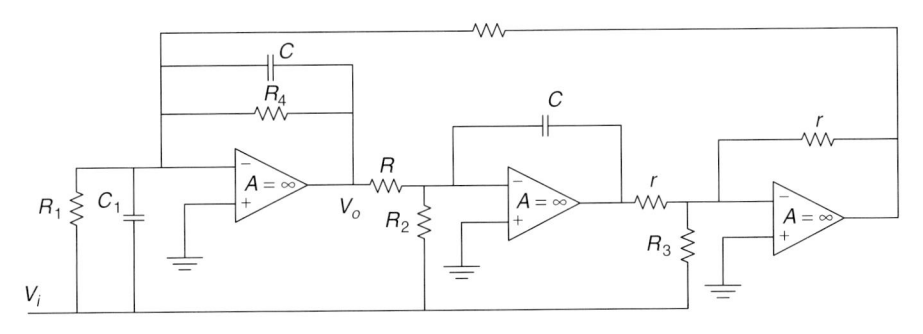

Figure 8.27 Tow–Thomas VM filter N employing three OAs.

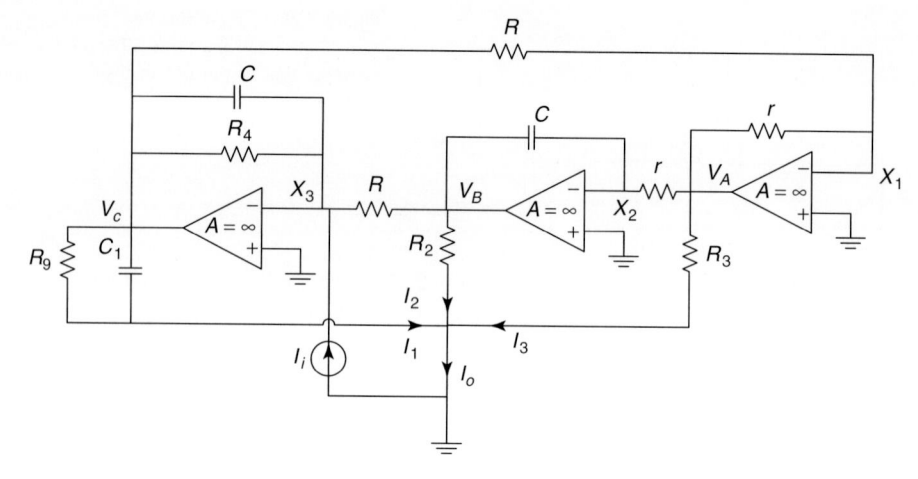

Figure 8.28 Filter whose CTF is the same as the VTF of Figure 8.27.

8.6.1.3 Ackerberg and Mossberg LP and BP Filters

As a final example, consider the Ackerberg and Mossberg filter, which can produce both LP and BP filter responses depending upon the choice of the output signal node, as shown in Figure 8.29 (Ackerberg and Mossberg, 1974). The two VTFs of the A & M filter are

$$\frac{V_{o1}}{V_i} = -\frac{s/RC_1}{s^2 + s\dfrac{1}{R_1 C_1} + \dfrac{r_1}{C_1 C_2 R_2 rr_2}} \tag{8.37a}$$

and

$$\frac{V_{o2}}{V_i} = -\frac{\dfrac{r_1}{C_1 C_2 Rrr_2}}{s^2 + s\dfrac{1}{R_1 C_1} + \dfrac{r_1}{C_1 C_2 R_2 rr_2}} \tag{8.37b}$$

Figure 8.30 shows the corresponding CTF network N_C obtained directly from the VM filter by reversing the input/output ports of each of the VCVS elements in

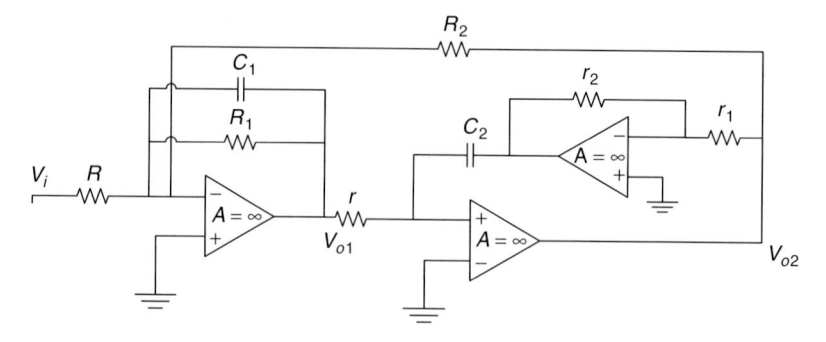

Figure 8.29 Ackerberg and Mossberg filter realizing LP and BP functions.

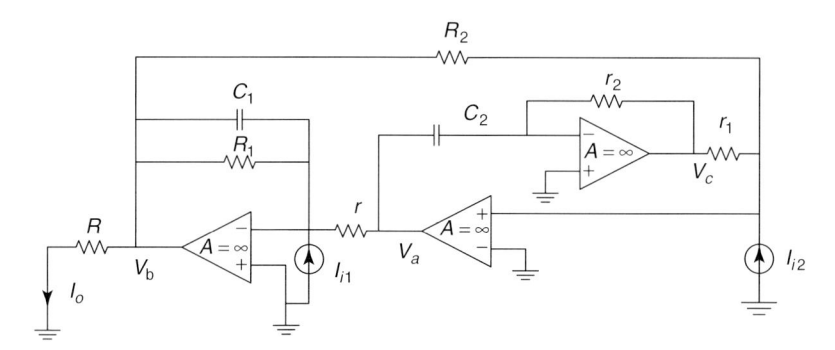

Figure 8.30 Filter whose CTFs are the same as the LP and BP VTFs of Figure 8.29.

Figure 8.29. It can easily be shown that the two CTFs are given by

$$\frac{I_o}{I_{i2}} = -\frac{\dfrac{r_1}{C_1 C_2 R r r_2}}{s^2 + \dfrac{1}{R_1 C_1} s + \dfrac{r_1}{C_1 C_2 r r_2 R_2}} \quad \text{with } I_{i1} = 0 \quad (8.38a)$$

and

$$\frac{I_o}{I_{i1}} = -\frac{\dfrac{1}{RC_1} s}{s^2 + \dfrac{1}{R_1 C_1} s + \dfrac{r_1}{C_1 C_2 r r_2 R_2}} \quad \text{with } I_{i2} = 0 \quad (8.38b)$$

which are, respectively, the same as the VTFs (V_{o2}/V_1) and (V_{o1}/V_1) of the VM filter of Figure 8.29.

Thus, we can obtain a CTF (VTF) filter from a VTF (CTF) filter that employs three-terminal ideal VCVSs by simply reversing the input/output terminals of each of the VCVSs in the latter. Similarly, we can conclude that if the VTF (CTF) filter consisted of only three-terminal CCCS elements, then we can obtain the corresponding CTF (VTF) filter by simply reversing the input/output terminals of the CCCSs in the former.

8.6.2
Effect of Finite Gain and Bandwidth of the OA on the Pole Frequency, and Pole Q

The analysis in the previous section has been made under the assumption of ideal passive components and ideal active devices (i.e., ideal OAs). However, it is important to study the effect of the nonideal characteristics of the active device. Since the active device used is the OA, one important deviation from its ideal characteristic arises because of the limited amplifier bandwidth. We now analyze the effect of finite bandwidth on the pole frequency and pole Q factor of the CTF filter and compare them with those of the corresponding VTF filter.

The analytical procedure is demonstrated for a multi-OA filter, namely, the A & M three-OA biquad. For the A & M filter shown in Figure 8.29, let us assume that each OA has a gain A. Then, the LP and BP VTFs are given by

$$\frac{V_{o2}}{V_i} = -\frac{(1/rR)}{D_V(s)} \tag{8.39}$$

and

$$\frac{V_{o1}}{V_i} = -\frac{\frac{1}{R}\left(\frac{\frac{1}{r}+sC_2}{A} + \frac{\frac{sC_2}{r_1}}{\frac{1}{r_2}+\frac{1/r_2+1/r_1}{A}}\right)}{D_V(s)} \tag{8.40}$$

where

$$D_V(s) = \frac{1}{R_2 r} + \left(\frac{\frac{1}{R}+\frac{1}{R_2}}{A} + \left(1+\frac{1}{A}\right)\left(\frac{1}{R_1}+sC_1\right)\right)\left(\frac{\frac{1}{r}+sC_2}{A} + \frac{\frac{sC_2}{r_1}}{\frac{1}{r_2}+\frac{1/r_2+1/r_1}{A}}\right) \tag{8.41}$$

The LP and BP CTFs for the filter shown in Figure 8.30 are obtained as

$$\frac{I_o}{I_{i2}} = -\frac{(1/rR)}{D_I(s)} \tag{8.42}$$

and

$$\frac{I_o}{I_{i1}} = -\frac{\frac{1}{R}\left(\frac{\frac{1}{R_2}+\frac{1}{r_1}}{A} + \frac{\frac{sC_2}{r_1}}{\frac{1}{r_2}+\frac{1/r_2+sC_2}{A}}\right)}{D_I(s)} \tag{8.43}$$

where

$$D_I(s) = \frac{1}{R_2 r} + \left(\frac{1}{Ar} + \left(1+\frac{1}{A}\right)\left(\frac{1}{R_1}+sC_1\right)\right)\left(\frac{\frac{1}{R_2}+\frac{1}{r_1}}{A} + \frac{\frac{sC_2}{r_1}}{\frac{1}{r_2}+\frac{1/r_2+sC_2}{A}}\right) \tag{8.44}$$

To evaluate the performance of the VM and CM filter, we use the one-pole model for the gain of the OA,

$$A = A(s) = \frac{A_o}{1+\frac{s}{\omega_p}} \tag{8.45a}$$

where ω_p is the pole frequency. For frequencies $\omega \gg \omega_p$, we may assume

$$A = A(s) \approx \frac{A_o \omega_p}{s} \approx \frac{\omega_t}{s} \tag{8.45b}$$

where $\omega_t \approx A_o \omega_p$ is the unity-gain bandwidth of the OA. We further assume that

$$R = R_2 = r = r_1 = r_2, \; C_1 = C_2 = C, \; R_1 = Q_o R \tag{8.46a}$$

and

$$\omega_o \ll \omega_t \tag{8.46b}$$

where ω_o and Q_o are the pole frequency and pole Q of the filter. With the above assumptions, and approximating $s^3 = -\omega_o^2 s, s^5 = \omega_o^4 s$, we may simplify $D_V(s)$ and $D_I(s)$. This leads to

$$\text{Pole-frequency deviation of the VTF: } \delta_V = \frac{\omega_{oa} - \omega_o}{\omega_o} \approx -\frac{3\omega_o}{2\omega_t} \tag{8.47}$$

$$\text{Pole- frequency deviation of the CTF: } \delta_I = \frac{\omega_{oa} - \omega_o}{\omega_o} \approx -\left(\frac{2\omega_o}{\omega_t} + \frac{\omega_o}{2Q_o\omega_t}\right) \tag{8.48}$$

$$\text{Pole-Q deviation of the VTF: } \eta_V = \frac{Q_{oa} - Q_o}{Q_o} \approx \frac{\omega_o}{2\omega_t} \tag{8.49}$$

$$\text{Pole-Q deviation of the CTF: } \eta_I = \frac{Q_{oa} - Q_o}{Q_o} \approx \frac{\omega_o}{2Q_o\omega_t} \tag{8.50}$$

where ω_{oa} and Q_{oa} are the realized values. Simulation results and calculations show that the performances of the CTF and VTF filters match very well, especially when the pole frequency of the filter is far smaller than that of the OA unity-gain bandwidth (Raut, Swamy, and Tian, 2007).

8.7
Switched-Current Techniques

In an SC system, signal voltages are sampled, converted to charge packets by the capacitors, and then reconverted to voltages by redistribution of the charges among a different set of capacitors. Existence of linear capacitors is necessary for processing the voltage signals in an SC system. In an SI system, the presence of linear capacitors is not needed. A current signal injected into a node develops a voltage on the parasitic capacitance at that node and is then transferred back to a current at another node by the simple current mirroring principle. Since switching is used, the current mirrors are named differently, that is, current copiers and dynamic current mirrors. Except for this difference, the operations in the SI systems are similar to those in SC systems and require similar building blocks such as addition, subtraction, multiplication, and delay. Like the SC systems, the SI system is attractive for implementation in an IC technology (Mohan, 2002). The operations of these basic building blocks in an SI system are discussed below. We will consider realizations using MOS and CMOS transistors.

8.7.1
Add, Subtract, and Multiply Operations

Consider the current mirror system shown in Figure 8.31. The inputs nodes marked + and − have low resistances because of the diode-connected transistors. They receive bidirectional input currents representing positive and negative signals, and hence, the input transistors are biased by currents I to ensure forward conduction (Hughes, Macbeth, and Pattullo, 1990). The output transistors M_4 and M_5 have a width-to-length ratio (or the aspect ratio), W/L, that is α times that of the input transistors M_2 and M_3. In the following, the notation $1:\alpha$ implies an aspect ratio of $1:\alpha$ between the input and output transistors. Other output transistors may be included with other aspect ratios to produce a fan out capability.

Both the inputs are current summing nodes and it is easily seen that $i_o = \alpha(\sum_{j=1}^{k} i_{pj} - \sum_{j=1}^{l} i_{nj})$.

In the above, and what follows, the upper case letters $I, J, \ldots$, are used to signify DC bias values, while the lower case letters $i, j, \ldots$, are used to signify the signal currents.

8.7.2
Switched-Current Memory Cell

Figures 8.32 and 8.33 show two possible memory cells. Clock phases ϕ_1 and ϕ_2 are defined by nonoverlapping voltages and it is assumed that the switches turn on when their control voltage is high. The circuit of Figure 8.32 is a simple current mirror with the switch S separating its input and output transistors. On phase ϕ_2 switch S is closed and both the oxide capacitances C_{ox1} and C_{ox2} are charged to V_{gs}, where

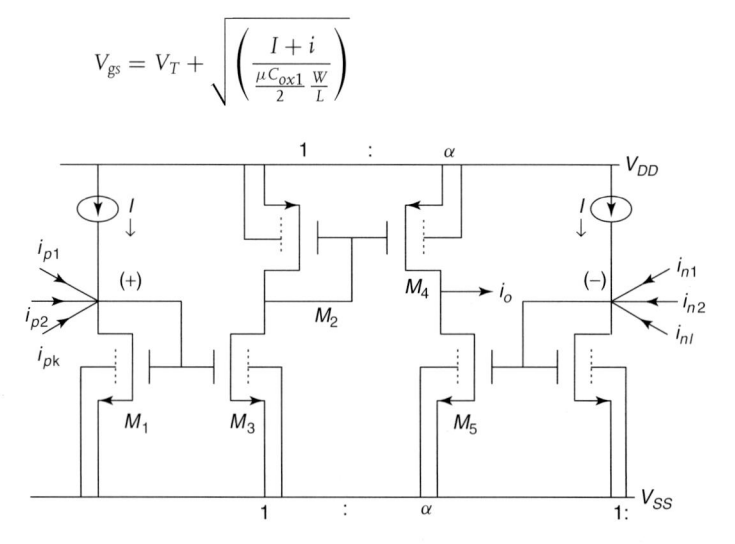

$$V_{gs} = V_T + \sqrt{\left(\frac{I+i}{\frac{\mu C_{ox1}}{2} \frac{W}{L}}\right)}$$

Figure 8.31 Current summing and differencing employing CMOS current mirrors.

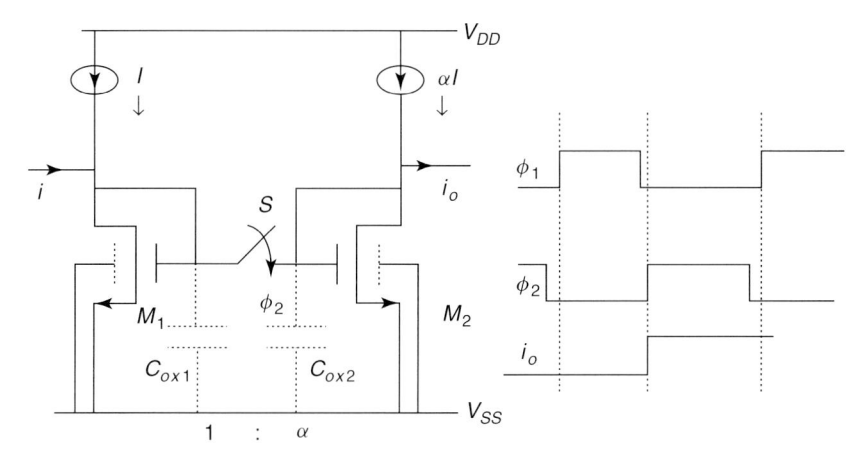

Figure 8.32 SI memory cell type 1.

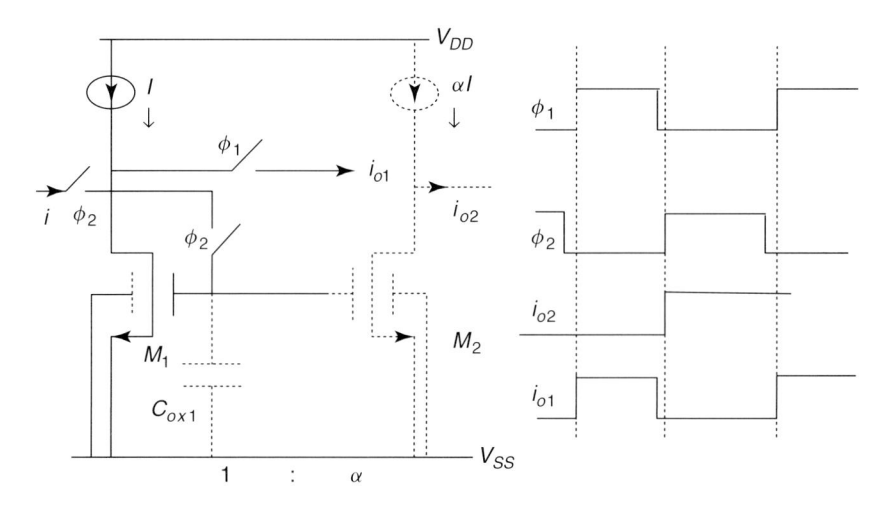

Figure 8.33 SI memory cell type 2.

By normal current mirror action $i_o = -\alpha i$, and the current i_o is available simultaneously with the input sample. On phase ϕ_1, switch S opens and isolates the input from the output. A voltage close to V_{gs} is held on C_{ox2} and sustains a current close to $-\alpha i$ at the output.

The arrangement in Figure 8.33 can achieve memory within a single transistor M_1. On phase ϕ_2, M_1 is diode connected and conducts a current $I + i$ and as before, V_{gs} is stored on the oxide capacitance. On phase ϕ_1, M_1 maintains its current $I + i$, and hence, $i_{o1} = -i$. To achieve scaling by a factor α, an extra output stage (M_2) can be used to make $i_{o2} = -\alpha i$. As M_1 is used alternately as an input diode and output transistor, i_{o1} is available only during phase ϕ_1. Output current i_{o2} is available for the whole period as with the other memory cell.

8.7.3
Switched-Current Delay Cell

The memory cells of the type shown in Figure 8.32 can be used to produce a unit delay. This will need a cascode of an NMOS and a PMOS transistor. The arrangement is shown in Figure 8.34.

The NMOS is clocked on phase ϕ_2 and the PMOS memory on phase ϕ_1. During phase ϕ_2 (the end of the $(n-1)$th period), the input current $i(n-1)$ together with the bias current I enters the input diode of the NMOS memory cell. During phase ϕ_1 (the start of the nth period), this current is stored in the NMOS memory, fed to the PMOS memory and, since switch S_2 is closed, a current $i_o(n)$ equal to $i(n-1)$ flows in the output. The input current, $i(n)$, does not propagate to the PMOS memory on this phase as switch S_1 is open. On the next phase, ϕ_2, $i_o(n)$ is sustained at the value $i(n-1)$, so the output clearly is the input signal delayed by one clock period.

8.7.4
Switched-Current Integrators

Figure 8.35 shows a simple SI integrator. It is formed from a delay circuit with two output stages weighted α and β, the output from the β weighted transistors being fed back to the input summing node. The signal, i_o, from the α weighted transistors is the integrator output signal. The system operates as described below. The output signal is established as soon as S_2 closes at the beginning of phase ϕ_1 and is held until the end of phase ϕ_2. Thus, it covers a time span of one clock period (duration of ϕ_1 plus duration of ϕ_2).

At the onset of ϕ_2, the $(n-1)$th clock period, the output signal is $i_o(n-1)$ and the value of i_f is $(\beta/\alpha)i_o(n-1)$. The input current is $i_1(n-1)$ and the total current in M_1 is

$$I_1 = I + i_1(n-1) + (\beta/\alpha)i_o(n-1) \tag{8.51}$$

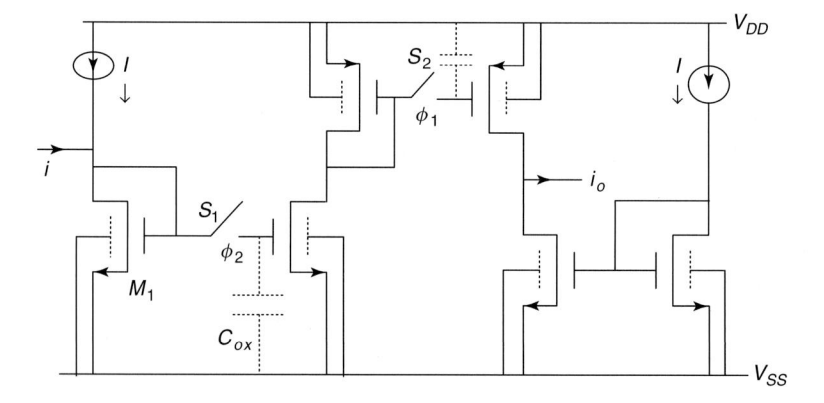

Figure 8.34 SI delay cell.

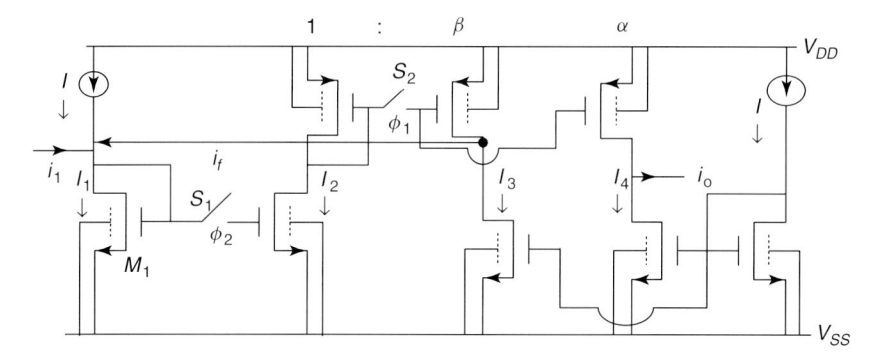

Figure 8.35 A simple SI integrator.

Since the switch S_1 is closed, $I_2 = I_1$. During the subsequent phase ϕ_1, the switch S_2 is closed, the currents i_1, i_o, and i_f change to $i_1(n), i_o(n)$, and $\beta i_o(n)/\alpha$ respectively. However, since S_1 is open, I_2 remains at its previous value, and since now S_2 is closed, it is mirrored on to the α and β transistors as I_3 and I_4. Thus,

$$I_4 = \alpha I_1 = \alpha \left[I + i_1(n-1) + \frac{\beta}{\alpha} i_o(n-1) \right] \tag{8.52}$$

and

$$i_o(n) = I_4 - \alpha I = \alpha i_1(n-1) + \beta i_o(n-1) \tag{8.53}$$

On taking the z-transform, we get

$$H_1(z) = I_o(z)/I_1(z) = \frac{\alpha z^{-1}}{1 - \beta z^{-1}} \tag{8.54}$$

where, $I(z)$ represents the z-transform of $i(n)$.

The above corresponds to an integrator associated with the forward Euler mapping between the analog and sampled-data domains, that is, $s \rightarrow (1 - z^{-1})/Tz^{-1}$.

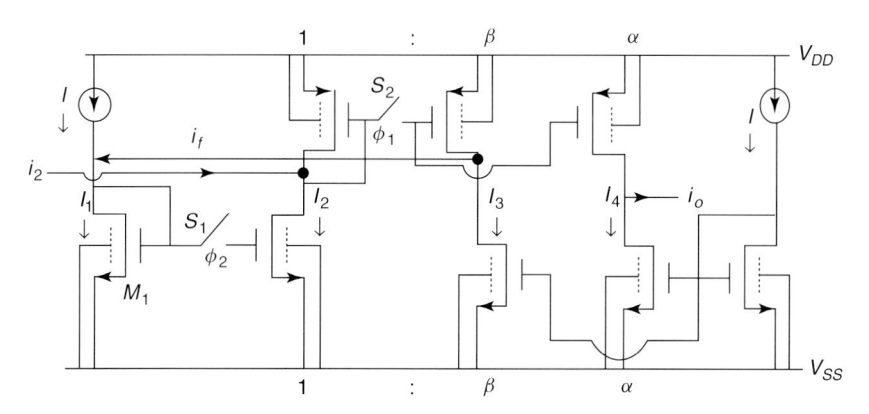

Figure 8.36 SI integrator with backward Euler mapping.

If $\beta = 1$, the integrator becomes a lossless integrator. Otherwise, the integrator is a lossy one.

In Figure 8.36 we show an integrator, which corresponds to a backward Euler mapping, that is, $s \to (1 - z^{-1})/T$. This is achieved by applying the input signal i_2 to the summing node at the input of the PMOS memory. In this case, it can be readily shown that

$$i_o(n) = \beta i_o(n-1) - \alpha i_2(n) \qquad (8.55)$$

On taking z-transform and simplifying, we get

$$H_2(z) = \frac{I_o(z)}{I_2(z)} = -\frac{\alpha}{1 - \beta z^{-1}} \qquad (8.56)$$

If $\beta = 1$, the integrator is lossless.

The integrator of Figure 8.37 has its input current, i_3, connected to the summing node of the NMOS memory on phase ϕ_2 and to the summing node of the PMOS memory on phase ϕ_1. We can show that

$$i_o(n) = \beta i_o(n-1) - \alpha[i_3(n) - i_3(n-1)] \qquad (8.57)$$

Expressed in the z-domain, we get

$$H_3(z) = \frac{I_o(z)}{I_3(z)} = -\alpha \frac{1 - z^{-1}}{1 - \beta z^{-1}} \qquad (8.58)$$

This corresponds to the damped integration of the differential-input signal, $i_3(n) - i_3(n-1)$. With $\beta = 1$, $H_3(z) = -\alpha$, and hence, the input signal is merely scaled and inverted.

8.7.5
Universal Switched-Current Integrator

The universal SI integrator can be obtained by combining the various operations as discussed in connection with the integrators in Figures 8.35–8.37. Figure 8.38

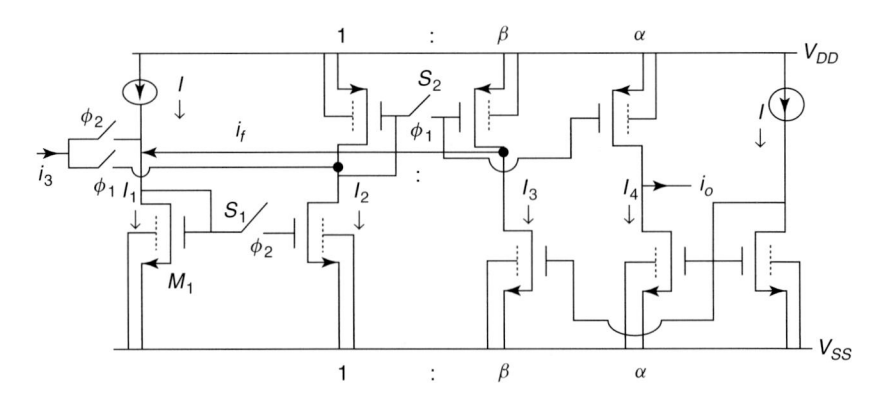

Figure 8.37 A lossy SI integrator building block.

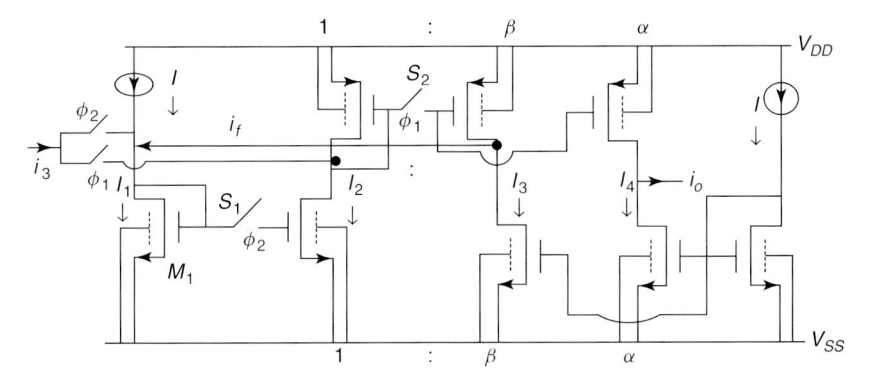

Figure 8.38 A universal SI integrator.

shows the universal SI integrator structure. For simplicity, the substrate terminals of the individual MOS transistors have not been shown in the diagram. The input current signals are weighted by factors α_1, α_2, and α_3 prior to application to the integrator system. This can be achieved by additional current mirror circuits preceding the integrator. By superposition,

$$i_o(n) = \beta i_o(n-1) + \alpha_1 i_1(n-1) - \alpha_2 i_2(n) - \alpha_3[i_3(n) - i_3(n-1)] \quad (8.59)$$

On taking z-transforms

$$I_o(z) = \alpha_1 \frac{z^{-1}}{1-\beta z^{-1}} I_1(z) - \alpha_2 \frac{1}{1-\beta z^{-1}} I_2(z) - \alpha_3 \frac{1-z^{-1}}{1-\beta z^{-1}} I_3(z) \quad (8.60)$$

From Eq. (8.60), we see if $i_2 = -i_1 = i$, $i_3 = 0$, and $\alpha_1 = \alpha_2 = \alpha_3$ then

$$H(z) = \frac{I_o(z)}{I(z)} = \alpha \frac{1+z^{-1}}{1-\beta z^{-1}}$$

This corresponds to the bilinear $s \leftrightarrow z$ transformed lossy (with $\beta \neq 1$) or lossless (with $\beta = 1$) SI integrator.

The SI integrators discussed above are good for illustrative purposes only and should be replaced by more accurate systems to reduce the effects due to the threshold voltage and gain mismatch, finite output impedance (due to r_{ds}), junction leakage, clock feedthrough, and so on. These problems have been addressed in the literature (Hughes, Bird, and Macbeth, 1989) and should be consulted for reliable system implementation in an IC technology.

8.8 Switched-Current Filters

Design of an SI filter is carried out in almost the same way as that of an SC filter. In SC filters, the coefficients of the transfer function can be related to the ratio of the capacitances. Similarly, in an SI filter the coefficients of the transfer function are related to the ratio of the area of the MOS transistors acting as current mirrors

8 Current-Mode Filters

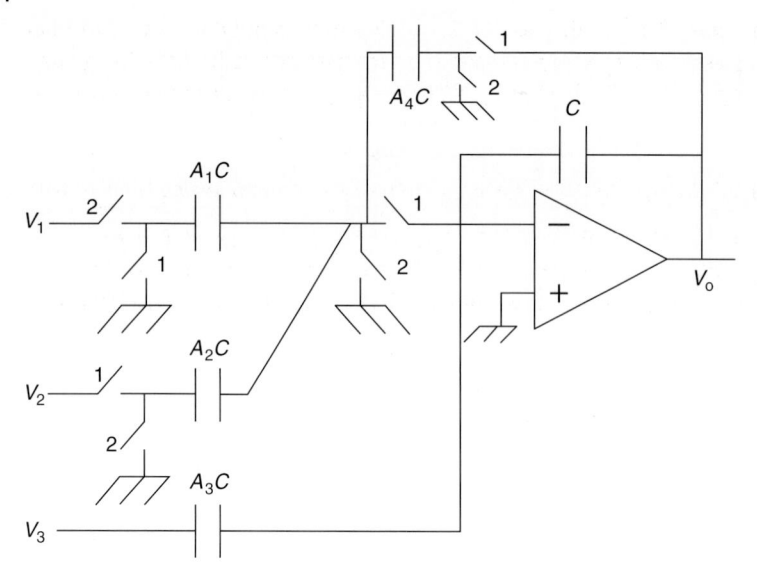

Figure 8.39 A generic SC network.

(or copiers). The preliminary phase of the design involves the same steps as the design of an SC filter (see Chapter 6). Since the SI filters are successors of SC filters, it is often convenient to start off the design of the SI filter from an associated SC filter, serving as the prototype. Once the SC filter structure is known, the SC integrators are simply replaced by corresponding SI integrators, one for one, to evolve the complete SI filter. To recapitulate, we consider a generic SC network (Figure 8.39) containing inverting and noninverting integration and a simple gain function. In Figure 8.39, the clock phases 1 and 2 are designated by numbers 1 and 2 respectively. Using the principle of analysis of SC networks, we can write

$$V_o^{(1)} = A_1 \frac{z^{-1/2}}{1-z^{-1}} V_1^{(2)} - A_2 \frac{1}{1-z^{-1}} V_2^{(1)} - A_3 V_3^{(1)} - A_4 \frac{1}{1-z^{-1}} V_o^{(1)} \qquad (8.61)$$

If V_1 is sampled in phase 1 and held thereafter, $V_1^{(2)} = z^{-1/2} V_1^{(1)}$. On collecting $V_o^{(1)}$ terms together, and simplifying, we finally get

$$V_o^{(1)} = \alpha_1 \frac{z^{-1}}{1-\beta z^{-1}} V_1^{(1)} - \alpha_2 \frac{1}{1-\beta z^{-1}} V_2^{(1)} - \alpha_3 \frac{1-z^{-1}}{1-\beta z^{-1}} V_3^{(1)} \qquad (8.62)$$

where

$$\alpha_1 = \frac{A_1}{1+A_4}, \quad \alpha_2 = \frac{A_2}{1+A_4}, \quad \alpha_3 = \frac{A_3}{1+A_4}, \text{ and } \beta = \frac{1}{1+A_4} \qquad (8.63)$$

On comparing the terms in Eq. (8.62) with those in Eqs. (8.54), (8.56), and (8.58), we can easily understand that a noninverting integration path in the SC network will correspond to an SI integrator of the type shown in Figure 8.35 (forward Euler). Similarly, it can be shown that an inverting integrator path in the SC network will correspond to an SI integrator of the type shown in Figure 8.36 (backward Euler) and an unswitched capacitor path will correspond to the differential SI integrator

8.8 Switched-Current Filters

shown in Figure 8.37 for the case of $\beta = 1$. We can thus build the entire SI filter network by associating an appropriate SI integrator with each of the integrating paths in the prototype SC filter network. The design procedure is now demonstrated with a simple example.

Example 8.3. We now consider the design of the second-order analog BP filter with a transfer function given by $H(s) = \dfrac{2027.9s}{s^2 + 641.28s + 1.0528 \times 10^8}$, considered earlier in Example 6.1. Using a clock frequency of 8 kHz, the sampled-data transfer function, under bilinear transformation and prewarping, becomes (see Example 6.1)

$$H(z) = \frac{9.192 \times 10^{-2}(1 - z^{-2})}{1 - 0.5521z^{-1} + 0.9418z^{-2}}$$

As a first step, the design has to be implemented using SC networks. Using the structure of Figure 6.20, a possible SC filter design is as shown in Figure 8.40 (see Example 6.3).
The capacitance values are

$$A = B = D = 1, I = 0.09192, E = 0.0582, C = 1.3897, G = 1, H = 1, \text{ and } J = 0.908$$

Upon examining the SC network around OA1, we see that the capacitors G and C correspond to inverting integration, the capacitor H to a noninverting integration, and E to a simple phase inversion. Similarly for OA2, the capacitors A and J correspond to noninverting integration, and capacitor I to an inverting integration.

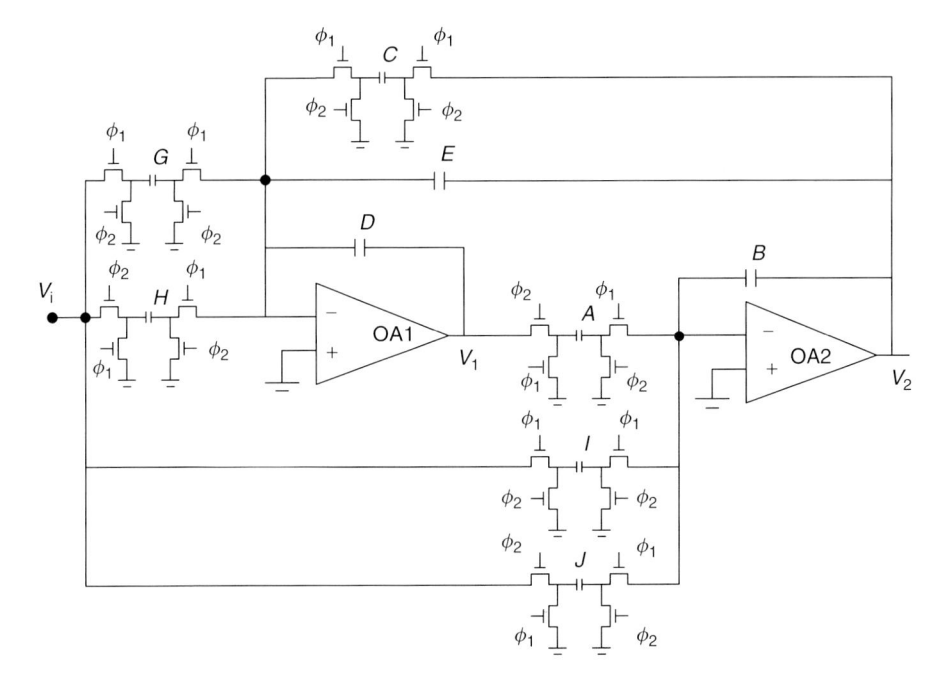

Figure 8.40 A standard building block for a second-order SC filter.

Using the standard analysis procedure for SC filter circuits, we can write (note: $\phi_1 \to$ phase (1), $\phi_2 \to$ phase (2)):

$$V_1^{(1)} = -\frac{1}{1-z^{-1}}V_i^{(1)} + \frac{z^{-\frac{1}{2}}}{1-z^{-1}}V_i^{(2)} - \frac{1.3897}{1-z^{-1}}V_2^{(1)} - 0.0582V_2^{(1)} \qquad (8.64)$$

and

$$V_2^{(1)} = \frac{z^{-\frac{1}{2}}}{1-z^{-1}}V_1^{(2)} - \frac{0.09192}{1-z^{-1}}V_i^{(1)} + \frac{0.9081z^{-\frac{1}{2}}}{1-z^{-1}}V_i^{(2)} \qquad (8.65)$$

Using the sample-hold conditions on V_i and V_1 (i.e., $V_i^{(2)} = z^{-1/2}V_i^{(1)}$, $V_1^{(2)} = z^{-1/2}V_1^{(1)}$), we can rewrite the above equations as:

$$V_1^{(1)} = -V_i^{(1)} - \frac{1.3897}{1-z^{-1}}V_2^{(1)} - 0.0582V_2^{(1)} \qquad (8.66)$$

and

$$V_2^{(1)} = \frac{z^{-1}}{1-z^{-1}}V_1^{(1)} - \frac{0.09192}{1-z^{-1}}V_i^{(1)} + \frac{0.9081z^{-1}}{1-z^{-1}}V_i^{(1)} \qquad (8.67)$$

The first and third terms in Eq. (8.66) can be implemented by the SI integrator of Figure 8.37 (with $\beta = 1$), the second terms in Eqs. (8.66) and (8.67) by the integrator of Figure 8.36 (with $\beta = 1$), while the first and third terms of Eq. (8.67) can be implemented by the integrator of Figure 8.35 (with $\beta = 1$). The signal input to these integrators will, however, be current signals i_i, i_1, and i_2 analogous to V_i, V_1, and V_2 in Figure 8.40.

Practice Problems

8.1 Derive the [z] of the current–controlled voltage source (CCVS) built using CCII and shown in row H in Table 8.2.

8.2 Derive the [y] of the negative impedance inverter (NII) built using CCII and shown in row K in Table 8.2.

8.3 Consider the CM filter shown in Figure P8.3 (Elwan and Soliman, 1996).
(a) Show that the CTF $\left(\frac{I_{o1}}{I_i}\right)$ gives an LP output

$$\frac{I_{o1}}{I_i} = \frac{1}{C_1 C_2 R_1 R_2 s^2 + \left(\frac{C_2 R_1 R_2}{R}\right)s + 1}$$

while the CTF $\left(\frac{I_{o2}}{I_i}\right)$ gives a BP output

$$\frac{I_{o2}}{I_i} = -\frac{C_2 R_1 s}{C_1 C_2 R_1 R_2 s^2 + \left(\frac{C_2 R_1 R_2}{R}\right)s + 1}$$

(b) Find the pole Q and the pole frequency of the filters and their sensitivities w.r.t. the various passive elements.

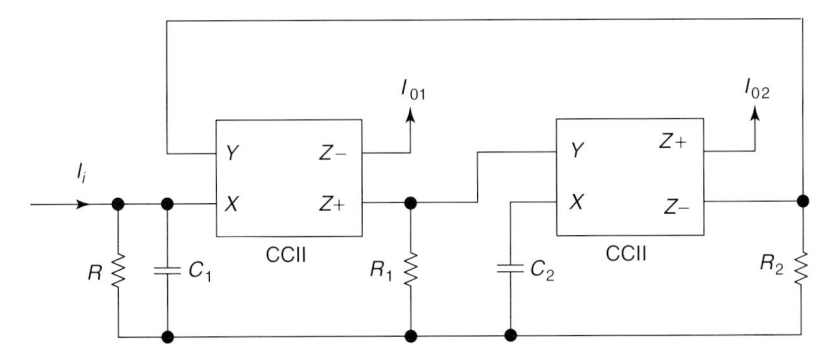

Figure P8.3

(c) Also, find the absolute value of the gain of the BP filter at its pole frequency.

8.4 Find the capacitive dual of the circuit of Figure 5.3 and show that its CTF is the same as the VTF of the original circuit, as given by Eq. (5.12).

8.5 Find the transpose of the circuit of Figure 5.3 and show that its reverse CTF is the same as the VTF of the original, as given by Eq. (5.12).

8.6 Consider Problems 5.2 and 5.3. Find the corresponding CM design for each case using transposition. Verify your design using a CAD tool.

8.7 Consider Problems 5.10–5.13. Provide a current-mode alternative design for each case. Verify your design using a CAD tool.

8.8 The Tow–Thomas VM filter N_V employing three infinite-gain OAs is shown in Figure 8.27.

(a) Show that its VTF is given by

$$\frac{V_o}{V_i} = -\frac{\frac{C_1}{C}s^2 + \frac{1}{RC}\left(\frac{R}{R_1} - \frac{r}{R_3}\right)s + \frac{1}{C^2 R R_2}}{s^2 + \frac{1}{CR_4}s + \frac{1}{C^2 R^2}}$$

(b) Find the corresponding CTF network N_C that also uses infinite-gain OAs and show that its CTF is the same as the VTF of the VM filter.

8.9 Find the transpose of the circuit of Figure P5.35, and find the expression for the reverse CTF of the transpose. Show that this CTF is the same as the VTF of the circuit of Figure P5.35.

8.10 The biquad shown in Figure P5.36 may be considered to be a single-input three-output network. Find the transpose of this circuit that will be a three-input single-output network. Find the CTF matrix of the transpose network. How is this matrix related to the VTF matrix of the original network?

8.11 (a) Show that the three-input four-output VM network of Figure P8.11a proposed in Sanchez-Sinencio, Geiger, and Nevarez-Lozano, (1998) and the four-input three-output CM network of Figure 8.11b proposed in Sun and Fidler (1996) are transposes of each other (Tang, 2004).

296 | 8 Current-Mode Filters

(a)

(b)

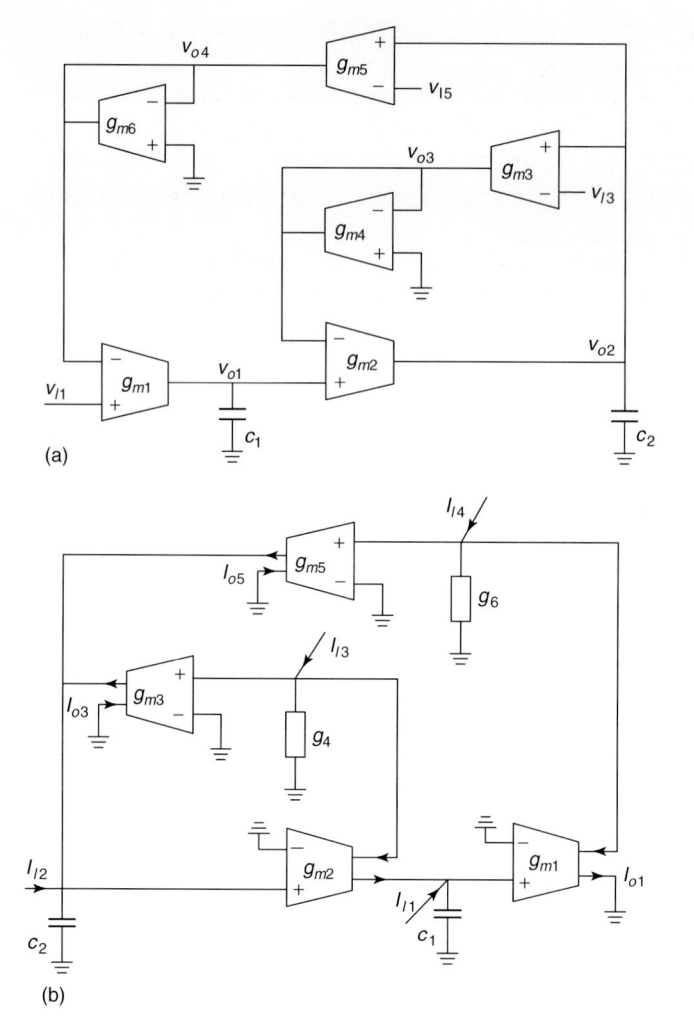

Figure P8.11

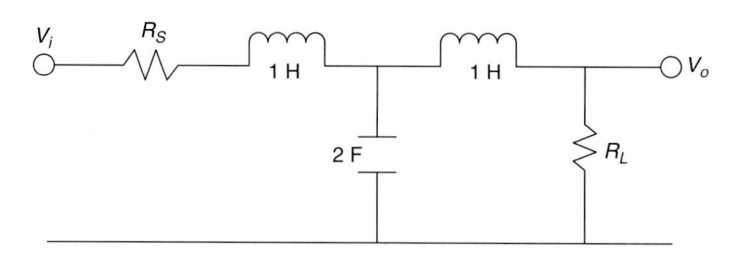

Figure P8.12

8.12 Figure P8.12 shows a doubly terminated third-order Butterworth ladder LP filter with load and source resistances of 1 Ω each.

(a) Obtain a leapfrog structure simulating the ladder for realizing $T_v(s) = \frac{V_O}{V_S}$ using OTAs with $g_m = 50$ μΩ; the cutoff frequency of the simulated LP filter should be 5 MHz.

(b) Using this leapfrog structure, obtain the corresponding CM leapfrog structure.

8.13 Consider Problems 6.2 and 6.3. Provide an SI design for each case.

8.14 Consider Problems 6.9–6.12. Provide an SI design for each case.

9
Implementation of Analog Integrated Circuit Filters

In the preceding chapters, we have discussed various topics including the mathematical background, basic theoretical approach, and analysis and design techniques related to analog filters. With the advancement in semiconductor technology (especially, submicron CMOS process), the trend of implementation of filters has moved toward microminiaturization. Historically, the first step in the scenario was to eliminate bulky inductors. In the 1960s, this was achieved by using active-RC techniques which employ hybrid IC technology. In the 1970s, SC technique facilitated the implementation of analog filters using monolithic CMOS IC technology. While the early goal was to use this technique for filters at low frequencies (less than 30 kHz), presently with the availability of submicron (less than 1 μm gate length) CMOS, and metal-semiconductor field-effect transistor (such as the gallium arsenide field-effect transistor), technologies, completely integrated circuit filters in the tens to hundreds of megahertz range have become possible. Switched-current techniques do not even need any capacitances to implement the filter. Again, with the availability of small on-chip inductors and the goal for implementing high-frequency monolithic filters, the notion of eliminating inductors is being discarded nowadays. Thus, while the basic principles of design of analog filters have not changed, the platform on which these are built as actual hardware has undergone progressive changes toward IC technology.

In this chapter, we introduce the basic principles of implementation of IC analog filters. The thrust is now toward elimination of even the resistors, which are essential elements in active-RC filters, the reason being the requirement of a large substrate area for a large-valued resistor. For the same reason, large capacitors are also to be avoided. In most cases, only transistors and small-valued capacitors are to be employed. The active devices, such as OA, OTA, and CC, are of course built from transistors, so that the entire filter can be implemented through a given IC technology. If large-valued R or C becomes essential, special electronic circuit techniques that are amenable to the IC environment are to be used. One such technique is to multiply the value of an element by utilizing the Miller effect. While saving active devices (OAs) was the motto during the era of hybrid-RC-active filters, OA and other active devices are used liberally to constrain the implementation to the platform of IC technology. In an IC technology, transistors and hence active devices such as OA, OTA, and CC are readily available and quite often lead to a saving of substrate area. The price paid is, however, somewhat increased DC power consumption. With the invention of various low-voltage and low-power techniques, together with the advent of current-mode techniques, the overall power consumption could be kept quite low despite the deployment of an increased number of transistors and active devices. In what follows, we briefly mention the active devices that are available in an IC technology. We discuss considerations related to the implementation of resistance and capacitance in a typical CMOS IC technology. Examples of analog filter implemented in a known IC technology are provided at the end.

9.1
Active Devices for Analog IC Filters

In Chapter 5, we have introduced the principal active devices that may be used to implement analog filters. These are OAs, OTAs, and CCs. For IC filter implementation, the same devices can be used. The emphasis is on devices with a large linear range of operation, wide bandwidth, and low-power consumption. Implementation of such special devices is the subject of analog IC design and shall not be pursued here. The interested reader may refer to several excellent books that are available on this subject (Geiger, Allen, and Strader, 1990; Gray and Meyer, 1993; Laker and Sansen, 1994; Johns and Martin, 1997; Baker, Li, and Boyce, 1998). For a preliminary design of the filter, the active device is assumed to have ideal characteristics. For high-performance design, the nonidealities of the active device are to be considered, and most often, the analysis and redesign are carried out with the aid of numerical and network simulation programs, such as MATLAB and SPICE.

9.2
Passive Devices for IC Filters

The passive devices are capacitors, resistors, and switches. The principal considerations are linearity of operation, good absolute accuracy in the design values, low sensitivity to process variations, and small substrate area requirement. In IC environment, it is difficult to hold the absolute values of R and C with good accuracy, but tracking the ratio of the values can be far more accurate. Hence, it is preferable to realize network functions as a ratio of component values, to ensure better accuracy.

9.2.1
Resistance

In Chapter 5, we have already discussed the various possible forms of semiconductor resistances that can be used in an IC environment. For filter design, the resistances are mostly realized from transistors, since they occupy small area and the value

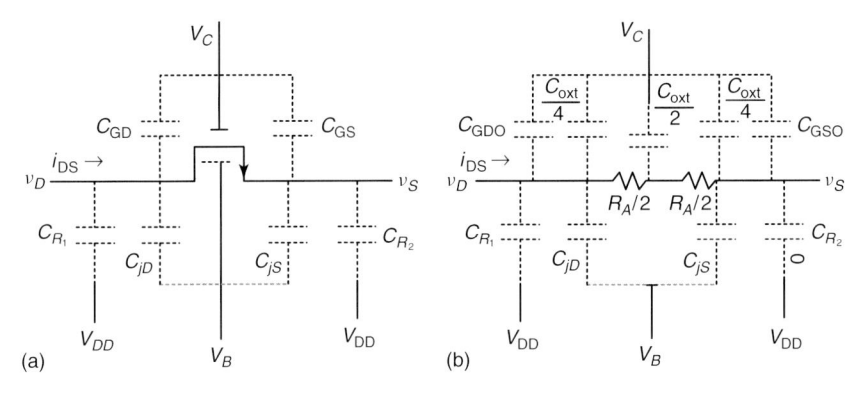

Figure 9.1 An MOSFET resistance (MOS-R) with parasitic capacitances. (a) Symbol and (b) equivalent circuit.

of the resistance can be controlled. In CMOS technology, the MOSFET operating in the linear region functions as an excellent resistance with values between few hundred ohms to few kilo ohms. The resistance value is nonlinearly dependent upon the ohmic drop across the transistor, and special techniques have been proposed to reduce the nonlinearity. The nonlinearity can be appreciated from the I–V equation of the MOSFET, shown in Figure 9.1a, with the equivalent circuit in Figure 9.1b (Laker and Sansen, 1994). The capacitances shown in broken lines are parasitic capacitances. The gate-control voltage V_C can be used to tune the value of the realized resistance. The I–V equation for an NMOS is given by (Laker and Sansen, 1994).

$$i_{DS} = \beta \left[(V_C - V_T)(v_D - v_S) - \frac{1}{2}\{(v_D)^2 - (v_S)^2\} \right] \tag{9.1}$$

where

$\beta = \mu C_{ox}(W/L)$

μ = electron mobility in the channel,

L = length of the channel,

W = width of the channel,

C_{ox} = gate capacitance per unit area, and

V_T = threshold voltage of the MOSFET (i.e., the minimum voltage by which V_C is greater than v_D and v_S for the device to operate in the triode region).

Typically V_T is about 1 V or less and μC_{ox} is about 100 µA V^{-2} or more. For small signal applications, $v_D - v_s \ll V_C - V_T$ and

$$i_{DS} = \beta(V_C - V_T)(v_D - v_S),$$

or

$$\frac{v_D - v_S}{i_{DS}} = R_A = \frac{1}{\beta(V_c - V_T)} = \frac{(L/W)}{\mu C_{ox}(V_c - V_T)} \tag{9.2}$$

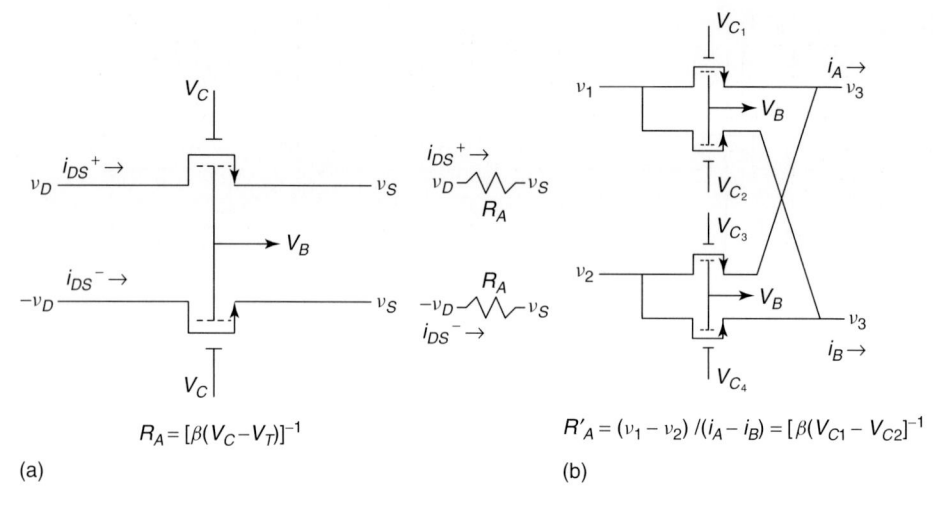

Figure 9.2 Floating linearized MOSFET resistors with (a) balanced input signals and (b) arbitrary input signals.

For linearity over a broader range of v_{DS}, special arrangement is required. Two such configurations (Tsividis, Banu, and Khoury, 1986; Ismail, Smith, and Beale, 1988) are shown in Figures 9.2a and 9.2b, along with the values of the resistances realized.

It should be noted that the source ends of the transistors are returned to the same signal potential. Further, in Figure 9.2a, the input signal is required to be balanced differential. In Figure 9.2b, the realized resistance R_A is a differential resistance. Thus, these will be employed in the realization of filters, where such special signaling conditions exist.

9.2.2 Switch

A transistor can be easily used as a switch by feeding a large positive (negative) signal at the base relative to the emitter for an npn-BJT (for a pnp-BJT), or at the gate relative to the source for an NMOS (for a PMOS). When the switching signal is high, the transistor has a low resistance R_{ON} between the collector and emitter terminals in the BJT, or between the drain and the source of the MOSFET. When the switching signal is low, the transistor is cut off and the OFF resistance, R_{OFF}, becomes very high. For linear circuits and systems, such as a filter, it is desirable that the switch operates linearly for a wide range of signal voltage across it. In this respect, MOS devices are preferred in analog filters. Figures 9.3a–9.3c show the schematic and the equivalent circuits for a typical MOSFET switch under OFF and ON conditions. The parasitic capacitances, ever present and important in an IC environment, are shown by broken lines. These produce undesirable effects like nonlinear distortion and clock feedthrough to the signal paths and the DC bias

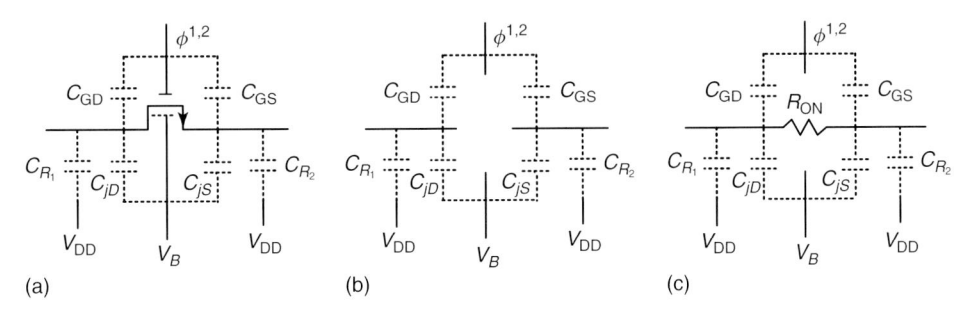

Figure 9.3 MOSFET switch with parasitic capacitances.
(a) Symbol, (b) OFF state, and (c) ON state.

lines. The clock feedthrough to the signal path produces a small DC offset voltage. The ON resistance is dependent upon the level of the clock signal, proportional to the length L and inversely proportional to the width W of the MOSFET. In SC filter applications, R_{ON} is to be chosen so that $R_{ON} C_{Tmx} \ll T/2$, where T is the clock period and C_{Tmx} is the sum of the highest design capacitance and the associated parasitic capacitance. This will ensure complete transfer of charge during a fraction of the ON period of the clock signal.

9.3
Preferred Architecture for IC Filters

An important problem in the IC environment is the coupling of noise signal from nearby circuit nodes and especially from the neighboring digital subsystems in an analog–digital mixed-mode VLSI system. Such noise signals constitute common-mode signals, which affect all nearby signal nodes. One way to reduce (or eliminate) the effect of such signals would be to subtract such signals without, however, affecting the strength of the primary information-bearing signal. This is very conveniently achieved by using differential architecture in analog system implementation. Apart from canceling the common-mode noise, differential signaling provides other advantages such as canceling even-order harmonics, doubling the signal swing, and so on. The price paid, however, is doubling of the number of components (i.e., doubling of the substrate area), doubling of the DC power consumption, and doubling of the noise power contribution. The source signal should be available in a balanced differential mode. For systems that are very sensitive to power and/or cost, differential architecture may not be a preferred approach. Conventional single-ended structures are to be used in such cases.

9.3.1
OA-Based Filters with Differential Structure

The method to derive a differential architecture from a given single-ended OA-based structure consists of the following steps (Schaumann, Ghausi, and Laker, 1990):

1) Form a mirror-image of the nominal single-ended structure about a plane through the ground node.
2) Divide the gain of all the active devices by 2.
3) Change the sign of the gain, the signs at the terminals, and the terminal signals in all the mirrored active elements.
4) Merge the resulting pair into a single balanced differential-in, differential-out (DIDO) device. The common ground node remains concealed as a common-mode plane.

To illustrate the procedure, we will consider the case of the inverting OA-integrator assuming the OA to be ideal with infinite open-loop gain. The various steps involved are shown in Figures 9.4a–9.4d. The input–output relation in a DIDO voltage amplifier is depicted in Figure 9.4e.

It should be noticed that, in the differential structure, both the inverting and noninverting integrator outputs are available. Hence, if in the original single-ended structure a unity-gain inverting amplifier is used to get a noninverting integrator, then in the differential structure, the unity-gain inverting amplifier is redundant and is replaced by cross-coupling of the dual outputs. This is illustrated in Figures 9.5a–9.5c considering a noninverting integrator circuit.

9.3.1.1 First-Order Filter Transfer Functions

In Chapter 5 (Table 5.1), we introduced several structures to realize first-order transfer functions using an OA. Some of the structures can be readily converted to the differential form, adopting the guidelines discussed above. This is illustrated in Table 9.1, wherein linearized MOS resistors (see Figure 9.2a) are used.

9.3.1.2 Second-Order Filter Transfer Functions

The technique of implementing first-order differential structures can be easily extended to differential structures around OAs, implementing second-order filter functions. In the early era of active-RC filters, SAB filters were popular because of the savings on the cost of the active device (viz., the OA). With the progress in the IC technology, this consideration is discarded in favor of the flexibility, tunability, and versatility afforded by multi-OA-architecture filters based on state-variable realizations. However, it should be mentioned that the SAB structures, especially the Sallen and Key unity-gain LP filters, still find use in antialiasing and reconstruction filters for active SC filters.

Figure 9.6a depicts a typical Tow–Thomas second-order filter network using three OAs, and RC elements (see Chapter 5). The outputs V_{BP} and V_{LP} are given by

$$\frac{V_{BP}}{V_i} = \frac{\frac{s}{R_1 C_1}}{s^2 + \frac{s}{R_1 C_1} + \frac{1}{R_2 R_3 C_1 C_2}} \tag{9.3}$$

and

$$\frac{V_{LP}}{V_i} = \frac{\frac{1}{R_2 R_4 C_1 C_2}}{s^2 + \frac{s}{R_1 C_1} + \frac{1}{R_2 R_3 C_1 C_2}} \tag{9.4}$$

9.3 Preferred Architecture for IC Filters

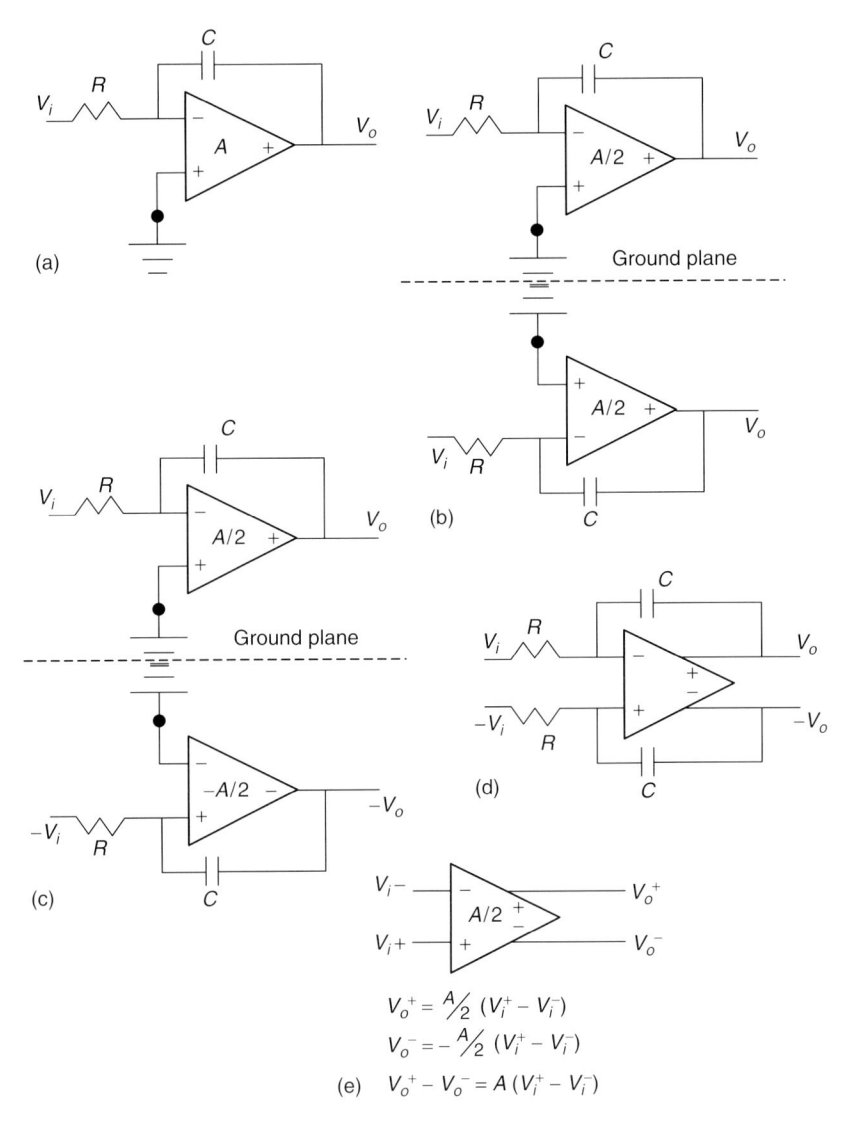

Figure 9.4 Procedure to convert a single-ended structure to a differential structure. (a) The original circuit, (b) mirroring about the ground plane, (c) changing the sign of the gain, the signs at the terminals, and the terminal signals in the mirrored circuit, (d) merging the mirror images, and (e) input–output equations of a differential-in, differential-out (DIDO) amplifier of gain A/2.

$$V_o^+ = \tfrac{A}{2}(V_i^+ - V_i^-)$$
$$V_o^- = -\tfrac{A}{2}(V_i^+ - V_i^-)$$
$$V_o^+ - V_o^- = A(V_i^+ - V_i^-)$$

A differential version of the same is given in Figure 9.6b. In the differential structure, *crossing of the wire* technique is used to accomplish the function of sign inversion. Thus, the OA used simply for sign inversion (viz., the inverting amplifier) is not required, as mentioned previously. Figure 9.6c shows the corresponding MOS-R C structure using the linearized MOS-R of Figure 9.2a.

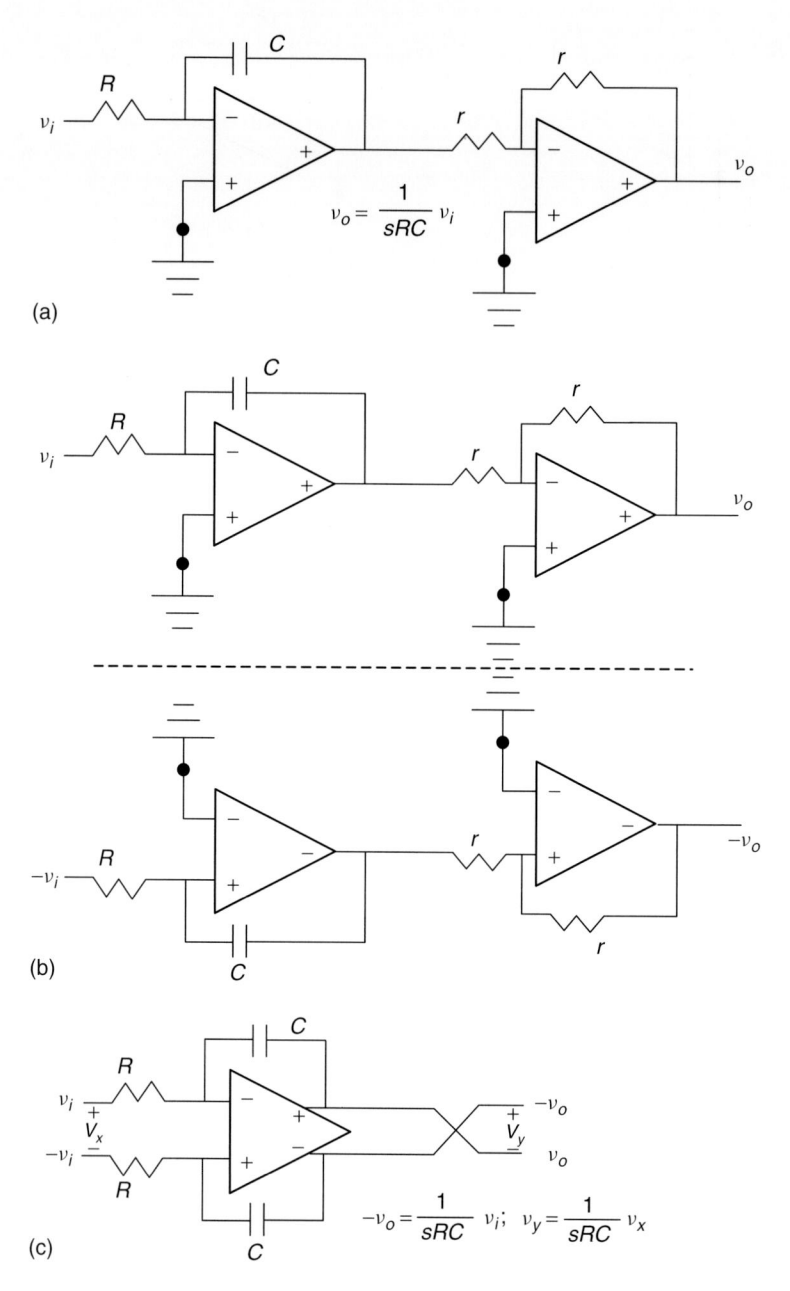

Figure 9.5 Various steps involved in converting a single-ended noninverting integrator to a differential structure.

9.3 Preferred Architecture for IC Filters | 307

Table 9.1 OA-based differential MOS-R C structures derived from the associated single-ended RC structures.

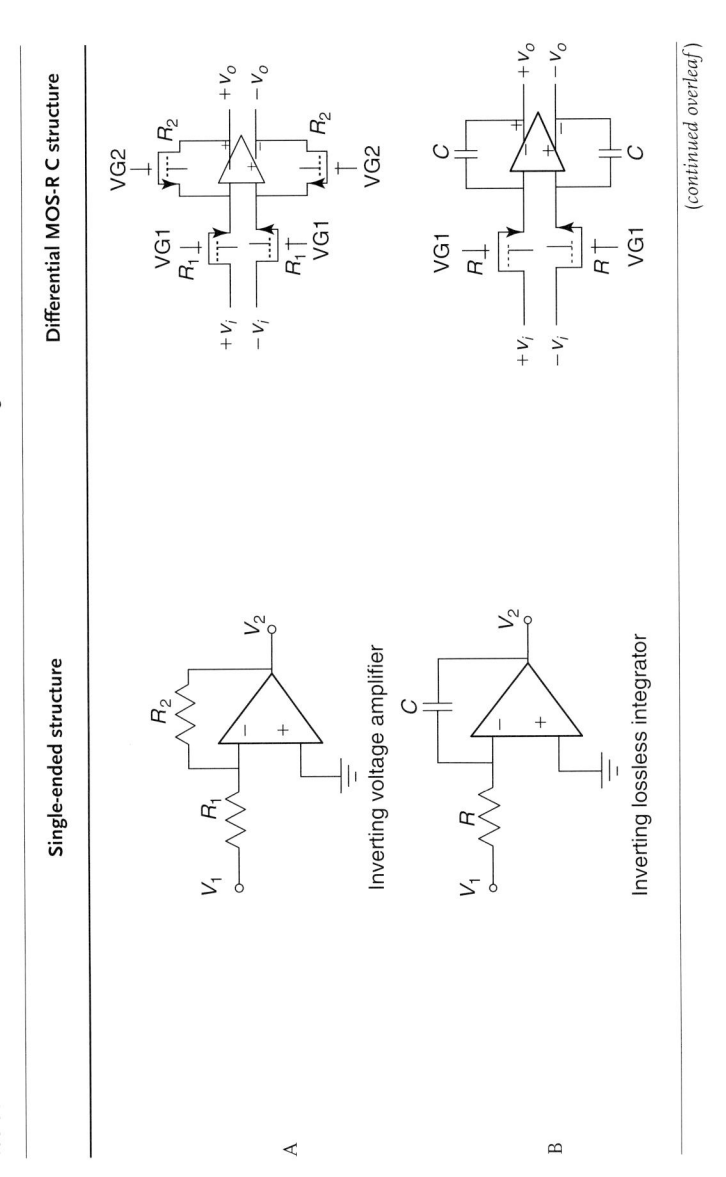

	Single-ended structure	Differential MOS-R C structure
A	Inverting voltage amplifier	
B	Inverting lossless integrator	

(continued overleaf)

Table 9.1 (continued).

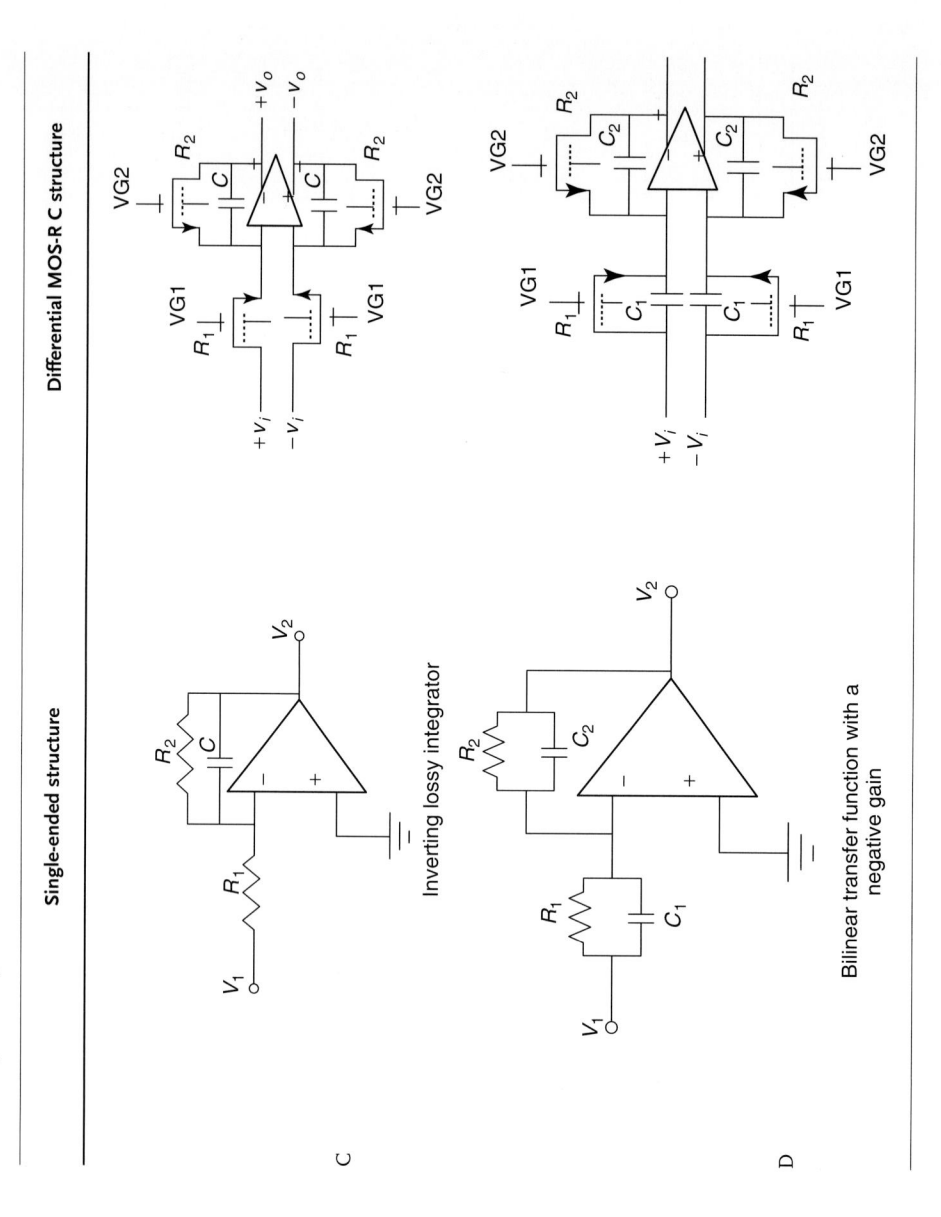

Single-ended structure	Differential MOS-R C structure
C — Inverting lossy integrator	
D — Bilinear transfer function with a negative gain	

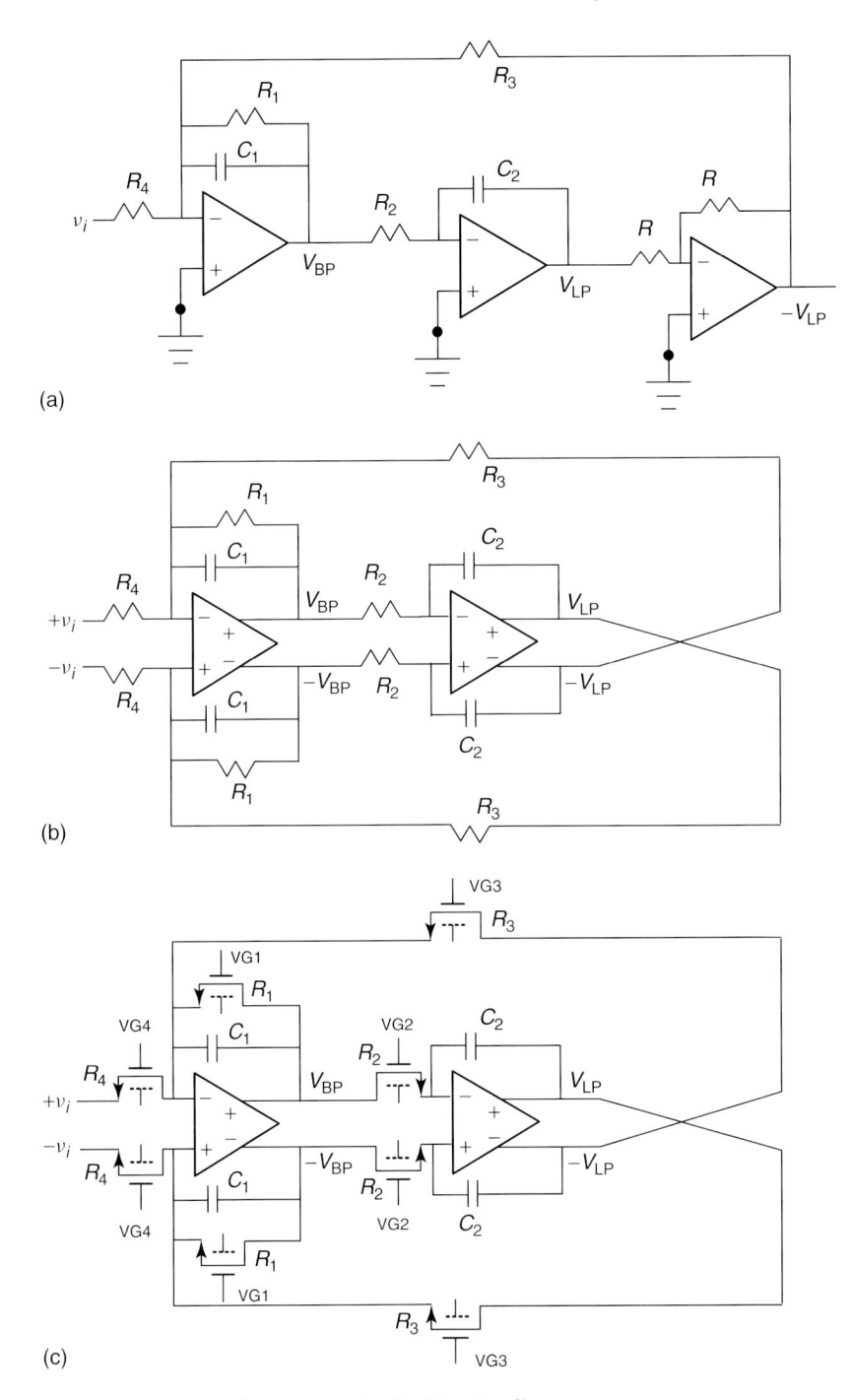

Figure 9.6 (a) Tow–Thomas second-order RC-active filter, (b) the corresponding differential configuration, and (c) the associated MOS-R C integrated circuit filter.

Similar structures can be developed for the various OA-based second-order RC filters introduced in Chapter 5.

9.3.2
OTA-Based Filters with Differential Structures

A single-ended OTA-based structure can be easily converted to the differential form by following a procedure similar to that used for OA-based structures. Since OTAs are used in conjunction with capacitors to implement a filtering function (i.e., OTA-C filters or g_m–C filters), it is necessary to know how to derive the differential structure for a capacitor (grounded or floating) used in a single-ended OTA-C filter. The procedure is as follows:

1) Form a mirror-image of the nominal single-ended structure about a plane through the ground node.
2) Change the signs of the gains (i.e., g_m's), the signs at the terminals, and the signs of the signals attached to all the mirrored OTAs.
3) Change each grounded capacitor C to $2C$.
4) Leave all the floating capacitors unchanged.
5) Merge the resulting pair with inverting and noninverting gains into a single balanced DIDO OTA.
6) Any two equal-valued grounded capacitors running off from a pair of nodes may be merged into one floating capacitor of one-half the value of either of the grounded capacitors, and connected between the same pair of nodes. The trade-off is a reduction in the substrate area versus an increased influence of the parasitic capacitances.

9.3.2.1 First-Order Filter Transfer Functions

Table 9.2 shows examples of differential structures corresponding to some of the OTA-based first-order transfer functions previously introduced in Chapter 5 (Table 5.10).

For purposes of illustration, the various steps involved in converting the single-ended OTA-based lossy integrator (Row C of Table 9.2) to the corresponding differential structure are shown in Figures 9.7a–9.7d. It may be mentioned that both Figures 9.7c and 9.7d are valid differential structures.

As explained in connection with SC filters (Chapter 6), an integrated capacitor is invariably associated with parasitic capacitances to ground from both the top and the bottom plates (Figure 6.8b, Chapter 6). In the structure of Figure 9.7c, the parasitic capacitance from the bottom plate (shown with curved line in Figure 9.8a) can be grounded, while in Figure 9.7d, it is not possible. If a floating capacitor, such as in Figure 9.7d, must be used to save the substrate area, then the capacitor C can be laid out as a parallel combination of two capacitors, each of value $C/2$, with the bottom plates inverted relative to each other, as indicated in Figure 9.8b (Schaumann and Van Valkenburg, 2001). This will preserve the symmetry and balance of the differential configuration.

Table 9.2 OTA-based differential structures derived from the associated single-ended structures.

	Single-ended structure	Differential structure	Remarks
A			$\frac{V_o}{V_i} = \frac{g_m}{sC}$ Voltage integrator
B			$\frac{I_o}{I_i} = \frac{g_m}{sC_i}$ Current integrator
C			$\frac{V_o}{V_i} = \frac{g_{m1}}{sC+g_{m2}}$ Lossy voltage integrator
D			$\frac{V_o}{V_i} = \frac{sC_1+g_{m1}}{s(C_1+C_2)+g_{m2}}$ First-order section

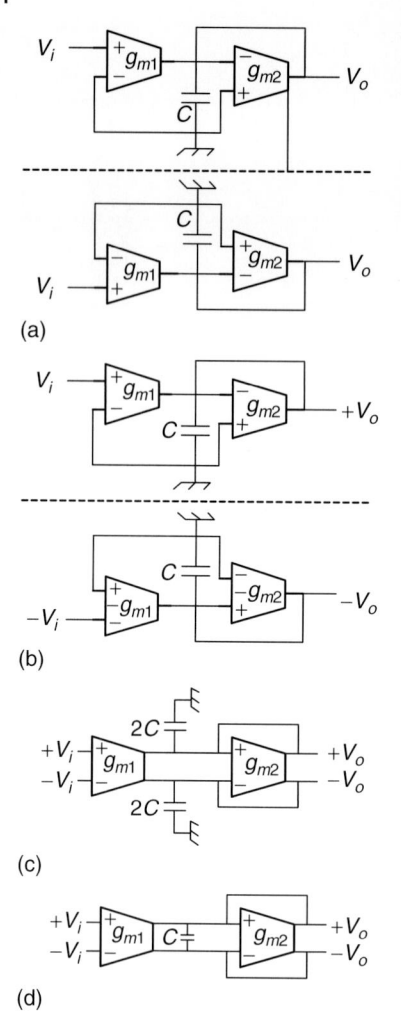

Figure 9.7 Various steps involved in deriving the differential structure for the lossy integrator of Table 9.2. (a) Mirroring around the ground plane, (b) changing the signs of the signals, terminals, and of the g_m's in the mirrored OTAs, (c) merging the mirrored parts, and (d) replacing the grounded capacitors by a floating capacitance in the merged structure.

In IC implementation of OTA-C (or, g_m–C) filters, parasitic capacitances as discussed above are to be carefully considered while creating the design capacitor layouts and the design capacitance values need to be compensated to account for the extra capacitances arising out of the parasitic components.

9.3.2.2 Second-Order Filter Transfer Functions

We now present an example of converting a single-ended OTA-C filter that provides both the LP and BP functions (Schaumann and Van Valkenburg, 2001). Figure 9.9a presents the single-ended OTA-C filter and Figure 9.9b depicts the differential version. The LP transfer function is given by

$$\frac{V_o}{V_1} = \frac{g_{m1}g_{m3}}{s^2 C_1 C_2 + sC_2 g_{m2} + g_{m3}g_{m4}} \quad (9.5)$$

9.3 Preferred Architecture for IC Filters

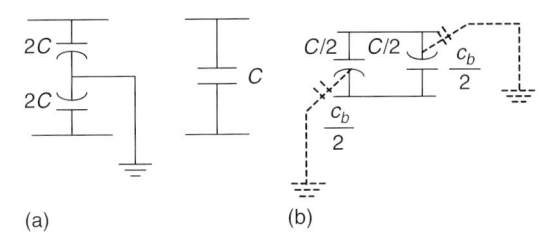

Figure 9.8 (a) Bottom-plate parasitic capacitance grounded and (b) bottom-plate parasitic capacitance distributed in a differential configuration.

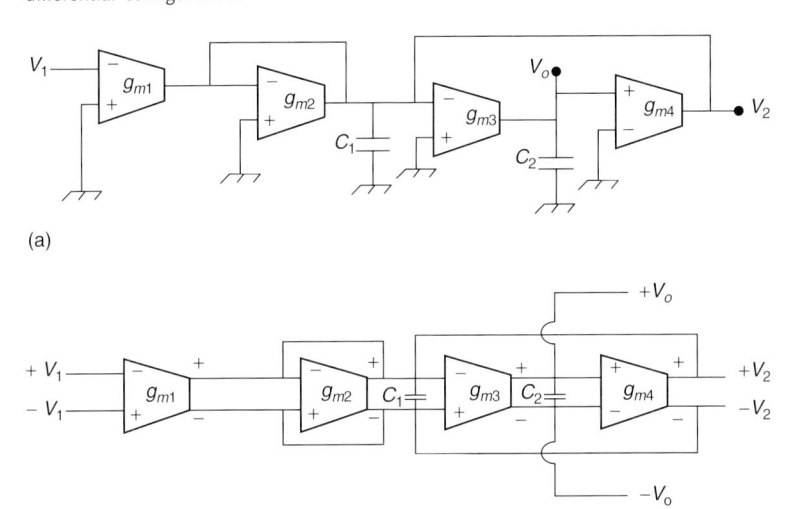

Figure 9.9 (a) Single-ended second-order filter with OTAs and (b) the corresponding differential version.

and the BP function by

$$\frac{V_2}{V_1} = \frac{-sC_2 g_{m1}}{s^2 C_1 C_2 + s C_2 g_{m2} + g_{m3} g_{m4}} \tag{9.6}$$

To take note of the parasitic elements in the differential version, we must recognize that each OTA has parasitic capacitances at its input and output, and a finite output resistance. For simplicity, we assume that all the OTAs in Figure 9.9a have the same values for these parameters, and designate them by C_i, C_o, and r_o. In this case, the capacitor C_1 will be in parallel with $3r_o$, $3C_o$ (due to the OTAs with g_{m1}, g_{m2}, g_{m4}), and $2C_i$ (due to OTAs with g_{m2}, g_{m3}), while the capacitor C_2 will be in parallel with r_o, C_o, and C_i. The parasitic components will lead to the equivalent admittances

$$sC_1 \rightarrow s(C_1 + 3C_o + 2C_i) + 3g_o = sC_{1\text{eff}} + 3g_o, sC_2 \rightarrow s(C_2 + C_o + C_i) + g_o$$
$$= sC_{2\text{eff}} + g_o$$

where $g_o = 1/r_o$. As a result, the transfer function expression also changes. In particular, the denominator expression becomes

$$s^2 C_{1\text{eff}} C_{2\text{eff}} + s[C_{2\text{eff}}(g_{m2} + 3g_o) + C_{1\text{eff}} g_o] + g_{m3} g_{m4} + g_{m2} g_o + 3g_o^2.$$

The capacitance compensation technique, discussed under the case of the first-order transfer function realization, can be adopted for the second-order transfer function as well. Thus, the design capacitance C_1 can be created (i.e., laid out by the IC fabrication tool) as $C_1' = C_1 - 3C_o - 2C_i$. To eliminate the effect of the parasitic resistance r_o, negative resistance insertion technique can be used (Szczepansky, Jakusz, and Schaumann, 1997; Patel and Raut, 2008).

9.4
Examples of Integrated Circuit Filters

9.4.1
A Low-Voltage, Very Wideband OTA-C Filter in CMOS Technology

We first present the design of a low-voltage, fully differential, very wideband OTA-C filter using a 0.18-µm CMOS technology (Li and Raut, 2003). The filter is of order 6, with a CHEB response having 1-dB ripple through the passband of 2–400 MHz.

The task of design starts with choosing an OTA with a good performance in terms of bandwidth, output resistance, dynamic range and the transconductance value measured relative to DC power consumption, and semiconductor wafer area required for implementation. To compare the overall performance of several OTAs, the following metric is used:

$$S_i|_{i=1,2,3,\ldots} = \frac{g_{mi} + \text{BW}_i + \text{DR}_i + R_{oi}}{P_{DC}} + \frac{g_{mi} + \text{BW}_i + \text{DR}_i + R_{oi}}{A}$$

where the numerator symbols g_{mi}, BW_i, DR_i, and R_{oi} represent the transconductance, bandwidth, dynamic range, and output resistance, respectively, and the denominator symbols P_{DC} and A represent the DC power and wafer area, respectively. Obviously, the ith OTA with the highest value of S_i will be the best candidate to be employed.

According to the above criterion, the OTA shown in Figure 9.10 is used (Szczepansky, Jakusz, and Schaumann, 1997) for the design of the filter. Multiple-loop feedback topology discussed in Section 7.4.1 is used for the implementation. The overall system is shown in Figure 9.11. This exploits the differential configuration which is preferred for IC filter implementation. It should be observed that the structure of Figure 9.11 is nothing but the differential version of the FLF single-ended structure of Figure 7.26, with $n = 3$ and $T_1(s) = T_2(s) = T_3(s) = T(s)$. The amplifier in Figure 9.11 performs the summation of the signals and corresponds to the summer in Figure 7.26. As a consequence, the transfer function of the structure of

9.4 Examples of Integrated Circuit Filters | 315

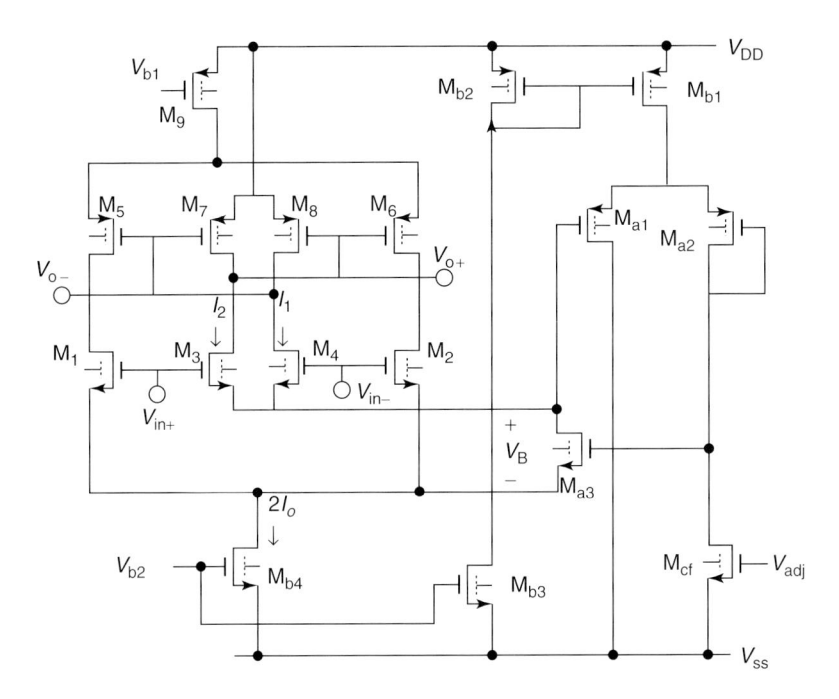

Figure 9.10 The OTA used in the design of the filter (© IEEE, 1997).

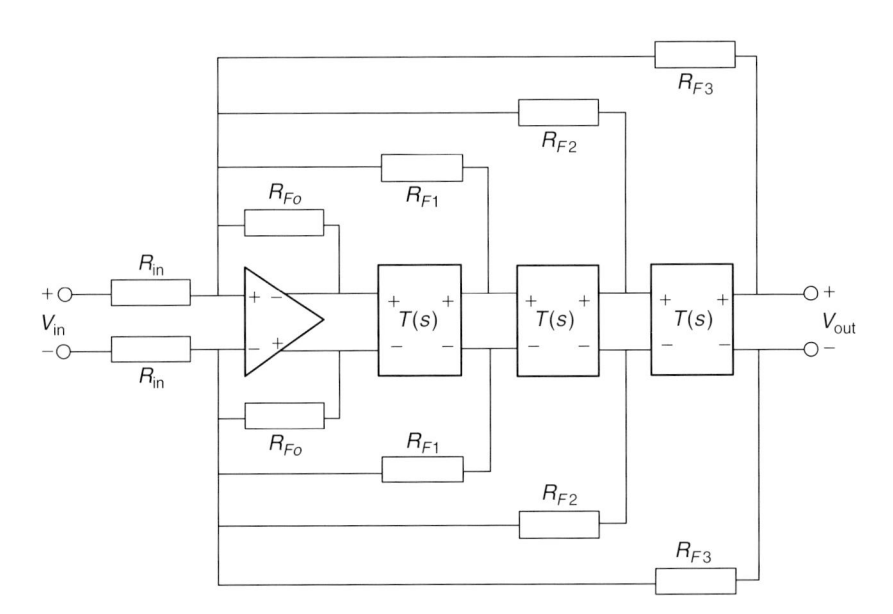

Figure 9.11 Sixth-order BPF using the MLF topology.

Figure 9.11 is given by

$$H(s) = \frac{V_{out}}{V_{in}} = -\frac{K(T(s))^3}{1 + F_1 T(s) + F_2(T(s))^2 + F_3(T(s))^3} \quad (9.7)$$

where $K = \frac{R_{Fo}}{R_{in}}$ and $F_i = \frac{R_{Fo}}{R_{Fi}}, i = 1, 2, 3$.

The following procedure is now followed for the design of the required BP filter.

From the specifications for the BP filter to be designed, the filter band-edges are at 2 and 400 MHz; hence, we can determine the system Q to be

$$Q = \frac{\omega_0}{BW} = \frac{\sqrt{2(400)}}{400 - 2} = 0.071 \quad (9.8)$$

For a sixth-order CHEB BP filter, the corresponding normalized CHEB LP transfer function is (see Appendix A)

$$H_N(s) = \frac{0.491}{s^3 + 0.988s^2 + 1.238s + 0.491} \quad (9.9)$$

As was done in Section 7.4.1 earlier, we first assume the transfer function $T(s)$ of each of the blocks in Figure 9.10 to be that of a lossless integrator, that is, $T(s) = \frac{1}{s+\alpha}$. Then from Eq. (9.7), we have

$$H(s) = \frac{K}{(s + \alpha)^3 + F_1(s + \alpha)^2 + F_2(s + \alpha) + F_3} \quad (9.10)$$

Following the procedure in Section 7.4.1, we now compare Eq. (9.10) with Eq. (9.9) to obtain the following relations:

$$K = 0.491$$
$$F_1 = 0.988 - 3\alpha$$
$$F_2 = 1.238 - 2F_1\alpha - 3\alpha^2 \quad (9.11)$$

and

$$F_3 = 0.491 - F_2\alpha - F_1\alpha^2 - \alpha^3$$

It is clear from Eq. (9.11) that α is arbitrary. Depending on the value of α, we can get the values of F_1, F_2, and F_3. For example, if $\alpha = 0.45$ then $F_1 = -0.362$, $F_2 = 0.9563$, and $F_3 = 0.0429$. For $\alpha = 0.25$, the corresponding values are 0.238, 0.9315 and 0.2276. The negative value for F_1 implies that *crossing the wire technique* is to be employed in the differential structure to achieve a reversal of the associated gain value. It should be pointed out that if α is chosen as 0.988/3, then $F_1 = 0$, that is, there is no feedback path F_1 corresponding to the case considered in Section 7.4.3. The lossy integrator $T(s) = \frac{1}{s+\alpha}$ can be implemented by a differential OTA-structure of the type shown in Table 9.2. Then, Figure 9.11 would realize the LP filter given by Eq. (9.9).

Now, we introduce the LP to BP transformation

$$s \to Q \frac{s^2 + 1}{s}$$

9.4 Examples of Integrated Circuit Filters

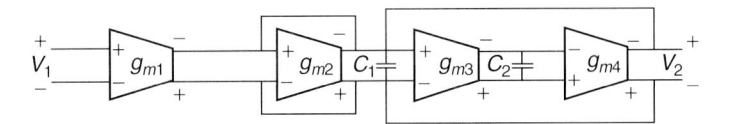

Figure 9.12 OTA-based biquad BP section.

Then,

$$(s+\alpha) \to \left(Q\frac{s^2+1}{s}+\alpha\right) = \frac{Q}{s}\left(s^2+\frac{\alpha}{Q}s+1\right) \quad (9.12)$$

Hence, each of the lossy integrators in the square boxes in Figure 9.11 is replaced by the normalized second-order transfer function

$$T_N(s) = \frac{s/Q}{s^2+(\alpha/Q)s+1} = \frac{s/(Q_p\alpha)}{s^2+(1/Q_p)s+1} = \frac{(H_0/Q_p)s}{s^2+(1/Q_p)s+1} \quad (9.13)$$

where $Q_p = \frac{Q}{\alpha}$ is the pole Q and $H_0 = \frac{1}{\alpha}$ is the gain at the center frequency of each of the second-order BP filters.

Now for the desired BP filter, Q has already been found to be 0.071. Choosing α to be 0.45, we have already calculated F_1, F_2, and F_3 to be $-0.362, 0.9563$, and 0.0429. Also, each of the individual second-order BP filters has a value of $Q_p = \frac{Q}{\alpha} = 0.158$, and the value of the gain at the center frequency is given by $H_0 = \frac{1}{\alpha} = 2.22$. We now apply frequency scaling to the normalized BP transfer function $T_N(s)$ by letting $s \to \frac{s}{\omega_p}$, where $\omega_p = 2\pi\sqrt{800} = 177.7 \times 10^6$ rad s^{-1}. Thus, the denormalized transfer of each of the second-order BP filters is given by

$$T(s) = \frac{H_0(\frac{\omega_p}{Q_p})s}{s^2+(\frac{\omega_p}{Q_p})s+\omega_p^2} \quad (9.14)$$

where $H_0 = 2.222$, $Q_p = 0.158$, and $\omega_p = 177.7 \times 10^6$ rad s^{-1}. We may realize each of these biquad BP filters by the OTA-based structure of Figure 9.9b. This is redrawn here for convenience as Figure 9.12. The BP transfer function $T(s)$ of the structure of Figure 9.12 is given by

$$T(s) = \frac{V_2}{V_1} = \frac{sC_2g_{m1}}{s^2C_1C_2+sC_2g_{m2}+g_{m3}g_{m4}} \quad (9.15)$$

The center frequency, the pole Q, and the gain at the center frequency are given by

$$\omega_p = \sqrt{\frac{g_{m3}g_{m4}}{C_1C_2}}, \quad Q_p = \frac{1}{g_{m2}}\sqrt{\frac{g_{m3}g_{m4}C_1}{C_2}}, \quad H_0 = \frac{g_{m1}}{g_{m2}} \quad (9.16)$$

For implementing the system in an IC technological process, one may choose C_1 and C_2 in the range 1–10 pF. Choosing $C_1 = C_2 = 2$ pF and setting $g_{m3} = g_{m4}$, one can determine the values of g_{m1}, g_{m2}, g_{m3}, and g_{m4} from Eq. (9.16) to be $g_{m3} = g_{m4} = 355.4$ μ℧, $g_{m2} = 2249$ μ℧ and $g_{m1} = 4997$ μ℧.

Table 9.3 Attenuation characteristics of the BP filter considered.

Frequency	f_o to $f_o \pm 100$ kHz	$f_o \pm 800$ kHz	$f_o \pm 1.6$ MHz	$f_o \pm 3$ MHz	$f_o \pm 6$ MHz
Attenuation (dB)	≤ 0.5	≥ 5	≥ 10	≥ 15	≥ 30

In order to implement the complete filter using OTA-based biquads, certain special arrangements are necessary. Since an OTA ideally behaves like a VCCS, and since $H(s)$ represents a VTF, each OTA-based biquad (as shown in Figure 9.12) has to be followed by a voltage buffer circuit to present a low output impedance to the following biquad section. For the differential configuration of Figure 9.11, each $T(s)$ consists of the biquad in Figure 9.12 with a pair of buffer circuits, one at each of the differential outputs. Similarly, the summing block in Figure 9.11 consists of a differential OTA appended with voltage buffer circuits.

The sixth-order filter has been implemented using the above considerations in a 0.18-μm CMOS technology, and for more details the reader is referred to (Li and Raut, 2003).

9.4.2
A Current-Mode Filter for Mobile Communication Application

We now present the case of a high-frequency, high-Q BP filter implemented using a BiCMOS technological process (Fabre et al., 1998). The filter can be used as an intermediate frequency filter in GSM cellular telephones. The attenuation characteristics for the BP for mobile communication application are shown in Table 9.3, where f_o is the center frequency of the BP filter. For GSM application, we assume f_o to be 85 MHz.

9.4.2.1 Filter Synthesis

To design the BP filter for the specifications of Table 9.3, we may consider the corresponding normalized LP filter for which $\omega_c = 1$, $\omega_s = \frac{6000}{100} = 60$, $A_p = 0.5\ dB$ and $A_a = 30\ dB$. If a MFM approximation (see Chapter 3) is used, the order of the normalized LP filter becomes two. For the pass-band loss of 0.5 dB (i.e., $\varepsilon = 0.3493$, the MFM approximation, following the procedure used in Example 3.6 (Chapter 3), Appendix, and the LP to BP transformation, leads to the frequency denormalized BP transfer function given by

$$H(S) = \frac{0.45209 \times 10^{13}}{s^4 + 30069s^3 + 0.57047 \times 10^{18} s^2 + 0.85768 \times 10^{24} s + 0.81357 \times 10^{35}}$$

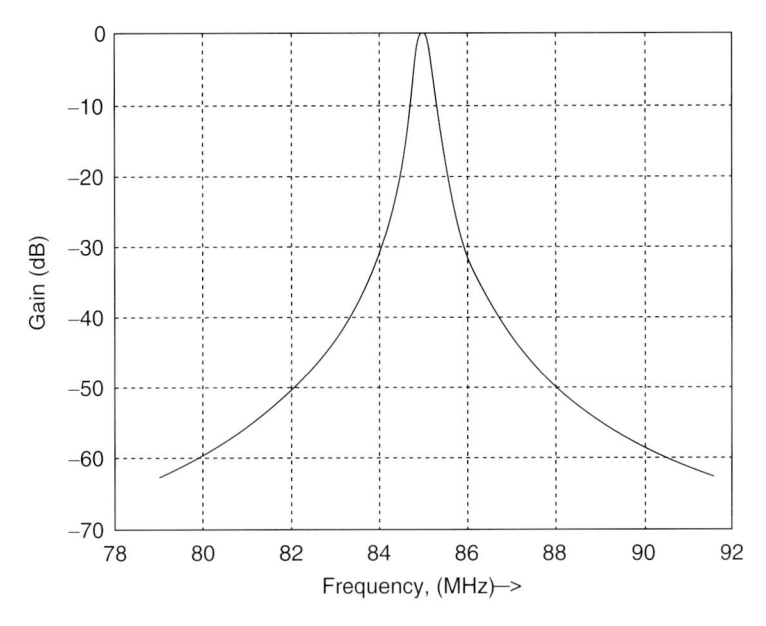

Figure 9.13 Theoretical magnitude response of H(s).

If the above function is realized as a cascade of two BP filter functions, the transfer function can be written as (using MAPLE program)

$$H(s) = T_1(s)T_2(s)$$
$$= \frac{21262s}{s^2 + 15056s + 0.28604 \times 10^{18}} \frac{21262s}{s^2 + 15013s + 0.2844 \times 10^{18}} \quad (9.17)$$

The above expressions represent two BP filters in cascade with

$$\omega_{o1} = 0.53483(10^9), \omega_{o2} = 0.53331(10^9), Q_1 = Q_2 = 355.22,$$
$$H_{o1} = 1.4122, H_{o2} = 1.4162 \quad (9.18)$$

where, H_{o1}, H_{o2} are the gains at the BP center frequencies. A plot of the magnitude response of the transfer function is shown in Figure 9.13. It may be observed that the attenuation in the stop band exceeds the given specification. This is very desirable in practice.

An implementation of the filter using a BiCMOS technological process has been reported in Fabre et al. (1998). Some details about the approach followed therein are given below.

9.4.2.2 Basic Building Block

The basic building block used for the implementation is a controlled CCII+ (CCCII+), as shown in the transistor level schematic of Figure 9.14a. The BJT devices in the BiCMOS process are used in the signal path to preserve high-frequency operation, while several MOS transistors are used to provide

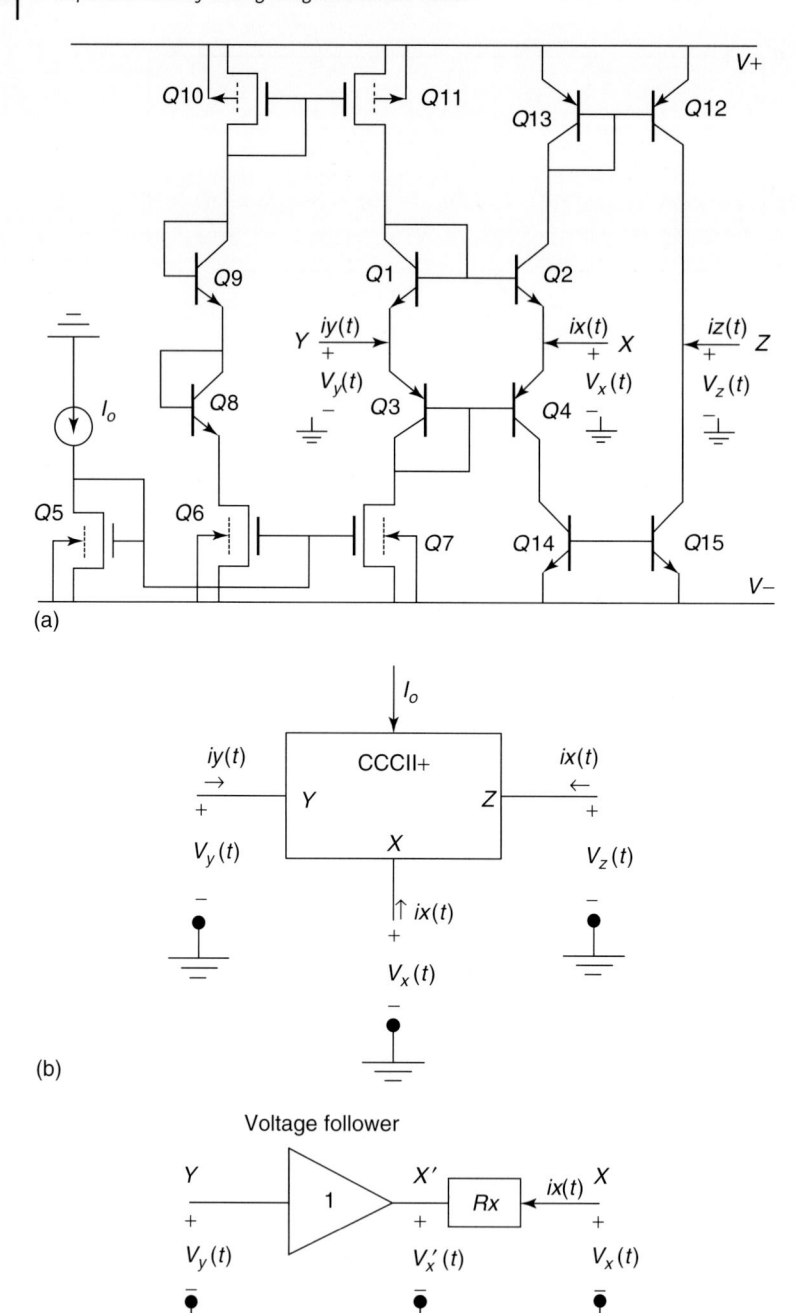

Figure 9.14 (a) Schematic of the CCCII+ in BiCMOS technology, (b) its symbol, and (c) equivalent circuit for AC operation. (adapted from Fabre et al., © IEEE, 1998.)

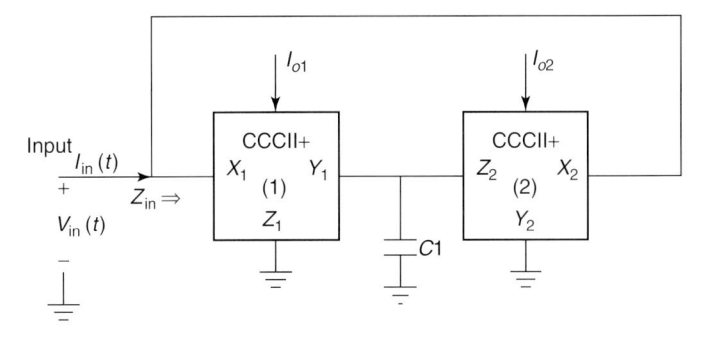

Figure 9.15 Block diagram for grounded inductance implementation. (adapted from Fabre et al., © IEEE, 1998.)

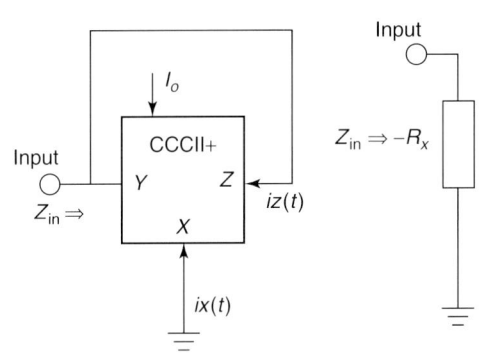

Figure 9.16 Generation of a negative resistance.

the control bias current I_o. The bias current can be changed to adjust the characteristics of the CCCII+. The CCCII+ is shown in its symbolic form in Figure 9.14b and the AC equivalent model is depicted in Figure 9.14c. The small signal resistance looking into the X terminal is approximately (from basic BJT principle) $V_T/2I_o = R_x$, and the signal voltage difference $v_{xy} = i_x R_x$.

9.4.2.3 Inductance and Negative Resistance

The CCCII+ can be configured to realize a grounded inductance, or a grounded resistance (positive or negative). These concepts have already been presented in Chapter 8. The associated configurations using the CCCII+ are shown in Figures 9.15 and 9.16. The equivalent circuit model for the realized inductance, together with the parasitic resistances, is shown in Figure 9.17. The negative resistance shown in Figure 9.16 can be used to compensate for the parasitic resistances in Figure 9.17.

9.4.2.4 Second-Order Elementary Band-Pass Filter Cell

Figure 9.18 shows the BP elementary cell realized with the CCCII+ building blocks. The CCCII+ blocks numbered as (1) and (2) realize the imperfect inductor with

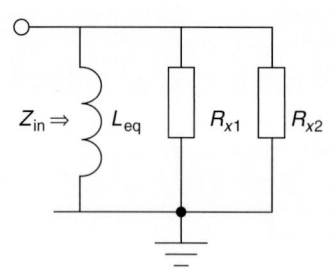

Figure 9.17 CCCII+ based inductance with parasitic resistances due to the two CCCII+ devices.

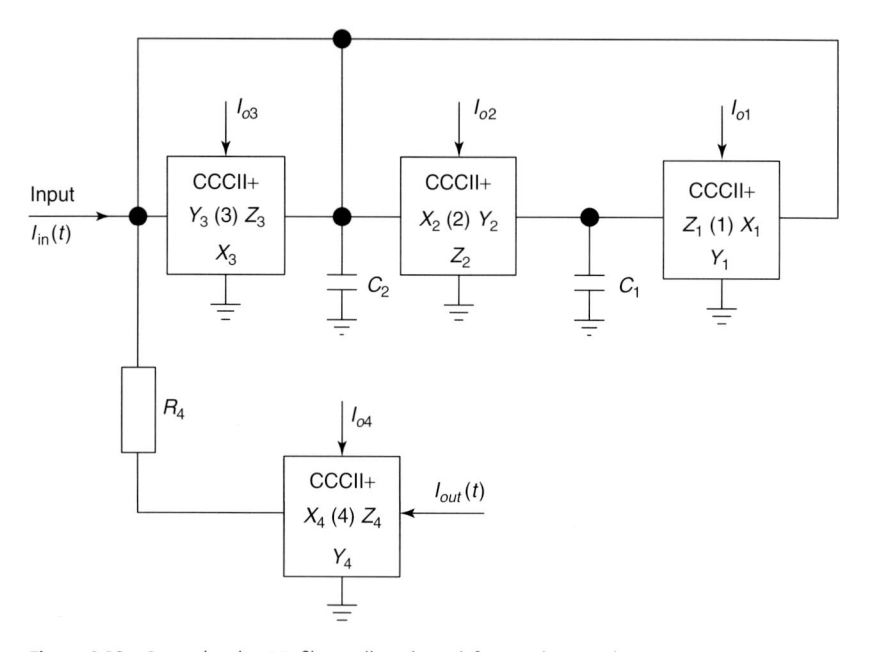

Figure 9.18 Second-order BP filter cell. (adapted from Fabre et al., 1998.)

the capacitor C_1. The capacitor C_2 is the resonating capacitor for the LC BP circuit. The conveyor numbered (3) produces a negative resistance to compensate for the parasitic resistance in the imperfect inductor. The resistance R_4 and the conveyor (4) form the output-current-sensing subsystem with an overall gain (I_{out}/I_{in}) close to unity at the center frequency f_o. The transfer function of the filter circuit is given by

$$\frac{I_{out}(s)}{I_{in}(s)} = \frac{\left(\frac{R_{x1}R_{x2}C_1}{R_4+R_{x4}}\right)s}{1+\left(\frac{R_{x1}R_{x2}C_1}{R_{eq}}\right)s+(R_{x1}R_{x2}C_1C_2)s^2} \quad (9.19)$$

where

$$\frac{1}{R_{eq}} = \left[\frac{1}{R_{x1}} + \frac{1}{R_{x2}} + \frac{1}{R_4+R_{x4}} - \frac{1}{R_{x3}}\right] \quad (9.20)$$

The filter parameters are then designable according to

$$\omega_o = \frac{1}{\sqrt{R_{x1} R_{x2} C_1 C_2}},$$

$$Q = \frac{R_{eq}}{\sqrt{R_{x1} R_{x2}}} \sqrt{\frac{C_2}{C_1}},$$

and

$$H_o = \frac{R_{eq}}{R_{x4} + R_4} \tag{9.21}$$

It can be seen that the above filter affords ω_o-tuning by varying R_{x1} or R_{x2} (which are controllable by the associated bias currents I_{o1} and I_{o2}), and Q-tuning by adjusting R_{eq} (by the bias current I_{o3} or by varying R_4). The parameters of the BP filter cells (namely, C_1, C_2, R_{x1}, R_{x2}, ...) are to be designed to match the coefficients of the transfer functions, $T_1(s)$ and $T_2(s)$, given by Eq. (9.17). For further details one may refer to the work of Fabre et al. (1998).

Practice Problems

9.1 For the Bainter band-reject biquad of Problem 5.13 of Chapter 5, find the corresponding differential structure and evaluate its VTF $\frac{V_o}{V_s}$.

9.2 For the OTA-biquad shown in Figure P5.36, find the corresponding differential structure and determine the VTFs $\frac{V_{o1}}{V_i}$, $\frac{V_{o2}}{V_i}$, and $\frac{V_{o3}}{V_i}$.

9.3 Consider Problems 5.20–5.23, as well as the specifications of the filters in Problem 5.33. Provide an integrated circuit design solution for each case. You may choose any of the following strategies: (a) MOS-R C solution with CMOS OA, (b) OTA-C solution with CMOS OTA, or (c) SC solution in CMOS technology. Assume that the capacitor values are to be less than 20 pF in IC technology. For a CMOS technological process, you may assume that $\mu C_{ox} = 100$ µA V^{-2}, $|V_{TH}| = 0.8$ V, $V_{DD} = |-V_{SS}| = 2.5$ V. The feature size (i.e., minimum value of W or L of the MOSFET) of the process can be taken as 0.5 µm. Verify your designs using appropriate CAD simulation programs.

Appendices

Appendix A

A.1
Denominator Polynomial D(s) for the Butterworth Filter Function of Order n, with Passband from 0 to 1 rad s^{-1}

Table A.1 gives the coefficients a_i of the denominator polynomial $D(s)$ of a Butterworth filter, where $D(s) = s^n + a_1 s^{n-1} + a_2 s^{n-2} + \cdots + a_{n-1} s + a_n$, while Table A.2 gives the polynomial $D(s)$ in factored form for n up to 6. For more extensive tables up to order 10, please refer to Weinberg (1962), Schaumann, Ghausi, and Laker (1990), Schaumann and Van Valkenburg (2001), Su (1996), and Huelsman (1993).

Table A.1 Coefficients of the polynomial $D(s)$ for a Butterworth filter function.

n	a_1	a_2	a_3	a_4	a_5	a_6
2	1.4142	1				
3	2.0000	2.0000	1			
4	2.6131	3.4142	2.6131	1		
5	3.2361	5.2361	5.2361	3.2361	1	
6	3.8637	7.4641	9.1416	7.4641	3.8637	1

Table A.2 Denominator $D(s)$ in factored form for a Butterworth filter function.

n	D(s)
1	$s+1$
2	$s^2 + \sqrt{2} s + 1$
3	$(s+1)(s^2 + s + 1)$
4	$(s^2 + 0.765s + 1)(s^2 + 1.848s + 1)$
5	$(s+1)[(s+0.3090)^2 + 0.9511^2][(s+0.8090)^2 + 0.5878^2]$
6	$[(s+0.2588)^2 + 0.9659^2][(s+0.7071)^2 + 0.7071^2][(s+0.9659)^2 + 0.2588^2]$

Modern Analog Filter Analysis and Design: A Practical Approach. Rabin Raut and M. N. S. Swamy
Copyright © 2010 WILEY-VCH Verlag GmbH & Co. KGaA, Weinheim
ISBN: 978-3-527-40766-8

Table A.3 Coefficients of the polynomial D(s) for a Chebyshev filter function.

Passband ripple A_p	N	a_1	a_2	a_3	a_4	a_5	a_6
0.5 dB $\varepsilon = 0.3493$	1	2.863					
	2	1.425	1.516				
	3	1.253	1.535	0.716			
	4	1.197	1.717	1.025	0.379		
	5	1.1725	1.9374	1.3096	0.7525	0.1789	
	6	1.1592	2.1718	1.5898	1.1719	0.4324	0.0948
1.0 dB $\varepsilon = 0.5089$	1	1.965					
	2	1.098	1.103				
	3	0.988	1.238	0.491			
	4	0.953	1.454	0.743	0.276		
	5	0.9368	1.6888	0.9744	0.5805	0.1228	
	6	0.9282	1.9308	1.2021	0.9393	0.3071	0.0689
2.0 dB $\varepsilon = 0.7648$	1	1.308					
	2	0.804	0.637				
	3	0.738	1.022	0.327			
	4	0.716	1.256	0.517	0.206		
	5	0.7065	1.4995	0.6935	0.4593	0.0817	
	6	0.7012	1.7459	0.8670	0.7715	0.2103	0.0514

A.2
Denominator Polynomial D(s) for the Chebyshev Filter Function of Order n, with Passband from 0 to 1 rad s^{-1}

Table A.3 gives the coefficients a_i of the denominator polynomial D(s) of a Chebyshev filter, where $D(s) = s^n + a_1 s^{n-1} + a_2 s^{n-2} + \cdots + a_{n-1} s + a_n$, while Table A.4 gives the polynomial D(s) in factored form for n up to 6 and for three values of the passband ripple A_p. For more extensive tables up to order 10 and for other values of A_p, refer to Weinberg (1962), Schaumann, Ghausi, and Laker (1990), Schaumann and Van Valkenburg (2001), Su (1996), and Huelsman (1993).

A.3
Denominator Polynomial D(s) for the Bessel Thomson Filter Function of Order n

Table A.5 gives the coefficients a_i of the denominator polynomial D(s) of a Bessel–Thomson filter, where $D(s) = s^n + a_1 s^{n-1} + a_2 s^{n-2} + \cdots + a_{n-1} s + a_n$, while Table A.6 gives the polynomial D(s) in factored form for n up to 6. For more extensive tables up to order 10, refer to Weinberg (1962), Schaumann, Ghausi, and Laker (1990), and Schaumann and Van Valkenburg (2001).

Table A.4 Denominator $D(s)$ in factored form for a Chebyshev filter function.

Passband ripple A_p	n	$D(s)$
0.5 dB $\varepsilon = 0.3493$	1	$s + 2.863$
	2	$s^2 + 1.425s + 1.516$
	3	$(s + 0.626)(s^2 + 0.626s + 1.142)$
	4	$(s^2 + 0.351s + 1.064)(s^2 + 0.845s + 0.356)$
	5	$(s + 0.3623)[(s + 0.1120)^2 + 1.0116^2][(s + 0.2931)^2 + 0.6252^2]$
	6	$[(s + 0.0777)^2 + 1.0085^2][(s + 0.2121)^2 + 0.7382^2][(s + 0.2898)^2 + 0.2702^2]$
1.0 dB $\varepsilon = 0.5089$	1	$s + 1.965$
	2	$s^2 + 1.098s + 1.103$
	3	$(s + 0.494)(s^2 + 0.490s + 0.994)$
	4	$(s^2 + 0.279s + 0.987)(s^2 + 0.674s + 0.279)$
	5	$(s + 0.2895)[(s + 0.0895)^2 + 0.9901^2][(s + 0.2342)^2 + 0.6119^2]$
	6	$[(s + 0.0622)^2 + 0.9934^2][(s + 0.1699)^2 + 0.7272^2][(s + 0.2321)^2 + 0.2662^2]$
2.0 dB $\varepsilon = 0.7648$	1	$s + 1.308$
	2	$s^2 + 0.804s + 0.637$
	3	$(s + 0.402)(s^2 + 0.369s + 0.886)$
	4	$(s^2 + 0.210s + 0.928)(s^2 + 0.506s + 0.221)$
	5	$(s + 0.2183)[(s + 0.0675)^2 + 0.9735^2][(s + 0.1766)^2 + 0.6016^2]$
	6	$[(s + 0.0470)^2 + 0.9817^2][(s + 0.1283)^2 + 0.7187^2][(s + 0.1753)^2 + 0.2630^2]$

Table A.5 Coefficients of the polynomial $D(s)$ for a Bessel–Thomson filter with a normalized delay of 1 s at DC.

n	a_1	a_2	a_3	a_4	a_5	a_6
1	1					
2	3	3				
3	6	15	15			
4	10	45	105	105		
5	15	105	420	945	945	
6	21	210	1260	4725	10395	10395

Table A.6 Denominator $D(s)$ in factored form for a Bessel–Thomson filter with a normalized delay of 1 s at DC.

n	$D(s)$
1	$s + 1$
2	$s^2 + 3s + 3$
3	$(s + 2.322)(s^2 + 3.678s + 6.460)$
4	$(s^2 + 5.792s + 9.140)(s^2 + 4.208s + 11.488)$
5	$(s + 3.6467)[(s + 3.3520)^2 + 1.7427^2][(s + 2.3247)^2 + 3.5710^2]$
6	$[(s + 4.2484)^2 + 0.8675^2][(s + 3.7356)^2 + 2.6263^2][(s + 2.5159)^2 + 4.4927^2]$

Table A.7 Parameters of a second-order elliptic function $H(s) = H\frac{s^2+a_0}{s^2+b_1 s+b_0}$.

ω_s	α_1 = 0.7	0.8	0.9	0.99	Coefficients
2.0	0.597 566	0.761 953	1.09 079	1.70 530	b_1
	0.748 566	0.889 100	1.21 614	3.39 116	b_0
	7.46 410	7.46 393	7.46 410	7.46 437	a_0
	0.070 208	0.095 295	0.146 639	0.449 766	$H = \alpha_2$
1.6	0.568 640	0.716 947	0.942 467	1.21 673	b_1
	0.780 727	0.923 621	1.24 863	3.01 139	b_0
	4.55 831	4.55 842	4.55 832	4.55 836	a_0
	0.119 892	0.162 086	0.246 530	0.654 022	$H = \alpha_2$
1.3	0.507 505	0.622 959	0.766 598	0.658 076	b_1
	0.835 122	0.975 687	1.27 415	2.34 888	b_0
	2.76 980	2.76 981	2.76 982	2.76 972	a_0
	0.211 054	0.281 803	0.414 008	0.839 569	$H = \alpha_2$
1.1	0.372 652	0.428 498	0.457 760	0.244 714	b_1
	0.916 613	1.03 128	1.23 375	1.63 605	b_0
	1.71 409	1.71 409	1.71 408	1.71 394	a_0
	0.374 317	0.481 308	0.647 782	0.945 011	$H = \alpha_2$

A.4
Transfer Functions for Several Second-, Third-, and Fourth-Order Elliptic Filters

The transfer function coefficients for several second-, third-, and fourth-order elliptic filters are given in Tables A.7–A.9, respectively, for a range of passband edge and stopband edge attenuations ($\alpha_1 = 10^{-A_p/20}, \alpha_2 = 10^{-A_s/20}$), and stopband to passband edge ratio ω_s. For more extensive tabulation, one may refer to Schaumann, Ghausi, and Laker (1990) and Huelsman (1993).

A.4 Transfer Functions for Several Second-, Third-, and Fourth-Order Elliptic Filters

Table A.8 Parameters of a third-order elliptic function $H(s) = H \dfrac{s^2+a_1}{(s+p)(s^2+b_{11}s+b_{10})}$.

	α_1				
ω_s	0.7	0.8	0.9	0.99	Coefficients
1.6	0.334131	0.429765	0.595349	1.29096	p
	0.243387	0.306340	0.404241	0.641363	b_{11}
	0.875469	0.925798	1.03480	1.62210	b_{10}
	3.22360	3.22359	3.22359	3.22359	a_1
	0.090743	0.123426	0.191108	0.649595	H
	0.021308	0.028976	0.044840	0.150824	α_2
1.4	0.353124	0.455659	0.635840	1.44703	p
	0.222806	0.278406	0.361387	0.514156	b_{11}
	0.890734	0.938912	1.04183	1.55605	b_{10}
	2.41363	2.41362	2.41363	2.41363	a_1
	0.130318	0.177253	0.274453	0.932875	H
	0.036758	0.049969	0.077236	0.254634	α_2
1.2	0.395725	0.514269	0.729373	1.84049	p
	0.181717	0.223182	0.278588	0.308389	b_{11}
	0.919159	0.962023	1.05044	1.41484	b_{10}
	1.69962	1.69962	1.69962	1.69962	a_1
	0.214008	0.291087	0.450785	1.53210	H
	0.077873	0.105649	0.162349	0.488077	α_2
1.1	0.448812	0.588202	0.850207	2.38157	p
	0.139154	0.167009	0.197936	0.164793	b_{11}
	0.945461	0.981243	1.05130	1.27550	b_{10}
	1.37033	1.37031	1.37031	1.37031	a_1
	0.309657	0.421193	0.652271	2.21688	H
	0.136715	0.184503	0.279160	0.702853	α_2

Table A.9 Parameters of a fourth-order elliptic function $H(s) = H\dfrac{(s^2+a_1)(s^2+a_2)}{(s^2+b_{11}s+b_{10})(s^2+b_{21}s+b_{20})}$.

ω_s	0.8	0.9	0.95	0.99	Coefficients
1.3	0.574 306	0.779 239	0.983 063	1.48 067	b_{11}
	0.326 332	0.428 090	0.563 827	1.07 349	b_{10}
	1.87 204	1.87 203	1.87 203	1.87 203	a_1
	0.139 842	0.182 294	0.218 409	0.277 321	b_{21}
	0.962 302	1.00 947	1.06 699	1.23 958	b_{20}
	8.09 589	8.09 589	8.09 613	8.09 589	a_2
	0.016 576	0.025 661	0.037 702	0.086 921	$H = \alpha_2$
1.2	0.589 557	0.799 091	1.00 707	1.49 416	b_{11}
	0.363 967	0.479 903	0.637 238	1.23 106	b_{10}
	1.57 240	1.57 240	1.57 240	1.57 242	a_1
	0.120 623	0.155 159	0.182 889	0.218 090	b_{21}
	0.969 765	1.01 180	1.06 243	1.20 610	b_{20}
	6.22 423	6.22 422	6.22 421	6.22 434	a_2
	0.028 851	0.044 651	0.065 716	0.150 187	$H = \alpha_2$
1.1	0.611 017	0.822 969	1.02 740	1.43 445	b_{11}
	0.442 216	0.587 138	0.785 841	1.52 732	b_{10}
	1.29 041	1.29 041	1.29 090	1.29 092	a_1
	0.088 880	0.111 155	0.126 428	0.132 384	b_{21}
	0.981 121	1.01 394	1.05 200	1.14 901	b_{20}
	4.34 613	4.34 581	4.34 973	4.34 993	a_2
	0.061 824	0.095 443	0.139 862	0.309 376	$H = \alpha_2$
1.05	0.621 079	0.825 168	1.00 616	1.24 184	b_{11}
	0.532 447	0.709 059	0.949 099	1.76 639	b_{10}
	1.15 363	1.15 363	1.15 362	1.15 362	a_1
	0.062 131	0.075 144	0.081 657	0.073 511	b_{21}
	0.989 514	1.01 374	1.04 043	1.09 961	b_{20}
	3.31 250	3.31 240	3.31 238	3.31 266	a_2
	0.110 293	0.169 268	0.245 492	0.503 140	$H = \alpha_2$

(Column group header: α_1)

Appendix B

B.1
Bessel Thomson Filter Magnitude Error Calculations (MATLAB Program)

```
%BT filter magnitude error calculations
            nj = 1;
            for n = 3:15;
                nn = n-2;

            for m = 1:45
% variable of the BT polynomial is set
                y = 0.1*m;
                x = 0 +y*i;
% calculations pertaining to ck begins
                n2 = 2*n;

                n1 = n +1;
                blast = 0;
                for k = 1:n1
     k1 = k-1;% k1 goes from 0 to n
 k2 = n2-k1;% means 2n-k1, k1 from 0 to n
 k3 = n-k1;% means n-k1, k1 from 0 to n
            % (2n-k1) ! loop begins
                fac1 = 1;
                for l1 = 1:k2
                facx = fac1*l1;
                fac1 = facx;
           end;% (2n-k1) ! loop ends
                faca = fac1;
             %(n-k1) ! loop begins
                fac2 = 1;
                for l2 = 1:k3
                    h = ck;
                end;
              % if loop ends
       % BT polynomial is calculated
              % as an iterative sum
              bsum = blast +ck*x^k1;
                blast = bsum;
                end;
              % k-loop ends
                bsum = blast;
       % the BT transfer func value for a
       % given 'n' and 'x' calculated
                hs = h /bsum;
                hsm = abs(hs);
                hsdb = -20*log10(hsm);
                ax(m) = y;
                ay(nn,m) = hsdb;
                end;
             %m-loop ends
                end;
             % n-loop ends
             kk = linspace(1,13,13)
             plot(ax,ay(kk,:),'w')
             axis([0 4.5 0 6])
                grid
        xlabel('normalized frequency -->')
        ylabel('magnitude error in dB')
             text(1.8,4.5,'n = 3')
```

(continued)

```
            facy = fac2 *12;
            fac2 = facy;
        end;%(n-k1)! loop ends
            facb = fac2;
        %k! loop begins
            fac3 = 1;
            for l3 = 1:k1
            facz = fac3 *l3;
            fac3 = facz;
        end;%(k1!) loop ends
            facc = fac3;
        %ck is computed
ck = faca /(facb *facc *2 ^k3);
            If k = = nj
    %(contd. On next column)
```

```
            text(2.4,4.5,'4')
            text(2.6,4,'5')
            text(2.9,3.8,'6')
            text(3.1,3.7,'7')
            text(3.3,3.5,'8')
            text(3.5,3.4,'9')
            text(3.6,3.3,'10')
            text(3.7,3.1,'11')
            text(4.0,3.4,'12')
            text(4.1,3.2,'13')
            text(3.9,2.7,'14')
            text(4.1,2.6,'15')
        end
```

B.2
Bessel Thomson Filter Delay Error Calculations (MATLAB Program)

```
        %BT filter delay calculations
    % variable of the BT polynomial is set
            for nn = 1:13;
            n = nn +2;
        % m-loop begins
            for m = 1:100
            x = 0.1 *m;
    % calculations pertaining to Ck begins
            n2 = 2 *n;
            n1 = n +1;
        bev1 = 0; bod1 = 0;
        evp1 = 0;odp1 = 0;
            Kev = 1;
        % k-loop begins
            for k = 1:n1
    k1 = k-1;% k1 goes from 0 to n
k2 = n2-k1;% means 2n-k1, k1 from 0 to n
k3 = n-k1;% means n-k1, k1 from 0 to n
        % (2n-k1)! loop begins
            fac1 = 1;
            for l1 = 1:k2
            facx = fac1 *l1;
            fac1 = facx;
        end;% (2n-k1)! loop ends
            faca = fac1;
        %(n-k1)! loop begins
            kev = kev-1;
            else
        bod = bod1-ck *(i) ^k *x ^k1;
            bod1 = real(bod);
        odp = odp1-k1 *ck *(i) ^k *x ^(k1-1);
            odp1 = real(odp);
            kev = kev +1;
            end
        % if-else loop ends
            end
        % k-loop ends
            x1 = bev1;
            x2 = evp1;
            y1 = bod1;
            y2 = odp1;
        anum = x1 *y2-y1 *x2;
        den = x1 ^2 +y1 ^2;
        der = anum /den;
        er = (1-der) /der;
        ax(m) = x;
        ay(nn,m) = er *100;
            End
        % m-loop ends
            res1 = ax;
            res2 = ay;
            end
```

(continued)

```
        fac2 = 1;
          for l2 = 1:k3
         facy = fac2 *l2;
            fac2 = facy;
       end;%(n-k1) ! loop ends
            facb = fac2;
         %k ! loop begins
            fac3 = 1;
          for l3 = 1:k1
         facz = fac3 *l3;
            fac3 = facz;
       end;%(k1 !) loop ends
            facc = fac3;
         %ck is computed
    ck = faca /(facb *facc *2 ^k3);
    % even part of BT polynomial is
    calculated as an iterative sum
            If kev = = 1
     bev = bev1 +ck *(i) ^k1 *x ^k1;
          bev1 = real(bev);
  evp = evp1 +k1 *ck *(i) ^k1 *x ^(k1-1);
          evp1 = real(evp);
        %(contd. On next column)
```

```
         % n-loop ends
            ax = res1;
            ay = res2;
         plot(ax,ay,'w-')
         axis([0 10 0 5])
            grid
     xlabel('normalized fequency -->')
         ylabel('% delay error')
         text(0.8,4.3,'n = 3')
         text(2,4.3,'4')
         text(2.8,4.3,'5')
         text(3.8,4.3,'6')
         text(4.8,4.3,'7')
         text(5.7,4.3,'8')
         text(6.6,4.3,'9')
         text(7.3,4.1,'10')
         text(8.1,3.9,'11')
         text(9,3.7,'12')
         text(9.5,1.7,'13')
         text(9.5,.4,'14')
         text(9.5,.1,'15')
            end
```

Appendix C

C.1
Element Values for All-Pole Single-Resistance-Terminated Low-Pass Lossless Ladder Filters

In this section, we present the element values for all-pole LP single-resistance-terminated lossless ladder filters with Butterworth, Chebyshev, and Bessel–Thomson approximations. Figures C.1a and C.1b correspond to voltage-driven structures for even- and odd-order filters, respectively, while Figures C.1c and C.1d to current-driven structures for even- and odd-order filters, respectively. The element values for several orders are given in Table C.1 for the Butterworth and Chebyshev filters, and Table C.2 for the Bessel–Thomson filter. For other values of n and ripple factors, one may refer to Weinberg (1962) and Huelsman (1993).

C.2
Element Values for All-Pole Double-Resistance-Terminated Low-Pass Lossless Ladder Filters

In this section, we present the element values for all-pole LP double-resistance-terminated lossless ladder filters with Butterworth, Chebyshev, and Bessel–Thomson approximations. Figures C.2a and C.2b correspond to structures for even- and odd-order filters, respectively, while c and d are alternate structures for even- and odd-order filters, respectively. The element values for several orders are given in Table C.3 for the Butterworth and Chebyshev filters, and Table C.4 for the Bessel–Thomson filter. For other values of n and ripple factors in the case of Chebyshev filters, one may refer to Weinberg (1962) and Huelsman (1993).

Appendix C

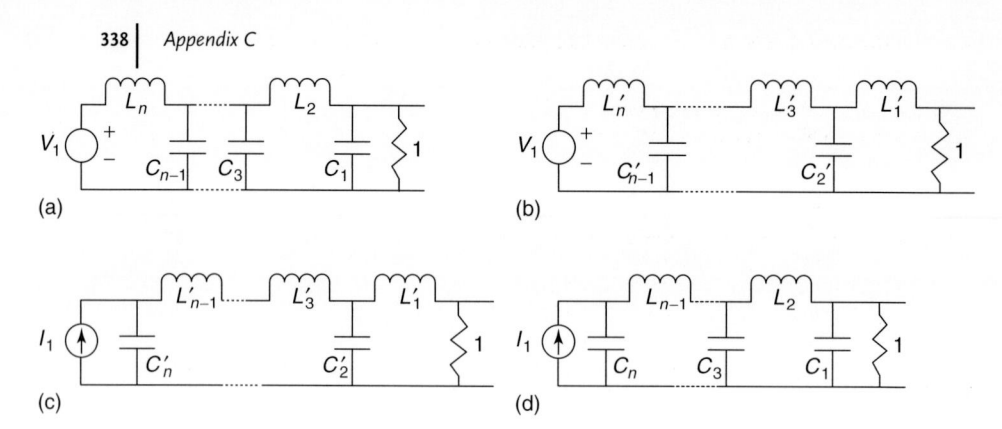

Figure C.1 Low-pass single-resistance-terminated lossless all-pole ladder filters; (a) voltage-driven, even-order; (b) voltage-driven, odd order, (c) current-driven even-order, (d) current-driven odd order.

Table C.1 Element values for singly terminated lossless Butterworth and Chebyshev filters (bandwidth normalized to 1 rad s^{-1}).

Order n	C_1	L_2	C_3	L_4	C_5	L_6
Butterworth filter						
2	0.7071	1.4142				
3	0.5000	1.3333	1.5000			
4	0.3827	1.0824	1.5772	1.5307		
5	0.3090	0.8944	1.3820	1.6944	1.5451	
6	0.2588	0.7579	1.2016	1.5529	1.7593	1.5529
Chebyshev filter (0.5 dB ripple)						
2	0.7014	0.9403				
3	0.7981	1.3001	1.3465			
4	0.8352	1.3916	1.7279	1.3138		
5	0.8529	1.4291	1.8142	1.6426	1.5388	
6	0.8627	1.4483	1.8494	1.7101	1.9018	1.4042
Chebyshev filter (1.0 dB ripple)						
2	0.9110	0.9957				
3	1.0118	1.3332	1.5088			
4	1.0495	1.4126	1.9093	1.2817		
5	1.0674	1.4441	1.9938	1.5908	1.6652	
6	1.0773	1.4601	2.0270	1.6507	2.0491	1.3457
n	L'_1	C'_2	L'_3	C'_4	L'_5	C'_6

C.2 Element Values for All-Pole Double-Resistance-Terminated Low-Pass Lossless Ladder Filters

Table C.2 Element values for singly terminated Bessel–Thomson filter with a delay of 1 s at DC.

Order n	C_1	L_2	C_3	L_4	C_5	L_6
2	0.3333	1.0000				
3	0.1667	0.4800	0.8333			
4	0.1000	0.2899	0.4627	0.7101		
5	0.0667	0.1948	0.3103	0.4215	0.6231	
6	0.0476	0.1400	0.2246	0.3005	0.3821	0.5595

n	L'_1	C'_2	L'_3	C'_4	L'_5	C'_6

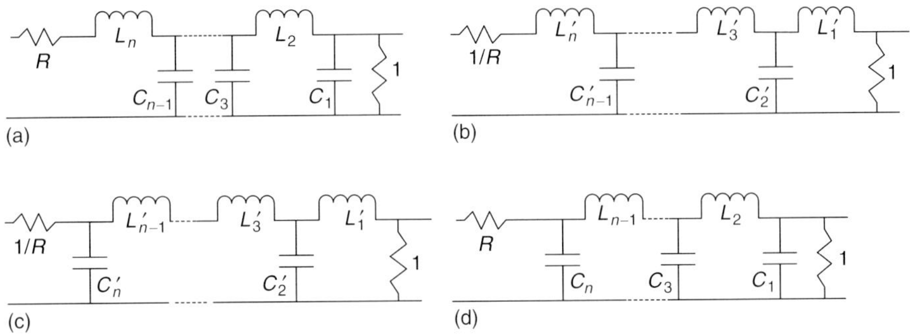

Figure C.2 Low-pass double-resistance-terminated lossless all-pole ladder filters.

Table C.3 Element values for doubly terminated lossless Butterworth and Chebyshev filters (bandwidth normalized to 1 rad s^{-1}).

n	C_1	L_2	C_3	L_4	C_5	L_6	C_7
Butterworth filter							
2	1.4142	1.4142					
3	1.0000	2.0000	1.0000				
4	0.7654	1.8478	1.8478	0.7654			
5	0.6180	1.6180	2.0000	1.6180	0.6180		
6	0.5176	1.4142	1.9319	1.9319	1.4142	0.5176	
Chebyshev filter (0.5 dB ripple)							
3	1.5963	1.0967	1.5963				
5	1.7058	1.2296	2.5408	1.2296	1.7058		
7	1.7373	1.2582	2.6383	1.3443	2.6383	1.2582	1.7373
Chebyshev filter (1.0 dB ripple)							
3	2.0236	0.9941	2.0236				
5	2.1349	1.0911	3.0009	1.0911	2.1349		
7	2.1666	1.1115	3.0936	1.1735	3.0936	1.1115	2.1666

n	L'_1	C'_2	L'_3	C'_4	L'_5	C'_6	L'_7

Table C.4 Element values for singly terminated Bessel–Thomson filter with a delay of 1 s at DC.

Order n	C_1	L_2	C_3	L_4	C_5	L_6
2	1.5774	0.4226				
3	1.2550	0.5528	0.1922			
4	1.0598	0.5116	0.3181	0.1104		
5	0.9303	0.4577	0.3312	0.2090	0.0718	
6	0.8377	0.4116	0.3158	0.2364	0.1480	0.0505
n	L'_1	C'_2	L'_3	C'_4	L'_5	C'_6

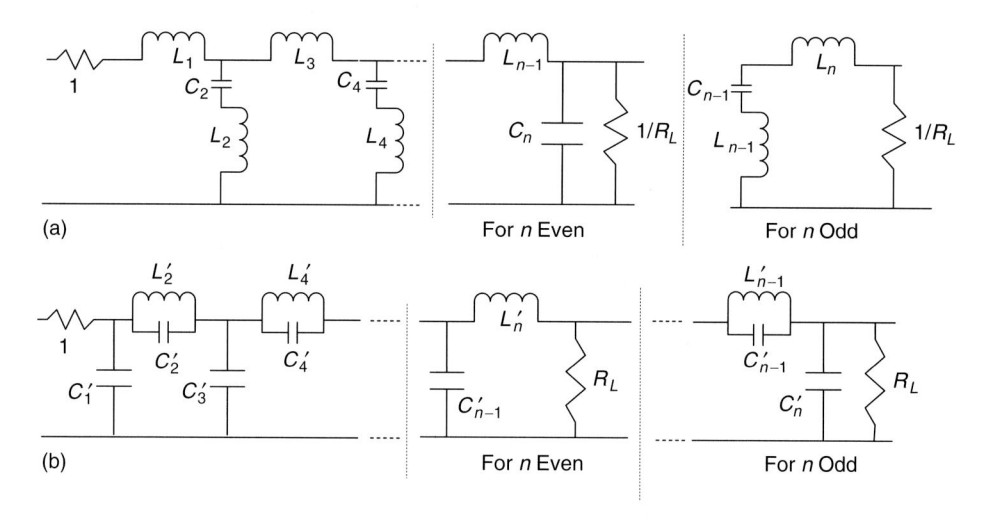

Figure C.3 Elliptic LP double-resistance-terminated lossless filter structures.

C.3
Element Values for Elliptic Double-Resistance-Terminated Low-Pass Lossless Ladder Filters

In this section, we present the element values for elliptic LP double-resistance-terminated lossless ladder filters. Figure C.3a shows the structures for both even- and odd-order filters while Figure C.3b gives alternate structures for the filters.

Table C.5 provides element values for odd orders (3 and 5) and for even order (4) for the case when the response at infinite frequency is forced to zero by adopting a modified expression for the even order transfer function. In this transfer function, the denominator is of degree n while the numerator degree is forced to be $n - 2$. In this case, the load and source resistances are *equal* with a value of 1 Ω each. The modified elliptic function has the form

C.3 Element Values for Elliptic Double-Resistance-Terminated Low-Pass Lossless Ladder Filters

Table C.5 Element values for the elliptic filter of Figure C.3a and b.

n	ω_s	A_a	L_1	C_2	L_2	L_3	C_4	L_4	L_5	See Figure C.3a
3	1.05	1.748	0.3555	0.15 374	5.39 596	0.3555				
	1.1	3.374	0.44 626	0.26 993	2.70 353	0.44 626				
	1.2	6.691	0.57 336	0.4498	1.30 805	0.57 336				
4	1.05	3.284	0.00 442	0.17 221	4.93 764	1.01 224	0.84 445	(Passband ripple 0.1 dB)		
	1.1	6.478	0.17 279	0.32 758	2.30 986	1.04 894	0.89 415			
	1.2	12.085	0.37 139	0.56 638	1.09 294	1.11 938	0.92 440			
5	1.05	13.841	0.70 813	0.76 630	0.73 572	1.12 761	0.20 138	4.38 116	0.04 985	
	1.1	20.050	0.81 296	0.92 418	0.49 338	1.22 445	0.37 193	2.13 500	0.29 125	
	1.2	28.303	0.91 441	1.06 516	0.31 628	1.38 201	0.60 131	1.09 329	0.52 974	
3	1.05	8.134	1.05 507	0.25 223	3.28 904	1.05 507				
	1.1	11.480	1.22 525	0.37 471	1.94 752	1.22 525				
	1.2	16.209	1.42 450	0.52 544	1.11 977	1.42 450				
4	1.05	11.322	0.63 708	0.35 277	2.41 039	1.11 522	1.39 953	(Passband ripple 1.0 dB)		
	1.1	15.942	0.80 935	0.54 042	1.40 015	1.18 107	1.45 001			
	1.2	22.293	1.00 329	0.77 733	0.79 634	1.26 621	1.49 217			
5	1.05	24.134	1.56 191	0.67 560	0.83 449	1.55 460	0.26 584	3.31 881	0.88 528	
	1.1	30.471	1.69 691	0.77 511	0.58 827	1.79 892	0.39 922	1.98 907	1.12 109	
	1.2	38.757	1.82 812	0.87 005	0.38 720	2.09 095	0.56 347	1.16 672	1.38 094	
n	ω_s	A_a	C_1'	L_2'	C_2'	C_3'	L_4'	C_4'	C_5'	See Figure C.3b

Table C.6 Alternate values for the elliptic filters of Figure C.3a and b.

n	ω_s	A_a	L_1	C_2	L_2	L_3	C_4	L_4	L_5	C_6	See Figure C.3a
4	1.05	4.485	0.15780	0.18091	4.73822	1.20743	0.82637				
	1.10	8.308	0.33411	0.33438	2.28333	1.26881	0.84827				
	1.20	14.387	0.53773	0.55478	1.12558	1.36980	0.85261				
							(Passband ripple 0.1 dB) $R_L = 0.73781\,\Omega$				
6	1.05	20.307	0.57153	0.65752	1.01346	0.92972	0.32584	2.72744	1.03524	0.88809	
	1.10	27.889	0.70783	0.81703	0.67992	1.10484	0.51890	1.54640	1.19779	0.88523	
	1.20	37.827	0.84244	0.98082	0.43111	1.32791	0.75659	0.88144	1.37708	0.87992	
4	1.05	13.243	0.95111	0.26779	3.20104	1.90749	0.80699				
	1.10	18.140	1.16239	0.39958	1.91077	2.05228	0.80907				
	1.20	24.700	1.40135	0.56068	1.11374	2.23453	0.80633				
							(Passband ripple 1.0 dB) $R_L = 0.37598\,\Omega$				
6	1.05	30.730	1.40432	0.58067	1.14761	1.37588	0.31837	2.79144	1.79883	0.82259	
	1.10	38.342	1.56906	0.69149	0.80335	1.66832	0.45609	1.75937	1.99786	0.82076	
	1.20	48.285	1.73631	0.80659	0.52424	2.01190	0.62218	1.07185	2.22816	0.81822	

n	ω_s	A_a	C'_1	L'_2	C'_2	C'_3	L'_4	C'_4	C'_5	L'_6	See Figure C.3b

C.3 Element Values for Elliptic Double-Resistance-Terminated Low-Pass Lossless Ladder Filters

$$H_N(s) = \frac{H_c \prod_{i=2}^{n/2} (s^2 + \Omega_i^2)}{a_0 + a_1 s + \cdots + a_{n-1} s + a_n s^n} \qquad (C.3.1)$$

More details on this are available in Huelsman (1993).

Table C.6 provides alternate set of element values for even orders (4 and 6), where the modified elliptic transfer function has the same form as in Eq. (C.3.1), but the values of Ω_i are slightly different. The source and load resistances for this alternate case are *unequal* with $R_s = 1\,\Omega$.

In each of these tables only two values are used for A_p, namely, 0.1 and 1.0 dB. For a more exhaustive set of tables, refer to Huelsman (1993).

References

Abuelma'atti, M.T. and Shabra, A.M. (1966) A novel current-conveyor-based universal current-mode filter. *Microelectr. J.*, **27**, 471–475.

Ackerberg, D. and Mossberg, K. (1974) A versatile active RC building block with inherent compensation for the finite bandwidth of the amplifier. *IEEE Trans. Circuits Syst.*, **CAS-21**, 75–78.

Al-Hashimi, B.M. and Fidler, J.K. (1988) New VCT-based active filter configurations and their applications. IEE Saraga Colloquium on Electronic Filters, London, pp. 6/1–6/4.

Allen, P.E. and Sanchez-Sinencio, E. (1984) *Switched Capacitor Circuits*, Van Nostrand Reinhold Company, New York.

Altun, M. and Kuntman, H. (2008) Design of fully differential current mode operational amplifier with improved input-output impedances and its filter applications. *AEÜ-Int. J. Electron. Commun.*, **62**, 239–244.

Antoniou, A. (2006) *Digital Signal Processing: Signals, Systems and Filters*, McGraw-Hill.

Antoniou, A. (1967) Gyrators using operational amplifiers. *Electron. Lett.*, **3**, 350–352.

Antoniou, A. (1969) Realization of gyrators using operational amplifiers and their use in RC-active network synthesis. *Proc. Inst. Elec. Eng.*, **116**, 1838–1850.

Assi, A., Sawan, M., and Raut, R. (1996) A new VCT for analog IC applications. *Proc. Eighth International Conference on Microelectronics*, Cairo, pp. 169–172.

Assi, A., Sawan, M., and Raut, R. (1997) A fully differential and tunable CMOS current mode opamap based on transimpedance-transconductance technique. *Proc. IEEE 40th Midwest Symp. Circuits and Systems*, Sacramento, pp. 168–171.

Bainter, J.R. (1975) Active filter has stable notch, and response can be regulated. *Electronics*, **48**, 115–117.

Baker, R.J., Li, H.W., and Boyce, D.E. (1998) *CMOS Circuit Design, Layout and Simulation*, IEEE Press.

Bermudez, J.C.M. and Bhattacharyya, B.B. (1982) Parasitic insensitive toggle-switched capacitor and its application to switched-capacitor networks. *Electron. Lett.*, **18**, 734–736.

Bhattacharyya, B.B. and Swamy, M.N.S. (1971) Network transposition and its application in synthesis. *IEEE Trans. Circuit Theory*, **CT-18**, 394–397.

Bhattacharyya, B.B., Sundaramurthy, M., and Swamy, M.N.S. (1981) Systematic generation of canonic sinusoidal RC-active oscillators. *IEE Proc. Part G*, **128**, 114–126.

Biolek, D., Hancioglu, E., and Keskin, A.U. (2008) High-performance current-differencing transconductance amplifier and its application in precision current-mode realization. *AEÜ-Int. J. Electron. Commun.*, **62**, 92–96.

Bobrow, L.S. (1965) On active RC synthesis using an operational amplifier. *Proc. IEEE*, **53**, 1648–1649.

Bruton, L.T. (1969) Network transfer functions using the concept of frequency dependent negative resistance. *IEEE Trans. Circuit Theory*, **CT-16**, 406–408.

Bruton, L.T. (1980) *RC-Active Circuits Theory and Design*, Prentice Hall, Inc.

Modern Analog Filter Analysis and Design: A Practical Approach. Rabin Raut and M. N. S. Swamy
Copyright © 2010 WILEY-VCH Verlag GmbH & Co. KGaA, Weinheim
ISBN: 978-3-527-40766-8

Bruun, E. (1991) A differential-input, differential-output current mode operational amplifier. *Int. J. Electron.*, **71**, 1047–1056.

Bruun, E. (1994) A high-speed CMOS current opamp for very low supply voltage operation. *Proc. IEEE Int. Symp. Circuits and Systems*, London, pp. 509–512.

Carlosena, A. and Moschytz, G.S. (1993) Nullators and norators in voltage to current mode transformations. *Int. J. Circuit Theory Appl.*, **21**, 421–424.

Chang, C.M., Soliman, A.M., and Swamy, M.N.S. (2007) Analytical synthesis of low sensitivity high-order voltage-mode DDCC and FDCCII-grounded R and C all pass filter structures. *IEEE Trans. Circuits Syst. - Part I*, **54**, 1430–1443.

Chang, C.-M. (1991) Current mode all-pass/notch and bandpass filter using single CCII. *Electron. Lett.*, **27**, 1812–1813.

Chang, C.-M. (1993) Universal active current filter with single input and three outputs using CCIIs. *Electron. Lett.*, **29**, 1932–1933.

Chen, W.-K. (1990) *Linear Networks and Systems: Algorithms and Computer-Aided Implementations*, 2nd edn, vol. 1, World Scientific Publishing Company Pvt. Ltd.

Chen, W.H. (1964) *Linear Network Design and Synthesis*, McGraw-Hill, New York.

Chen, W.-K. (1986) *Passive and Active Filter, Theory and Implementations*, John Wiley & Sons, Inc., New York.

Chen, W.-K. (Editor-in-Chief) (1995) *The Circuits and Filters Handbook*, Chapter 76, CRC Press.

Chen, H.P. (2009a) Versatile universal voltage mode filter employing DDCCs. *AEÜ-Int. J. Electron. Commun.*, **62**, 78–82.

Chen, H.P. (2009b) Single FDCCII-based universal voltage-mode filter. *AEÜ-Int. J. Electron. Commun.*, **63**, 713–719.

Cheng, K.H. and Wang, H.C. (1997) Design of current mode operational amplifier with differential input and differential output. *Proc. IEEE Int. Symp. Circuits and Systems*, Hong Kong, pp. 153–156.

Chiu, W.Y. and Horng, J.W. (2007) High-input and low-output impedance voltage-mode universal biquadratic filter using DDCCS. *IEEE Trans. Circuits Syst.- Part II*, **54**, 649–652.

Christian, E. and Eisermann, E. (1977) *Filter Design Tables and Graphs*, Transmission Networks International, Knightdale.

Daggett, K.E. and Vlach, J. (1969) Sensitivity-compensated active networks. *IEEE Trans. Circuit Theory*, **CT-16**, 416–422.

Darlington, S. (1939) Synthesis of reactance 4-poles which produce prescribed insertion loss characteristics. *J. Math. Phys.*, **18**, 257–353.

Deliyanis, T. (1968) High Q-factor circuit with reduced sensitivity. *Electron. Lett.*, **4**, 577.

Deliyanis, T., Sun, Y., and Fidler, J.K. (1999) *Continuous-time Active Filter Design*, CRC Press.

Director, S.W. and Rohrer, R.A. (1969) The generalized adjoint network and network sensitivities. *IEEE Trans. Circuit Theory*, **16**, 318–323.

Elwan, H.O. and Soliman, A.M. (1996) A novel CMOS current conveyor realization with an electronically tunable current mode filter suitable for VLSI. *IEEE Trans. Circuits Syst. - Part II*, **43**, 663–670.

Fabre, A. and Alami, M. (1995) Universal current-mode biquad implemented from 2nd generation current conveyors. *IEEE Trans. Circuits Syst. - Part I*, **42**, 383–385.

Fabre, A., Saaid, O., Wiest, F., and Boucheron, C. (1998) High-frequency high-Q BiCMOS current-mode bandpass filter and mobile communication application. *IEEE J. Solid-State Circuits*, **33**, 614–625.

Fan, S.C., Gregorian, R., Temes, G.C., and Zomorodi, M. (1980) Switched capacitor filters using unity gain buffers. *Proc. IEEE Int. Symp. Circuits and Systems*, Houston, pp. 334–337.

Fettweis, A. (1979a) Basic principles of switched capacitor filters using voltage inverter switches. *AEÜ*, **33**, 13–19.

Fettweis, A. (1979b) Switched capacitor filters using voltage inverter switches: further design principles. *AEÜ*, **33**, 107–114.

Fettweis, A., Herbst, D., Hoefflinger, B., Pandel, J., and Schweer, R. (1980) MOS switched-capacitor filters using voltage inverter switches. *IEEE Trans. Circuits Syst.*, **CAS-27**, 527–538.

Fleischer, P. and Tow, J. (1973) Design formulas for biquad active filters using three operational amplifiers. *Proc. IEEE*, **61**, pp. 662–663.

Fleischer, P.E. and Laker, K.R. (1979) A family of active switched-capacitor biquad building blocks. *Bell Syst. Tech. J.*, **58**, 2235–2269.

Fried, D.L. (1972) Analog sampled data filters. *IEEE J. Solid-State Circuits*, **SC-7**, 302–304.

Friend, J.J. (1970) A single operational amplifier biquadratic filter section. *Proc. IEEE Int. Symp. Circuit Theory*, Georgia, pp. 179–180.

Friend, J.J., Harris, C.A., and Hilberman, D. (1975) STAR: an active biquadratic filter section. *IEEE Trans. Circuits Syst.*, **CAS-22**, 115–121.

Geiger, R.L., Allen, P.E., and Strader, N.R. (1990) *VLSI Design Techniques for Analog and Digital Circuits*, McGraw-Hill.

Geiger, R.L. and Sanchez-Sinencio, E. (1985) Active filter design using operational transconductance amplifiers – a tutorial. *IEEE Circuits Devices Mag.*, **1**, 20–32.

Ghausi, M.S. and Laker, K.R. (1981) *Modern Filter Design – Active RC and Switched Capacitor*, Prentice Hall, Inc.

Girling, F.E.J. and Good, F.F. (1970) Active filters 12: the leap-frog or active-ladder synthesis. *Wirel. World*, **76**, 341–345.

Gray, P.R. and Hodges, D.A. (1979) MOS switched capacitor filters. *Proc. IEEE*, **67**, 61–75.

Gray, P.R. and Meyer, R.G. (1993) *Analysis and Design of Analog Integrated Circuits*, 3rd edn, John Wiley & Sons, Inc.

Grebene, A.B. (1984) *Bipolar and MOS Analog Integrated Circuit Design*, John Wiley & Sons, Inc.

Gregorian, R. and Nicholson, W.E. (1979) CMOS switched capacitor filters for a PCM voice codec. *IEEE J. Solid-State Circuits*, **SC-14**, 970–980.

Hakim, S.S. (1965) Active RC filters using an amplifier as the active element. *Proc. IEE*, **112**, 901–912.

Hodges, D.A., Gray, P.R., and Brodersen, R.W. (1978) Potential of MOS technologies for analog integrated circuits. *IEEE J. Solid-State Circuits*, **SC-13**, 285–294.

Hokënek, E. and Moschytz, G.S. (1980) Analysis of general switched capacitor networks using indefinite admittance matrix. *Proc. IEE, Part G*, **127**, 21–33.

Huelsman, L.P. (1993) *Active and passive analog filter design – An introduction*, McGraw-Hill.

Hughes, J.B., Bird, N.C., and Macbeth, I.C. (1989) Switched currents- a new technique for analogue sampled-data signal processing. *IEEE Int. Symp. Circuits and Systems*, pp. 1584–1587.

Hughes, J.B., Macbeth, I.C., and Pattullo, D.M. (1990) Switched current filters. *Proc. IEE, Part G*, **137**, 156–162.

Hurtig, G. III (1973) U.S. Patent 3,720,881.

Hwang, Y.S., Wu, D.S., Chen, J.J., and Chou, W.S. (2008) Realization of current-mode high order filters employing multiple output OTAs. *AEÜ-Int. J. Electron. Commun.*, **62**, 299–303.

Ismail, M., Smith, S.V., and Beale, R.G. (1988) A new MOSFET-C universal filter structure for VLSI. *IEEE J. Solid-State Circuits*, **SC-23**, 183–194.

Jiraseree-amornkun, A. and Surakampotorn, W. (2008) Efficient implementation of tunable ladder filers using multi-output current controlled conveyors. *AEÜ- Int. J. Electron. Commun.*, **62**, 11–23.

Johns, D.A. and Martin, K. (1997) *Analog Integrated Circuit Design*, John Wiley & Sons, Inc.

Kerwin, W.J., Huelsman, L.P., and Newcomb, R.W. (1967) State variable synthesis for insensitive integrated circuit transfer functions. *IEEE J. Solid State Circuits*, **SC-2**, 87–92.

Knob, A. (1980) Novel strays insensitive switched-capacitor integrator realizing the bilinear Z-transform. *Electron. Lett.*, **16**, 173–174.

Kung, H.T. (1982) Why systolic architecture. *Computer*, **15**, 37–46.

Kuo, B.C. (1967) *Automatic Control System*, Prentice Hall.

Kurth, C.F. and Moschytz, G.S. (1979) Nodal analysis of switched capacitor networks. *IEEE Trans. Circuits Syst.*, **CAS-28**, 93–104.

Laker, K.R. and Sansen, W.M.C. (1994) *Design of Analog Integrated Circuits and Systems*, McGraw-Hill.

Laker, K.R. (1979) Equivalent circuits for the analysis and synthesis of switched capacitor networks. *Bell Syst. Tech. J.*, **58**, 729–769.

Laker, K.R. and Ghausi, M.S. (1980) Design of minimum sensitivity multiple-loop feedback bandpass active filters. *J. Franklin Inst.*, **310**, 51–64.

Li, R.D. and Raut, R. (2003) A Very Wide-Band OTA-C Filter in CMOS VLSI Technology. Proceedings of the 7th World Multiconference on Systemics, Cybernatics and Informatics (SCI 2003), Orlando, Fl., USA, paper#002396.pdf, pp. 1–6, (on CD ROM).

Malawka, J. and Ghausi, M.S. (1980) Second order function realization with switched capacitor network and unity gain amplifiers. *Proc. IEE, Part-G*, **127**, 187–190.

Mikhael, W.B. and Bhattacharyya, B.B. (1975) A practical design for insensitive RC-active filters. *IEEE Trans. Circuits Syst.*, **CAS-22**, 407–415.

Minaei, S., Kuntman, H., and Cicekoglu, O. (2000) A new high output impedance current-mode universal filter with single input and three outputs using dual output CCIIs. Proc. 7th IEEE Int. Conf. Electronics, Circuits and Systems, Beirut, Lebanon, pp. 379–382.

Mitra, S.K. (1967) Transfer function realization using RC one-ports and two grounded voltage amplifiers. Proc. *First Annual Princeton Conf.* on Information Sciences and Systems, Princeton University, March 1967, pp. 18–23.

Mitra, S.K. (1969) *Analysis and Synthesis of Linear Active Networks*, John Wiley & Sons, Inc.

Mohan, P.V.A. (2002) *Current Mode Analog VLSI Filters-Design and Implementation*, Birkhauser.

Mohan, P.V.A., Ramachandran, V., and Swamy, M.N.S. (1982) General stray-insensitive first-order active SC network. *Electron. Lett.*, **18**, 1–2.

Mohan, P.V.A., Ramachandran, V., and Swamy, M.N.S. (1995) *Switched Capacitor Filters: Theory, Analysis and Design*, Prentice Hall, London.

Moschytz, G.S. (1974) *Linear Integrated Networks, Fundamentals*, Van Nostrand Reinhold.

Moschytz, G.S. and Carlosena, A. (1994) A classification of current-mode single amplifier biquads based on a voltage-to-current transformation. *IEEE Trans. Circuits Syst. - Part II*, **41**, 151–156.

Nelin, B.D. (1983) Analysis of switched capacitor networks using general purpose circuit simulation programmes. *IEEE Trans. Circuits Syst.*, **CAS-30**, 43–48.

Nguyen, N.M. and Meyer, R.G. (1990) Si-IC-compatible inductors and LC passive filters. *IEEE J. Solid-State Circuits*, **25**, 1028–1031.

Orchard, H.J. (1966) Inductorless filters. *Electron. Lett.*, **2**, 224–225.

Patel, V. and Raut, R. (2008) A study on CMOS negative resistance circuits. Proc. *IEEE Canadian Conf. Electrical and Computer Engineering*, pp. 1.283–1.288.

Prasad, D., Bhaskar, D.R., and Singh, A.K. (2009) Universal current-mode biquad filter using dual output current differencing transconductance amplifier. *AEÜ-Int. J. Electron. Commun.*, **63**, 497–501.

Raut, R. (1992) A novel VCT for analog IC applications. *IEEE Trans. Circuits Syst. - Part II*, **39**, 882–883.

Raut, R. (1993) A voltage to current transducer (VCT) in CMOS technology for applications in analog VLSIC. *Proc. Canadian Conf. Electrical and Computer Engineering*, Vancouver, pp. 82–85.

Raut, R. and Daoud, N.S. (1993) Current-mode oscillator realization using a voltage-to-current transducer in CMOS technology. *Proc. IEE, Part-G*, **140**, 462–464.

Raut, R. (1984) Some design and analysis techniques for practical sc sampled data filters using unity gain amplifiers. Ph.D. Thesis, Concordia University, Montreal.

Raut, R. and Bhattacharyya, B.B. (1984a) A note on the analysis of multiphase SC networks containing operational amplifiers. *Proc. International Conf. on Computers, Systems and Signal Processing*, Bangalore, pp. 1400–1403.

Raut, R. and Bhattacharyya, B.B. (1984b) Design of parasitic tolerant switched-capacitor filters using unity-gain buffers. *Proc. IEE, Part-G*, **131**, 103–113.

Raut, R., Bhattacharyya, B.B., and Faruque, S.M. (1992) Systolic array architecture implementation of parasitic-insensitive switched-capacitor filters. *Proc. IEE, Part-G*, **139**, 384–394.

Raut, R. and Guo, N. (1997) Low power wide band voltage and current mode second order filters using wide band CMOS transimpedance network. *Proc. IEEE Int. Symp. Circuits and Systems*, Hong Kong, pp. 313–316.

Raut, R., Swamy, M.N.S., and Tian, N. (2007) Current mode filters using voltage amplifiers. *Circuits, Syst. Signal Process.*, **26**, 773–792.

Roberts, G.W. and Sedra, A.S. (1989) All current mode selective circuits. *Electron. Lett.*, **25**, 759–761.

Roberts, G.W. and Sedra, A.S. (1992) A general class of current amplifier-based biquadratic filter circuits. *IEEE Trans. Circuits Syst. - Part I*, **39**, 257–263.

Ruston, H. and Bordogna, J. (1971) *Electrical Networks: Functions, Filters, Analysis*, McGraw-Hill.

Sallen, R.P. and Key, E.L. (1955) A practical method of redesigning RC-active filters. *IEEE Trans. Circuit Theory*, **CT-2**, 74–85.

Sanchez-Sinencio, E., Geiger, R.L., and Nevarez-Lozano, H. (1998) Generation of continuous-time two integrator loop OTA filter structures. *IEEE Trans. Circuits Syst.*, **CAS-35**, 936–946.

Schaumann, R. and Van Valkenburg, M.E. (2001) *Design of Analog Filters*, Oxford University Press.

Schaumann, R., Ghausi, M.S., and Laker, K.R. (1990) *Design of Analog Filters: Passive, Active RC, and Switched Capacitor*, Prentice Hall, Englewood Cliffs.

Sedra, A.S., Smith, K.C. (1970) A second generation current conveyor and its applications. *IEEE Trans. Circuit Theory*, **CT-17**, 132–134.

Sedra, A.L. and Smith, K.C. (2004) *Microelectronic Circuits*, 5th edn, Oxford University Press.

Sedra, A.S. and Brackett, P.O. (1978) *Filter Theory and Design: Active and Passive*, Matrix Publishers, Inc.

Senani, R. (1992) New current-mode biquad filters. *Int. J. Electron.*, **73**, 735–742.

Soliman, A.M. (1996) Applications of current feedback operational amplifiers. *Analog Integr. Circ. Signal Process.*, **11**, 265–302.

Soliman, A.M. (1997) New current mode filters using current conveyors. *AEÜ-Int. J. Electron. Commun.*, **51**, 275–278.

Su, K.L. (1996) *Analog Filters*, Chapman & Hall.

Sun, Y. and Fidler, J.K. (1995) High-order current-mode continuous-time multiple output OTA capacitor filters. *IEE 15th SARAGA Colloquium on Digital and Analogue Filters and Filtering Systems*, London, 1995, pp. 8/1–8/6.

Sun, Y. and Fidler, J.K. (1996) Structure generation of current-mode two integrator loop dual output-OTA grounded capacitor filters. *IEEE Trans. Circuits Syst. - Part II*, **43**, 659–663.

Sun, Y. and Fidler, J.K. (1997) Structure generation and design of multiple loop feedback OTA-grounded capacitor filters. *IEEE Trans. Circuits Syst. - Part I*, **44**, 1–11.

Swamy, M.N.S. (1975) Modern Filter Design, Classroom Notes, Department of Electrical and Computer Engineering, Concordia University, Montreal.

Swamy, M.N.S., Bhusan, C., and Bhattacharyya, B.B. (1974) Generalized duals, generalized inverses and their applications. *J. Inst. Electron. Rad. Eng.*, **44**, 95–99.

Swamy, M.N.S., Bhusan, C., and Bhattacharyya, B.B. (1976) Generalized dual transposition and its applications. *J. Franklin Inst.*, **301**, 465–476.

Swamy, M.N.S. and Thulasiraman, K. (1981) *Graphs, Networks and Algorithms*, John Wiley-Interscience, New York.

Swamy, M.N.S., Bhusan, C., and Thulasiraman, K. (1972) Sensitivity invariants for active lumed/distributed networks. *Electron. Lett.*, **8**, 26–29.

Swamy, M.N.S., Raut, R., and Tang, Z. (2004) Generation of new OTA-C oscillator structures using network transposition. *Proc. IEEE 47th Midwest Symposium on Circuits and Systems*, Hiroshima, 1, pp. 73–76.

Swamy, M.N.S. and Raut, R. (2002) Realization of g_m-C current-mode filters from associated g_m-C voltage-mode filters. *Proc. IEEE 45th Midwest Symposium on Circuits and Systems*, Tulsa, Oklahoma, 2, pp. 625–628.

Szczepansky, S. Jakusz, J., and Schaumann, R. (1997) A linear fully balanced CMOS OTA for VHF filtering applications. *IEEE Trans. Circuits Syst. - Part II*, **44**, 174–187.

Tang, Z. (2004) Application of network transposition in the design of OTA-C filters and oscillators, M.A.Sc. Thesis, Concordia University.

Tangsrirat, W., Tanjaroen, W., and Pukkalamm, T. (2009) Current-mode multiphase sinusoidal oscillator using CDTA-based all pass sections. *AEÜ-Int. J. Electron. Commun.*, **62**, 616–622.

Temes, G.C. and Mitra, S.K. (1973) *Modern Filter Theory and Design*, John Wiley & Sons, Inc., New York.

Thomas, R.E. (1959) Technical Note No. 8, Circuit Theory Group, University of Illinois, Urbana.

Thomas, L.C. (1971a) The biquad: part I - some practical design considerations; part II-a multipurpose active filtering system. *IEEE Trans. Circuit Theory*, **CT-18**, 350–357.

Thomas, L.C. (1971b) The Biquad: part II - a multipurpose active filtering system. *IEEE Trans. Circuit Theory*, **CT-18**, 358–361.

Thomson, W.E (1949) Delay networks having maximally flat frequency characteristics. *Proc. IEE, (part 3)*, vol. 96, 487–490.

Tow, J. (1968) Active RC filters – a state space realization. *Proc. IEEE*, **56**, 87–92.

Tow, J. (1969) A step-by-step active filter design. *IEEE Spectr.*, **6**, 64–68.

Tsividis, Y., Banu, M., and Khoury, J. (1986) Continuous-time MOSFET-C filters in VLSI. *IEEE Trans. Circuits Syst.*, **CAS-33**, 125–140.

Toumazou, C., Lidgey, J., and Haigh, D. (eds) (1990) *Analogue IC Design – The Current-Mode Approach*, IEE Press, London.

Tu, S.H., Chang, C.M., Ross, N.J., and Swamy, M.N.S. (2007) Analytical synthesis of current-mode high-order single-ended-input OTA and equal-capacitor elliptic filter structures with the minimum number of components. *IEEE Trans. Circuits Syst. - Part I*, **54**, 2196–2210.

Van Valkenburg, M.E. (1960) *Introduction to Modern Network Synthesis*, John Wiley & Sons, Inc.

Vandewalle, J., DeMan, H.J., and Rabey, J. (1981) Time, frequency and z-Domain modified nodal analysis of switched capacitor networks. *IEEE Trans. Circuits Syst.*, **CAS-28**, 186–195.

Wang, C., Zhou, L., and Li, T. (2008) A new OTA-C current-mode biquad with single input and multiple outputs. *AEÜ-Int. J. Electron. Commun.*, **62**, 232–234.

Weinberg, L. (1962) *Network Analysis and Synthesis*, McGraw-Hill, New York.

Yanagisawa, T. (1957) RC active networks using current inversion type negative impedance converters. *IEEE Trans. Circuit Theory*, **CT-4**, 140–144.

Zverev, A.L. (1967) *Handbook of Filter Synthesis*, John Wiley & Sons, Inc.

Index

a

Åckerberg-Mossberg biquad *see under* biquad
active building blocks   28, 41, 147, 161, 265
active devices   4, 5, 28, 103, 124, 126, 143, 191, 192, 207, 214, 228, 268, 283, 299, 300, 304
active resistor *see under* resistor
active RC filters   1, 4, 28, 30, 43, 103, 217, 221, 223, 299, 304
adjoint network   268
admittance matrix   9, 13, 15, 18
alias   168
All-pass filter   53, 108, 155, 265
all-pole function   4, 56, 65, 85, 89, 230, 233
all-pole filter   88, 217
anti-aliasing filter   168
analog filters   4, 7, 28, 30, 148, 161, 299, 300, 302
Antoniou's GIC   209, 210, 214, 251
approximation   4, 5, 41
– magnitude *see* magnitude approximation
– phase *see* phase approximation
– constant-delay *see* constant delay approximation

b

band-pass filter   30, 31, 115, 138, 154–157, 160, 172, 206, 222, 265, 321
band-reject filter   4, 53
band-stop filter *see* band-reject filter
bandwidth   3, 71, 73, 76, 108, 168, 172, 285, 314
Bessel polynomial   75
Bessel-Thomson (BT) filters   5, 75–77, 328, 329, 333–335, 339, 340
bilinear transformation *see under* transformation
biquadratic filter 103–105, 187–178–189, *see also* biquads and second order filters

biquadratic function   104, 199, 229
biquads 103–105, *see also* biquadratic filters and second order filters
– Åckerberg-Mossberg   156, 282
– Fleischer–Tow   123, 273, 274
– KHN *see under* KHN state variable filter
– Mikhael-Bhattacharyya   126
– single amplifier   109–117
– Tow-Thomas   121, 132, 223, 281
brick-wall characteristics   41, 53
Bruton's transformation *see under* transformation
Butterworth approximation *see under* magnitude approximation
Butterworth filter   56, 58, 59, 91, 95, 96, 99, 100, 232, 236, 237, 327, 337, 338, 339

c

capacitive dual   45–47, 79, 269, 270, 295
cascade   16, 23, 46, 129, 193, 202, 220, 225, 227, 228, 229, 239, 245, 248, 250, 258, 270, 319
Cauer networks   86–88, 90, 95, 97, 100
CCII *see* Current conveyor II
CCCS *see* Current controlled current source
CCVS *see* Current controlled voltage source
chain matrix   14–16, 42, 45–49, 89, 208–211
chain parameters   14
Chebyshev approximation *see under* magnitude approximation
Chebyshev filter   65, 95, 328, 329, 337, 338, 339
clock feed-through   174, 291, 302
clock frequency   162, 163, 167, 187
CMOS   2, 3, 5, 144, 147, 264, 266, 285, 318, 323

Modern Analog Filter Analysis and Design: A Practical Approach. Rabin Raut and M. N. S. Swamy
Copyright © 2010 WILEY-VCH Verlag GmbH & Co. KGaA, Weinheim
ISBN: 978-3-527-40766-8

Index

component simulation 207–209, 211–213, 215
component transformation 96
constant delay approximation 75–77
constrained networks 4, 7, 24–27, 110, 117
continued fraction (CF) expansion 86, 87, 90, 91
continuous-time domain 207, 251
continuous-time filters 2, 5, 138, 150
controlled sources 7, 16, 17, 24, 48, 52, 268, 280
current controlled current source (CCCS) 7, 16, 17, 18, 28, 31, 42, 48, 49, 79, 80, 113, 114, 256, 258, 262, 271, 272, 280, 283
current controlled voltage source (CCVS) 7, 16, 17, 18, 48, 49, 79, 80, 262, 294
current conveyor II (CCII) 3, 7, 31, 32, 114, 151, 257, 261–266, 294, 320–322
controlled currect conveyor II (CCCII) 114, 319–322
current amplifier 261, 269, 270, 271
current copier 285
current mirror 32, 149, 256, 285, 287, 291
current OA 114, 151
current-mode (CM) filters 3, 33, 150, 255–294, 345–347
current transfer function (CTF) 3, 12, 32, 36, 43, 46, 47, 48, 49, 52, 80, 84, 85, 93, 214, 265, 267, 268, 271, 272, 273, 274, 275, 276, 278, 279, 281, 282, 283, 284, 285
current transfer function (CTF) matrix 271, 276, 277, 295
CTF matrix *see* current transfer function (CTF) matrix
cut-off frequency 53, 54, 60, 69, 90, 91, 92, 221, 225, 226

d

D-network 201, 202
delay 53, 54, 68, 74, 75
delay and add 200–203
delay equalizers 53, 55, 77
differential architecture 5, 303
differential structure 303–307, 310–312, 316, 323
differential input OTA 277
diffused resistor *see under* resistor
digital filters 3, 192, 200
discrete-time equations 163, 175, 178
differential-input single-output (DISO)-OTA 277–279
doubly-terminated LC ladder 4, 92–100, 214

doubly-terminated network 50, 51, 83, 93, 252
driving point admittance 12, 86, 110
driving point function 48, 271
driving point impedance (DPI) 12, 42, 46, 50, 86, 94, 208, 210, 211, 214, 268
dual one-port 44, 45
dual networks 5, 44, 45, 270, 271, *also see* capacitive, resistive and generalized duals
dual transpose 50
dual two-port 4, 45, 47, 49, 50, 51, 79, 85, 94, 97, 100, 269, 270, 271, *also see* capacitive dual
dynamic current mirrors 285
dynamic range equalization 147, 191, 198, 207, 228, 229, 314

e

elliptic approximation *see under* magnitude approximation
elliptic filter 67, 68, 223, 249, 330, 311, 332
epitaxial and ion-implanted resistors 143
equivalent networks 16, 30
equal-ripple (equi-ripple) 62

f

feature size 147, 323
feed-forward technique 122
FDNR 208, 211, 213–217, 250, 252, 253
filter functions 5, 41, 122, 304, 319
floating inductor 208–210, 212, 215
Fleischer–Tow *see under* biquad
forward open circuit voltage gain 15
forward short circuit current gain 15
forward transconductance 15
frequency-dependent negative resistor *see* FDNR
frequency domain characterization 7, 163, 167
frequency scaling 51, 62, 64, 96
frequency transformations *see under* transformation
– low-pass to band-pass *see under* transformation
– low-pass to high-pass *see under* transformation
– low-pass to band-reject (bandstop) *see under* transformation

g

gain-bandwidth (GB) product 132, 134, 147, 284, 285
– effects of finite GB of an OA 283

generalized dual 45, 46, 47, 49, 255, 268, 269, *also see* dual two-ports and capacitive dual
generalized immittance converter 210, 211
general network theorems 41
g_m-C filter *see* OTA-C filter
grounded inductance 212, 321
gyrator 138, 208–210, 214, 239, 255, 263

h
high-pass filter 4, 53, 108, 115, 160
higher order active filters 207–251
Hurwitz polynomial 89
hybrid technology 141

i
ideal operational amplifier (OA) 101, 105, 128, 131, 132, 134, 224, 226, 280
immittance 210, 211
impedance converter 7, 17, 18, 33, 49, 80, 210, 268
impedance inverter 7, 17, 18, 48, 49, 138, 208, 263, 268, 271, 294, *see also* gyrator
impedance matrix 11, 14, 15, 48
impedance scaling 41–43
impedance transformation *see under* transformation
indefinite admittance matrix (IAM) 18–23, 39, 41
inductance simulation 5, 208, 210, 239
integrators
– ideal 241
– inverting 30, 181, 183, 184, 196, 220, 231, 233, 234, 246, 247, 292
Lossy 29, 106, 177, 181, 220, 221, 230, 231, 233, 260, 308, 310, 312, 316
Noninverting 181, 183, 196, 220, 304, 306
inverse Chebyshev approximation *see under* magnitude approximation
inverse follow the leader feedback (IFLF)
– structure 238
inverse network 44, 45, *see also* dual two-port network

k
KHN state variable filter 119–121, 127, 128

l
ladder filter 33, 101, 212, 214, 215, 222, 224, 239, 245, 250–252
leapfrog structure 217, 220, 221, 242, 243, 278, 279, 297
linear phase approximation *see* constant delay approximation
linear phase 75, 78

loop analysis 7–11
lossless digital integrator (LDI)
– transformation *see under* transformation
lossless integrator 106, 177, 179, 230, 290, 307, 316
lossless (LC) ladder filter 50, 88, 90, 92, 100, 207, 208, 212, 213, 214, 217, 245, 250, 337–341
lossless ladder simulation 207, 212, 213–216, 217–225, 239–245, 250
lossless network 80, 83, 89, 93, 94
lossy integrator 29, 106, 177, 181, 220, 230, 231, 233, 260, 308, 310, 312, 316
inverting 29, 106, 177, 181, 220, 221, 232, 233, 236, 308
Noninverting 177, 181, 186, 220
low-pass (LP) filter 4, 31, 33, 35, 154–156, 160, 204, 206, 240, 253, 337, 340
low-pass to band-pass transformation *see under* transformation
low-pass to high-pass transformation *see under* transformation
low-pass to band-reject (bandstop)
– transformation *see under* transformation

m
magnitude approximation 54–69
– Butterworth 56
– Chebyshev 60–65
– – maximally flat magnitude (MFM) 55–60
– elliptic 65–68
– inverse Chebyshev 68
MATLAB 4, 5, 58, 59, 64, 65, 68, 76, 171, 173, 300
maximum power transfer 50
mesh analysis *see* loop analysis
Mikhael-Bhattacharyya biquad *see under* biquad
mimimum sensitivity feedback structure 238
MFM approximation *see under* magnitude approximation
modified leap-frog (MLF) structure 238
modular approach 191–194
monolithic inductor 3, 147
MOS switch 175, 184
MOS capacitor 144–146, 175
multi-in single-output (MISO) system 122, 123
multiple amplifier biquad 117–128
multiple feedback 5, 207–237

n

negative impedance converter (NIC)   17, 18
– current inverting   17, 21, 22, 80, 263, 267
– voltage inverting   17, 267
negative impedance inverter (NII)   17, 18, 263, 294
network function   7, 12, 18, 138, 139, 154, 190, 213, 217
network transposition *see* transposition
nodal analysis   8, 9, 11
noise signal   303
nonreciprocal network   7, 15, 16, 18, 48, 271
nodal admittance matrix   18, 19
noninverting amplifier   131, 132, 238
notch filter   55, 140, 265, 266
– highpass notch   108, 109, 157
– lowpass notch   108, 109, 157
– symmetric   108, 109
null filter *see* notch filter
nullator   268, 279, 280

o

one-pole model   147, 184
one-port network   12, 16, 44, 45, 85, 94
OA *see* operational amplifier
open circuit impedance parameters   14
operational amplifier   24–26, 28, 29, 30, 105–107, 147, 148, 258
operational simulation technique   5, 217, 219–224, 239, 241, 253
operational transconductance amplifier (OTA)   2, 30, 31, 135, 139, 149, 239, 241, 243
OTA-C filters   5, 138, 139, 159, 310, 312, 314

p

parasitic capacitances   2, 138, 145, 159, 161, 174–176, 204, 239, 245, 260, 270, 285, 301–303, 310, 312, 313
parasitic-insensitive structures *see under* switched-capacitor (SC) filter
passband   41, 50, 53, 54, 55, 56, 60, 61, 63, 67, 68, 69, 71, 73, 96, 227, 228
phase approximation   73–77
phase delay   74
pinched resistor *see under* resistor
pole frequency   104
pole zero pairing   228, 229
pole-Q   104
positive impedance converter (PIC)   17
positive impedance inverter (PII)   18, 138, 208, *see also* gyrator

prewarping   170–172, 205, 248, 250, 251, 293
primary resonator block (PRB)   237

q

Quality factor (Q)   71, 211, 285
Q–enhancement   134
Q_p-sensitivity *see under* sensitivity

r

rational function   29, 50, 66
RC:CR transformation *see under* transformation
reactance function   85, 86, 87, 89, 90
reactance network   85, 86
reciprocal network   15, 47, 48, 100
reconstruction filter   168
reflection coefficient   94
resistive dual 45, *also see* dual networks
resistor
– active   143
– diffused   142
– pinched   142
reversed two-port (network)   47, 48

s

Sallen and Key filters   110–116
sampled-data domain   164, 166, 170, 207, 250, 251, 289
sampled-data frequency   163, 168–170
sampled-data transfer function   167–169
– Frequency domain characteristics of   167
second-order filters 4, 7, 30, 31, 35, 52, 103–140, 157, 159, 161, 181, 187, 189, 191, 193, 195, 197, 207, 223, 225, 239, 253, 266, 278, 280, *see also* biquadratic filters and biquads
sensitivity   4, 124–127, 129, 154, 207, 213, 216, 227–229, 238, 267, 272, 300
Q_p-sensitivity   125, 126, 127, 267
ω_p-sensitivity   126, 127
shifted-companion feedback structure   235
short circuit admittance parameters   13
signal-flow graph technique   192–198
Silicon dioxide   144–146
simulation of LC ladder   217
single-amplifier biquad *see under* biquad
single-input dual-output (SIDO) OTA   277–279
single-input multi-output (SIMO)   121
singly terminated network   50, 51, 83, 84, 90–92
source transformation   9–11, 239, 240
state variable filter   119, 120, 126, 127

stopband   41, 53, 54, 55, 63, 66, 67, 68, 69, 71, 72, 170,
super-capacitors 213, 215, 217, *also see* FDNR
switched-capacitor (SC) filter   2, 149, 161–206
– parasitic insensitive (PI)   4, 162, 175–187, 245–250
– second-order   162, 192, 196
– higher-order   248, 249
swiched-current filters *see* switched-current techniques
switched-current techniques   285–294
switched-current memory   286
systolic array architecture   192

t

time-delay *see* delay
three-terminal two-port   16
threshold voltage   144, 291, 301
topological dual   45
Tow-Thomas biquad *see under* biquad
transadmittance function (TAF)   12, 52, 84, 92, 118
transconductance   51, 135, 149, 260
transconductance amplifier   150, 258
transconductance-C filter *see* OTA-C filter
transformation
– bilinear (s→z)   161, 169–173
– Bruton's   214
– Frequency   69–71, 114
– impedance   42–43, 113–114, 214, 217
– lossless digital integrator (LDI)   166
– low-pass to band-pass   71, 88, 237, 316
– low-pass to high-pass   69–70, 212
– low-pass to band-reject (bandstop)   73
– RC:CR   79, 113–115
transimedance amplifier   258
transimedance function (TIF)   12, 15, 52, 84, 89, 93
transition band   53, 54, 67, 68, 69, 72, 168
transmission parameters *see* chain parameters
transmission zero   66, 88, 89, 128, 215, 223, 230, 233, 237
transposed networks   4, 48, 50, 268, 271, 273, 280

transposition   3, 5, 267, 268, 271–273, 275, 277, 279, 295
two-port networks   4, 7, 12, 13, 15–17, 22, 37, 41, 42, 45, 48, 268
two-port parameters   4, 13–18

u

unity gain amplifier (UGA)   105, 199–204

v

VCCS *see* Voltage controlled current source
VCVS *see* Voltage controlled voltage source
VLSI   2, 303
voltage amplifier   25–29, 105, 109, 117, 131, 137, 185, 199, 269, 307
voltage controlled current source (VCCS)   7, 16, 17, 18, 28, 30, 48, 49, 79, 80, 262, 271
voltage controlled voltage source (VCVS)   7, 16–18, 24, 26, 28, 29, 34, 42, 48, 49, 79, 113, 114, 184–186, 217, 262, 268, 271, 272, 280–283
voltage follower   105
voltage mode (VM)   5, 103, 217, 225, 255, 267–279
voltage transfer function (VTF)   12, 32, 35, 39, 43, 46–49, 52, 80, 84, 85, 93, 214, 265, 267, 268, 271–283, 285, 294, 295
VTF *see* votage transfer function
voltage transfer function (VTF) matrix   271, 276, 295
VTF matrix *see* voltage transfer function (VTF) matrix

w

warping   170

y

y-parameters *see* short circuit admittance parameters

z

z-parameters *see* open circuit impedance parameters
z-transform   163–165, 178, 179